# CHEMICAL QUALITY
## of WATER and the
# HYDROLOGIC CYCLE

Robert C. Averett and Diane M. McKnight

Proceedings of the symposium entitled *Chemical Quality of Water and the Hydrologic Cycle*, at the 8th Rocky Mountain Regional Meeting of the American Chemical Society, Denver, Colorado, June 8–12, 1986.

# LEWIS PUBLISHERS, INC.

**Library of Congress Cataloging-in-Publication Data**

Chemical quality of water and the hydrologic cycle.

    Proceedings of a symposium held in Denver, Colo., June 8-12, 1986, as part of the 8th Rocky Mountain Regional Meeting of the American Chemical Society.
    Includes bibliographies and index.
    1. Hydrologic cycle--Congresses.
2. Water Quality--Congresses. 3. Water, underground--Quality--Congresses.
III. American Chemical Society. Rocky Mountain Regional Meeting (8th:1986: Denver, Colo.)
GB848.C43   1987     551.48    87-14349
ISBN 0-87371-081-9

*Second Printing 1988*

LEWIS PUBLISHERS, INC.
121 South Main Street, Chelsea, Michigan 48118

PRINTED IN THE UNITED STATES OF AMERICA

This book is dedicated to
our families
for their enthusiasm,
support, and tolerance.

# PREFACE

An earth-science paradigm is that of the hydrologic cycle, that complex process which describes how water in its several forms is circulated throughout the earth's hydrosphere. Recently, this paradigm has been renamed the "active paradigm of the hydrologic cycle," signifying that man, himself, influences both the quality and quantity of water that circulates throughout the earth's hydrosphere.

Man's influence on the quality of the world's water is today commonplace. Indeed, water quality is a world-wide problem and of world-wide concern. There are numerous papers or scientific investigations concerned with the physical, chemical, and biological properties of water from different points of the hydrologic cycle or with the specific processes responsible for these properties. However, it is difficult to find studies that relate these properties and ensuing changes, in a holistic manner, to the hydrologic cycle. Such was our purpose for this volume; to present research that emphasizes the changes that take place in water as it circulates throughout the atmosphere, the earth's surface, and below the earth's surface. Obviously, it is not possible in a single volume to include all aspects of the quality of water in this context. Indeed, it is difficult to even focus on a few areas of investigation.

We feel, however, that the chapters included in this volume provide a valuable illustration of some of the important changes that take place as water moves through the hydrologic cycle. The holistic perspective conveyed in this book may benefit the scientific community, as well as water users and managers, in the attempt to understand the influence of man on this renewable, circulating resource. Relating man's influence on water quality to the hydrologic cycle also provides a framework for addressing the complex problems included in the active paradigm.

We enjoyed putting this volume together and hope it will encourage and stimulate future research using more holistic approaches in viewing and studying the quality of the world's water.

Robert C. Averett
Diane M. McKnight

# ACKNOWLEDGMENTS

We express our appreciation to the authors for their contribution to the symposium and the volume. We also are indebted to the many scientists who provided critical and thorough reviews of the manuscripts. We thank Marvin Goldberg and others for organizing the 8th Rocky Mountain Regional Meeting of the American Chemical Society and for sponsoring the symposium. Debra Sites contributed substantially in the coordination of the symposium and in the review process for the manuscripts. We are especially grateful to Leah Wilson, Betty Pedersen, and Pat Griffith for editorial assistance and manuscript preparation. During the symposium and preparation of the volume, we were members of the U.S. Geological Survey. Our efforts were conducted without remuneration.

Robert C. Averett
Diane M. McKnight

**Robert C. Averett** is a hydrologist with the U.S. Geological Survey in Reston, Virginia. He received his PhD from Oregon State University in aquatic ecology and began his professional career in Montana working in the field of water pollution. In 1968 he joined the U.S. Geological Survey, where he has held a number of positions related to water quality research, program direction, and research management. He served as a member of the Interior Department's Redwood National Park Study team, and was U.S. Chairman for the International Joint Commission, Poplar River Water Quality Study Board, Saskatchewan-Montana. The author of over 50 scientific papers, and consultant for a popular book series, Dr. Averett has conducted water quality research throughout the western United States.

**Diane M. McKnight** is a hydrologist for the U.S. Geological Survey, Water Resources Division, and Assistant Research Advisor for the Ecology Discipline in the National Research Program of the Water Resources Division. She received her PhD in Civil Engineering/Water Resources from the Massachusetts Institute of Technology in 1979. She has studied chemical, hydrological, and biological processes involving algae, trace metals, and dissolved organic material in different aquatic environments, including a drinking water reservoir receiving $CuSO_4$ treatment, an ombrotrophic bog in New England, Spirit Lake at the base of Mt. St. Helens, and pristine and acid mine drainage streams and lakes in the Colorado Rocky Mountains. Along with G.R. Aiken, R.L. Wershaw, and P. MacCarthy, she is coeditor of a recent book sponsored by the International Humic Substances Committee, *Humic Substances in Soil, Sediment, and Water*. Dr. McKnight is currently conducting an interdisciplinary study of the biogeochemistry of autochthonous aquatic humic substances in polar desert lakes in the Dry Valleys of Antarctica.

Chemical Quality of Water and

the Hydrologic Cycle

Table of Contents

LIST OF FIGURES

# LIST OF TABLES

SECTION I

PRECIPITATION CHEMISTRY

# CHAPTER 1

# THE INFLUENCE OF FOREST CANOPIES ON THE CHEMICAL QUALITY OF WATER AND THE HYDROLOGIC CYCLE

L. J. Puckett, U.S. Geological Survey, Reston, Virginia

## ABSTRACT

The chemical quality of precipitation is significantly altered upon contact with the forest canopy. Coniferous canopies tend to increase dissolved solutes in throughfall to a greater degree than deciduous canopies. Changes in chemistry may result from natural biological processes or, in polluted airsheds, by interaction of precipitation with anthropogenic compounds which have been deposited onto the forest canopy. The net result is a significant increase in solution ionic strength and a 72 to 114% increase in the concentration of $SO_4^{2-}$. Such increases in $SO_4^{2-}$ deposition are linked through the "salt effect" to decreases in stream alkalinity, increases in acidity, and accelerated base cation leaching.

## INTRODUCTION

The forest canopy, by virtue of its position, acts as the interface between the airshed and the watershed. As a result almost all exogenous inputs to the watershed must come into contact with the canopy. During this period of contact with the canopy various biological, chemical, and physical interactions may occur which affect precipitation chemistry. Biological processes are the most difficult to quantify and, therefore, the least understood. Because of interactions with the biological processes it may be equally difficult to quantify the chemical and physical processes. In order to gain a more complete understanding of the chemical quality of water in the hydrologic cycle these biological, chemical, and physical interactions need to be considered. In this chapter the current literature relative to the processes which determine throughfall chemistry will be reviewed. Research results on throughfall and stream chemistry at the Mill Run watershed in northwestern Virginia will also be presented.

*Chemical Quality of Water and the Hydrologic Cycle*, Robert C Averett and Diane M. McKnight (Eds.) © 1987 Lewis Publishers, Inc., Chelsea, Michigan. Printed in the United States of America.

Physical Processes

Much of the change in throughfall chemistry may result from interactions with anthropogenic dry deposited compounds. Dry deposition includes both particulate and gaseous compounds and is highly dependent on turbulent transport from the atmosphere to the canopy. Once in the canopy, compounds must pass through a laminar flow boundary layer in order to contact canopy surfaces. In the less than 0.1-μm-diameter particle size class Brownian diffusion is the primary means for penetration of the boundary layer. However, this process contributes little to dry deposition. Particles in the 0.1 to 1.0-μm-diameter size class have no efficient means of dry deposition (Fowler, 1980a). These compounds, $(NH_4)_2SO_4$, $(NH_4)_3H(SO_4)_2$, $H_2SO_4$, $NH_4HSO_4$, and $NH_4NO_3$ are highly hygroscopic and are deposited primarily in wet deposition through their role as cloud condensation nuclei. For particles larger than 1.0 μm in diameter sedimentation and inertial impaction are the primary means for penetrating the boundary layer (Fowler, 1980a). The majority of these particles are soil dust, usually of an alkaline nature, although they also may contain adsorbed acidic gases and aerosols. Therefore, except for relatively large particles (i.e., greater than 5.0 μm), dry deposition of particles is an inefficient process.

Deposition of gases such as $SO_2$, $HNO_3$, and $NO_2$ is the most important dry deposition process (Fowler, 1980b). Gases penetrate the boundary layer by molecular diffusion, a process which is rate-limited by the concentration gradient, the diffusion coefficient, and surface resistance. The total resistance to gaseous deposition is the sum of canopy resistance, stomatal resistance, leaf and stem resistance, and soil and litter layer resistance (Fowler, 1980a). Under most forest canopies the soil and litter layer resistance component is trivial, further reducing the problem to consideration of the stomatal component and the leaf and stem component. Published resistance values and deposition velocities (Fowler, 1980a; Grennfelt et al., 1980) indicate that stomatal uptake of gases is much greater than cuticular uptake. However, because stomata are closed a large amount of the time, the overall importance of the two pathways may be about equal. Adsorption and dissolution of gases occurs at a much faster rate to water than to dry surfaces. Therefore, resistance to gaseous deposition by a wet canopy is negligible (Grennfelt et al., 1980) facilitating rapid adsorption of $SO_2$, $HNO_3$, and $NO_2$ until equilibrium is achieved. For most temperate forests this is not considered a significant component of deposition.

Time between precipitation events is another important factor influencing throughfall chemistry (McColl and Bush, 1978). Dry deposition of gases and particles is a time dependent process; therefore, longer periods of time between events result in higher concentrations of some dissolved solutes in throughfall. Long dry periods also result in greater deposition of compounds onto the leaf surface from internal pools as a result of translocation through the cuticle (a wax-impregnated and coated protective surface layer) in the transpiration stream.

Wet deposition includes all forms of precipitation involving water: rain, sleet, snow, mist, and cloud condensation. Because of the physical processes involved, the effect of wet deposition on throughfall chemistry is a function of the amount of precipitation deposited, the rate of deposition (intensity), and the evaporative losses after coming into contact with canopy surfaces. The amount of rainfall is important in that throughfall, which is highly concentrated in dissolved ions during the early stages of a storm, becomes less concentrated as compounds are washed and leached from the canopy. Rain intensity may be a significant factor in determining throughfall chemistry (Tukey, 1970). As rainfall comes into contact with canopy surfaces, varying degrees of chemical equilibrium may be achieved at a rate dependent on the period of contact. High intensity events allow less time for equilibrium to be achieved; thus, throughfall concentrations are expected to be inversely related to intensity. However, the dilution effect may mask any effects of intensity.

## Biological and Chemical Processes

Forest canopies are biologically active substrates on which airshed-watershed interactions take place. Consequently, we might expect throughfall concentrations of biologically important solutes to be changed the most as a result of canopy interactions. In fact, the canopy acts as a sink for nitrogen ($NO_3^-$, $NO_2^-$, $NH_4^+$) and a source for $SO_4^{2-}$, $K^+$, $Ca^{2+}$, $Mg^{2+}$ and in small amounts $Na^+$ and $Cl^-$ (Tukey, 1970; Cole and Rapp, 1981). Depending on the type and proximity to emission sources, the canopy may serve as either a sink or source for $H^+$ (Cole and Rapp, 1981).

Biological uptake may take place either through the cuticle, or through stomata, or both. Stomatal uptake of gases is limited by the period during which the stomata are open and the rate at which the plant can metabolize the compound being absorbed (Fowler, 1980b). Uptake through the cuticle is more complex; adsorbed compounds and gases must pass through the wax layers of the cuticle. On some plants the cuticle is either discontinuous or nonexistent, facilitating absorption. Uptake is dependent on a continuous wet connection between the epidermal cells and the cuticle surface through microchannels having a diameter of 0.9 to 1.0 nm (Schonherr, 1976a, b). As the cuticle dries and shrinks, the wax layers in and on the cuticle either become more continuous and microchannels become blocked or their diameters decrease; thereby limiting diffusion and viscose flow through the cuticle (Schonherr and Schmidt, 1979). Also, during periods of low humidity the enhanced outward flow of water, due to intense evapotranspiration, may limit the influx of solutes. This also enhances movement of dissolved solutes to the leaf surface with the transpiration stream. Sulfur dioxide absorbed through stomata and oxidized internally to $SO_4^{2-}$ may be transported to the leaf surface in the transpiration stream (Bache, 1977; Johnson, 1984). The net result is that most of the dry deposited compounds remain on the leaf surface to be dissolved and washed off during subsequent rainfall.

During periods of high humidity or rain the liquid connection through the cuticle is reestablished and cuticular uptake may occur. Under these conditions uptake is controlled by the concentration gradient, the plant's ability to metabolize and/or translocate the compound (Schonherr, 1976a, b). For growth-limiting nutrients such as $NO_3^-$ and $NH_4^+$, or those rapidly incorporated into organic compounds in the leaf, the concentration gradient will favor uptake. During low-intensity precipitation or when leaves are wet from dew or mist, throughfall may become concentrated enough for uptake to occur.

Up to this point, biological uptake of compounds by the forest canopy has been discussed; however, leaching of compounds by precipitation is of equal importance. All of the major cations ($Ca^{2+}$, $Mg^{2+}$, $Na^+$, $K^+$) and anions ($Cl^-$, $NO_3^-$, $SO_4^{2-}$) can be extracted from leaves and needles (Tukey, 1970). Most cations are bound to exchange sites in the cuticle or free spaces in the leaf (Tukey, 1970, 1980). The cation exchange sites are primarily $-COO^-$ groups and significantly outnumber anion exchange sites (Schonherr, 1976a). The cuticle behaves as a polyelectrolyte with a zero point of charge at about pH 3.0 (Schonherr and Huber, 1977). In the pH 6.0 to 9.0 range, the cation exchange sites are highly selective for divalent ions, especially $Ca^{2+}$, however, in the pH 3.0 to 6.0 range, this selectivity is not as strong (Schonherr and Huber, 1977). Although the pH of the transpiration stream has not been determined, it, like most other internal plant solutions, is probably circumneutral, favoring saturation of exchange sites with divalent cations. The exchange capacity of the cuticle decreases with decreasing pH as the $-COO^-$ sites become protonated (Schonherr and Huber, 1977). During precipitation events, many of which are in the pH 3.0 to 4.0 range in the eastern United States and western Europe, displacement of cations from these exchange sites is quite pronounced. Because of the stronger selectivity of the exchanger for divalent cations, monovalent cations are leached most readily.

As with uptake, the size of the microchannel pore diameter (0.9–1.0 nm) may limit leaching of ions. Movement of ions through a fully hydrated cuticle is limited by their hydrated diameter (McFarlane and Berry, 1974; Schonherr, 1976a). Relative ion permeability based on hydrated ionic diameter (Table 1) is $NH_4^+>K^+>Na^+>Ca^{2+}>Mg^{2+}$ for cations and $Cl^- = NO_3^->SO_4^{2-}$ for anions.

Actual leaching rates also are dependent on the relative concentrations of the ions in vegetation (Tukey, 1970). Because concentrations of $NH_4^+$ and $NO_3^-$ are negligible in foliage (Johnson, 1984), it is unlikely that either could be leached by rainfall. For similar reasons significant net increases of $Na^+$ and $Cl^-$ in throughfall also may be attributed to washoff of dry deposited compounds. Although $Mg^{2+}$ is found in greater than trace quantities in foliage it is primarily bound in the chlorophyll molecule. This, in addition to its large hydrated ionic diameter (0.8 nm), limits the leachability of this ion.

Table 1.  Hydrated Ionic Diameters for the Major Cations and Anions
          in Throughfall.  Data are from Kielland (1937).

| Ion | Diameter (nm) |
|-----|---------------|
| Cations | |
| $NH_4^+$ | 0.25 |
| $K^+$ | 0.3 |
| $Na^+$ | 0.4-0.45 |
| $Ca^{2+}$ | 0.6 |
| $Mg^{2+}$ | 0.8 |
| Anions | |
| $Cl^-$ | 0.3 |
| $NO_3^-$ | 0.3 |
| $SO_4^{2-}$ | 0.4 |

The remaining ions, $K^+$, $Ca^{2+}$, and $SO_4^{2-}$, have potentially large
internal and external sources.  Potassium is recognized as the most
mobile ion in deciduous foliage (Eaton et al., 1973; Mollitor and
Berg, 1980; Cole and Rapp, 1981) and plays an important role in
regulating the opening and closing of stomates (Meyer et al.,
1983).   In  addition,  its  relatively  small  hydrated  diameter
(0.3 nm) and monovalent form make it potentially the most easily
leached ion.   In many throughfall studies, $K^+$ shows the greatest
net enrichment, especially under deciduous canopies (Cole and Rapp,
1981).   External  sources  include  windblown  primary  silicate
minerals  such  as  $KAlSi_3O_8$  as  well  as  eroded  foliar  structures
(trichomes in particular).  This makes it difficult to estimate how
much $K^+$ originates outside the watershed and is a net input, and
how much is resuspended plant and mineral components from within
the watershed.
     Much of the calcium in foliage is bound to cell walls where it
is responsible for maintaining structural integrity and also is
known to occur as calcium oxalate (Meyer et al., 1983).   Another
large fraction is bound to cation exchange sites of the cuticle as
discussed earlier.  This bound fraction is not as readily leached
as $K^+$ except at low pH (Schonherr and Huber, 1977).   The large
diameter  (>5.0 µm)  fraction  of  atmospheric  particulates  often
contain calcium (Lindberg et al., 1986), and dry deposition may
account for much of its net enrichment in throughfall.
     Sulfur in plants is incorporated into amino acids (cysteine,
cystine, methionine) and excess sulfur accumulates in foliage as
$SO_4^{2-}$  (Turner and Lambert, 1980).   The organically bound sulfur
occurs in plants in a molar ratio with nitrogen of 0.030, and
sulfur utilization is dependent on nitrogen availability (Turner

and Lambert, 1980; Johnson et al., 1982).  Plants receiving increased nitrogen inputs in rainfall are most likely accumulating
greater than the normal amounts of sulfur.  Still, excess sulfur,
obtained either as $SO_2$ or $SO_4^{2-}$ through the dry deposition processes described earlier, may be leached quite readily, both from
internal pools and the exterior surface of foliage (Lovett and
Lindberg, 1984; Johnson, 1984).


METHODS


## Site Description

     Precipitation and throughfall samples were collected in the
Mill Run watershed near Front Royal, Va., at latitude 38°52'30"N
and longitude 78°22'00"W (Fig. 1).  The Mill Run watershed lies in
a southwest to northeast trending, 365-ha stream valley between
Massanutten Mountain to the southeast and Little Crease Mountain to
the northwest.  Elevations in the watershed range from 353 to
690 m; the sample collection sites are at about 371 m.  The adjacent valley floors are at about 244 m on the northwest (Passage
Creek) and 183 m on the southeast (South Fork Shenandoah River).
     Predominant winds are out of the southwest and north, a pattern
that reflects the movement of weather systems across the State
(Crockett, 1971).  Mean annual precipitation for the 1951-80 period
at the nearby Woodstock climate station was 881.89 mm, making this
one of the driest parts of Virginia (Hayden, 1979).
     Woody vegetation in the watershed is dominated by mixed stands
of scarlet oak (*Quercus coccinea* Muench.), chestnut oak (*Q. prinus*
L.), white oak (*Q. alba* L.), and hickory (*Carya sp.*), but also includes black gum (*Nyssa sylvatica* Marsh.) and red maple (*Acer
rubrum* L.) as major subcanopy components.  White pine (*Pinus
strobus* L.), shortleaf pine (*P. echinata* Mill.), pitch pine (*P.
rigida* Mill.), and Virginia pine (*P. virginiana* Mill.) occur as
scattered individuals and isolated stands.  Tree ages established
by ring counts of increment cores indicate that the forest has not
been harvested since about 1880.


## Precipitation Sampling

     Collection of incident precipitation began in July 1982 in an
open field adjacent to the watershed. In May 1983 another precipitation collection site was established in a clearing in the watershed adjacent to the throughfall sites.  The original collector was
operated until December 1983 when it became necessary to dis-
continue the site due to excessive contamination from bird drop-
pings.  Samples were collected in an Aerochem Metric model 301
automatic wetfall-only sampler of the type used by the National
Atmospheric Deposition Program and the National Trends Network.
Precipitation quantity was measured with Belfort weight-recording
rain gages.

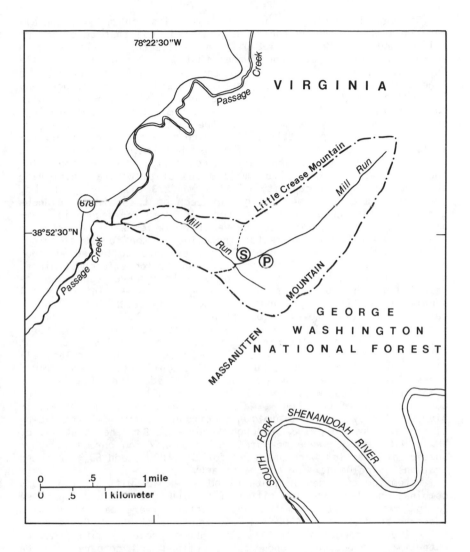

Figure 1.    Map showing the location of the Mill Run watershed.    The
            precipitation collection site is indicated as (P) and
            the stream sampling site as (S).

     Throughfall was collected in adjacent deciduous and coniferous
stands  in  paired,  automatic  wetfall-only  collectors.    Vegetation
directly  over  the  deciduous  throughfall  collectors  consisted  of
chestnut  oak  and  black  gum.    Over  the  coniferous  collectors  the
vegetation  was  white  pine.    Throughfall  volume  was  measured  with
graduated  plastic  rain  gages  attached  directly  to  the  wetfall-only
collector.    Sampling  of  throughfall  was  continuous  from  July  1982
through  October  1984  and  from  May  1985  through  November  1985.

Sequential samples of incident precipitation and throughfall were collected using an apparatus modified from that described by Liljestrand and Morgan (1981). The collector consisted of a 267-mm-diameter polyethylene funnel that was placed in a wetfall-only collector and connected to the sequential sampler with a length of Tygon tubing. The sequential sampler was constructed by connecting ten "Y" connectors in sequence with Tygon tubing with one arm of each "Y" leading to a 125-mL Nalgene polyethylene bottle. A second Tygon tube out of the bottle serves as a pressure relief line. As each bottle fills during an event, flow is diverted to the next bottle in line. The last bottle in line has a 1.0 L capacity and serves as an overflow reservoir. Each bottle holds the equivalent of about 2.5 mm (140 mL) of precipitation, giving the sampler a total capacity of 41.0 mm (2.3 L). During the May through November 1985 period, one of each of the paired deciduous and coniferous throughfall collectors was used for sequential sampling. Incident precipitation sequential samples were collected in a second wetfall-only collector installed in the forest clearing.

Precipitation and throughfall samples were collected as soon after an event as practical; however, due to logistic constraints, distance from the site, and lack of a field observer, some samples remained in the collectors for as long as 1 week. Initially all samples were filtered through 0.45-µm pore-size Nucleopore polycarbonate filters; however, beginning in May 1984, 0.2-µm filters were used. The 0.2-µm pore size is sufficient for cold sterilization, therefore further stabilizing the samples. Both an unacidified and an acidified sample were collected. Bottles, all glassware, and plasticware used during this study were triple rinsed with double-deionized water. In addition, bottles for acidified samples were soaked in 0.1 N HCl prior to rinsing. Concentrations of $Ca^{2+}$, $Mg^{2+}$, $Na^+$, and $K^+$ were determined on acidified samples by atomic absorption spectrophotometry with the exception that from July 1982 through November 1983 concentrations of $NH_4^+$, $Na^+$, and $K^+$ were measured using Dionex ion chromatography. Unacidified samples were analyzed for $Cl^-$, $NO_3^-$, and $SO_4^{2-}$ concentrations using Dionex ion chromatography.

Potentiometric measurements of pH were made with an Orion Ross combination electrode and either a Beckman pHI 21 or Orion Model 231 pH meter. The pH meter and electrode were calibrated with standard buffers at pH 4.00 and 7.00 and in a dilute $H_2SO_4$ solution ($H^+$ = 100 microequivalents$\cdot L^{-1}$) to insure proper liquid junction potential. Electrode slope was determined and compared to the theoretical Nernst value. Any electrode having a slope less than 95% of the theoretical Nernst value or which disagreed with the known pH of the dilute pH 4.00 standard by more than 0.05 units was reconditioned or replaced. The concentration of $H^+$ was then calculated as the antilog$_{10}$(-pH).

Rainfall measurements in millimeters were used to calculate rainfall volume in liters per square meter. Total deposition of each ion was then calculated by multiplying the concentration per liter by the sample volume in liters per square meter. Quarterly and annual volume-weighted means were calculated by dividing the total deposition of each ion by the total volume of precipitation or throughfall during the respective periods. Net deposition was calculated for each ion as the difference between the total throughfall deposition and the total rainfall deposition. The dataset of rainfall measurements also was used to classify each sample according to volume, duration, intensity, and total elapsed time between events.

## Streamflow

Water samples were collected from Mill Run approximately once a week and processed in the same manner as described for precipitation samples. Discharge was measured at 15-min intervals and used to calculate total discharge for the period that the water sample was assumed to represent. This period is defined as the time halfway back to the last sample date and halfway forward to the next sample date. The total discharge was then multiplied by the sample concentration of each ion to calculate total export. These values were then totaled for annual export estimates.

## Statistical Analyses

Net deposition of $SO_4{}^{2-}$ was regressed against the time (hours) between events and amount (cm) of rainfall using the SAS multiple regression procedure REG with the no-intercept option (SAS, 1982). The no-intercept equation, Net deposition = $b_1$ (dry period hours) + $b_2$ (rain amount cm), best represents the assumptions of the throughfall chemistry model; as the time between rainfall events and the amount of rainfall approach zero, so too will dry deposition and canopy leaching. Regression equations were determined for deciduous and coniferous throughfall for the growing season (May-October) and the dormant season (November-April) for the period July 1982 through November 1985. The resulting net regression coefficients provide an estimate of $SO_4{}^{2-}$ deposition in microequivalents $m^{-2}$ $hour^{-1}$ during the antecedent dry period and an estimate of canopy leaching in microequivalents $m^{-2}$ $cm^{-1}$ of rainfall.

## RESULTS AND DISCUSSION

## Seasonal Variability

Concentrations of the major ions in throughfall and rainfall vary during the year. This is especially true for those ions that are, in part, of anthropogenic origin (Table 2). Sulfate, $NO_3{}^{-}$, and $H^{+}$ all have two periods of high concentration. These peak periods are assumed to result from high fossil fuel consumption for heating in the winter quarter and cooling in the summer quarter. Relative to precipitation during the third quarter, there was an increase in $NO_3{}^{-}$ and $SO_4{}^{2-}$ concentrations in both deciduous and coniferous throughfall, and $H^{+}$ in coniferous throughfall. At the

Table 2.  Quarterly and Annual Volume-Weighted Mean Concentrations
μeq/L of Selected Ions in Coniferous (CT), Deciduous
(DT), and Precipitation (P) Throughfall during 1983 at
the Mill Run Watershed.

| Quarter | Type | pH | $H^+$ | $NO_3^-$ | $SO_4^{2-}$ |
|---------|------|------|-------|----------|-------------|
| 1 | CT | 4.03 | 93.2 | 64.3 | 113.4 |
|   | DT | 4.34 | 45.2 | 25.1 | 69.4 |
|   | P  | 4.45 | 35.3 | 18.5 | 20.6 |
| 2 | CT | 4.25 | 55.9 | 41.6 | 67.2 |
|   | DT | 4.49 | 32.5 | 24.1 | 61.9 |
|   | P  | 4.41 | 39.0 | 19.7 | 44.6 |
| 3 | CT | 3.87 | 135.8 | 81.5 | 190.1 |
|   | DT | 4.12 | 75.1 | 54.6 | 169.1 |
|   | P  | 3.99 | 103.2 | 34.0 | 102.7 |
| 4 | CT | 4.14 | 71.7 | 53.1 | 78.3 |
|   | DT | 4.63 | 23.4 | 12.7 | 55.6 |
|   | P  | 4.53 | 29.2 | 15.3 | 27.2 |
| Annual | CT | 4.10 | 78.9 | 54.5 | 97.5 |
|        | DT | 4.42 | 37.8 | 25.6 | 78.5 |
|        | P  | 4.33 | 47.1 | 20.7 | 45.4 |

same time there was a notable decrease of $H^+$ in deciduous throughfall.  This buffering effect of deciduous canopies has been
reported in most throughfall studies and has been attributed to $H^+$
exchange and Bronsted base reactions (Cronan and Reiners, 1983).
Failure of the deciduous stand to neutralize $H^+$ in the winter
.quarter points out the importance of leaves to this process and
indicates that the bark is not involved to any significant degree.
Still, deciduous bark is an effective substrate for dry deposition
and subsequent throughfall enrichment as indicated by the elevated
concentration of $SO_4^{2-}$ in deciduous throughfall during the winter
quarter.

On an annual basis, concentrations of $SO_4^{2-}$ were 114% greater
under the coniferous canopy and 72% greater under the deciduous
canopy relative to precipitation.  In addition, $SO_4^{2-}$ in coniferous throughfall was 25% more concentrated than deciduous throughfall.  Nitrate enrichment followed a similar pattern for coniferous
throughfall with 164% more than precipitation.  The rather small
$NO_3^-$ enrichment of deciduous throughfall (about 23%) reflects
uptake by the canopy.  Buffering of $H^+$ by the deciduous canopy was
important on an annual basis in spite of the abscence of this
process during the leaf-free period.

Deposition and Leaching

The net effect of the interaction of precipitation with the forest canopy can be seen best by examination of net deposition as calculated for the deciduous and coniferous stands (Table 3). Net deposition under the forest canopy reflects both washoff of compounds from foliar surfaces and leaching from internal pools. Based on the values in Table 3 the enrichment series for the deciduous canopy is $K^+ > Ca^{2+} > Mg^{2+} > Na^+$ and $SO_4^{2-} > NO_3^- > Cl^-$, and for the coniferous canopy $Ca^{2+} > H^+ > K^+ > Na^+ > Mg^{2+}$ and $SO_4^{2-} > NO_3^- > Cl^-$. Except for $Na^+$ the deciduous cation enrichment series is the same as that predicted based on hydrated ionic diameter. The anion enrichment series for both coniferous and deciduous throughfall is opposite that predicted but is reasonable when dry deposition and foliar concentration sources are considered. Coniferous throughfall enrichment of cations is the most anomalous series. The high $Ca^{2+}$ and $H^+$ values suggest greater dry deposition to the coniferous canopy of compounds such as $CaCO_3$, $H_2SO_4$, and $HNO_3$. Also the high $Na^+$ deposition suggests either a much more efficient scavenging of NaCl aerosols during the winter months or dry deposition of albite or some other $Na^+$ containing mineral. Dry deposition, therefore, is suggested as a source for much of the net deposition of several ions in throughfall.

Table 3.  Net Flux as Precipitation, and Deciduous (DT) and Coniferous (CT) Throughfall during 1983 to the Mill Run Watershed.  Units are Equivalents $ha^{-1}$.

[Negative values indicate net retention by the forest canopy]

| Ion | Rain | DT | CT |
|---|---|---|---|
| $H^+$ | 434.2 | -112.2 | 151.6 |
| $Ca^{2+}$ | 57.5 | 201.8 | 292.7 |
| $Mg^{2+}$ | 11.9 | 77.2 | 80.2 |
| $Na^+$ | 58.9 | 10.8 | 96.0 |
| $K^+$ | 27.5 | 266.6 | 149.6 |
| $NH_4^+$ | 145.1 | -61.2 | -8.3 |
| $Cl^-$ | 151.0 | 35.0 | 82.4 |
| $NO_3^-$ | 191.1 | 36.9 | 210.8 |
| $SO_4^{2-}$ | 418.3 | 230.1 | 314.4 |

Differentiation of the internal and external sources of $SO_4^{2-}$ in throughfall may be calculated from regression analyses, and researchers have approached this in different ways (Bache, 1977; Miller et al., 1976; Miller and Miller, 1980; Lakhani and Miller, 1980; Lovett and Lindberg, 1984). The method described earlier as used in this investigation is similar to that of Lovett and Lindberg (1984) with the exceptions that all types of samples were used, not just the single event samples, and the dataset extended over three years. Results of the multiple regression analyses are presented in Table 4 along with those of Lovett and Lindberg (1984). Although similar in magnitude and significance, the Mill Run results suggest a much greater dry deposition loading. At the Mill Run site the regression equations indicate that dry deposition to and leaching from the white pine canopy is about twice that of the deciduous canopy. Interestingly, the washoff component does not differ greatly between the growing season and the dormant season for either the pines or the hardwoods. Although this is not too surprising for the pines, which retain their foliage in the dormant season, it is surprising for the deciduous canopy. However, the deciduous tree bark may be as efficient a dry deposition substrate during the dormant season as the leaves during the growing season. Also, Lovett and Lindberg (1984) report similar results for white oak at Oak Ridge (Table 4). The internal leaching estimates reflect the result of the dormant versus the growing season comparison. As would be expected, this component is reduced considerably for both canopy types during the dormant season.

As another means of estimating the external washoff and internal leaching components, throughfall collected in sequential increments is examined. Figure 2 shows the net deposition of ions per square meter for discrete intervals (about 2.5 mm increments) in a coniferous and deciduous stand. The majority of washoff of dry deposited compounds occurs in the first three increments with the remainder attributable to leaching from internal pools. For $SO_4^{2-}$ the external washoff component may be calculated as the sum of the first three values and internal leaching as the sum of the remaining values. Obviously, some fraction of the first three increments also is internal leachate. To correct for this over-estimation the mean of the three throughfall increments assumed to represent internal leaching is subtracted from each throughfall increment assumed to represent external washoff. The total amount subtracted from external washoff is then added to the internal leachate value. Table 5 presents the external washoff and internal leachate components calculated from (1) the uncorrected sums (US) of the sequential throughfall samples, (2) the corrected sums (CS), and (3) the multiple regression estimates (REG). Both the CS and REG methods are in close agreement as to the relative proportions of external and internal souces of net $SO_4^{2-}$ enrichment of throughfall. The regression model has underestimated both components; however, the net regression coefficients are estimates of mean dry deposition and canopy leaching rates for the paired sites over four growing seasons and may be sensitive to fluctuations in weather and pollutant concentrations. Bache (1977) has provided evidence that dry deposition of $SO_4^{2-}$ to the canopy is controlled by the atmospheric concentration of $SO_2$. The effect on

Table 4.    Growing and Dormant Season Net $SO_4{}^{2-}$ Deposition Regression Coefficients for White Pine (WP) and Chestnut Oak/ Black Gum (CO/BG) Throughfall at the Mill Run Watershed and Chestnut Oak (CO) and White Oak (WO) Throughfall at the Oak Ridge Site.   Regressions are of the Form:   Net $SO_4{}^{2-} = b_1(TIME) + b_2(CM)$.   The Standard Error, $R^2$, and Significance Level also are Given.   Units are Microequivalents $m^{-2}$ $h^{-1}$ for the $b_1$ Coefficient and Microequivalents $m^{-2}$ $cm^{-1}$ of Rainfall for the $b_2$ Coefficient.

[NS, not significant; *, P<0.05; **, P<0.01; ***, P<0.001.   Oak Ridge Data are from Lovett and Lindberg (1984)]

| Site | $b_1$(SE) | $b_2$(SE) | $R^2$ | N |
|------|-----------|-----------|-------|---|
| Growing season | | | | |
| WP | 3.52 (0.52)*** | 210 (38)*** | 0.76 | 105 |
| CO/BG | 1.63 (0.46)*** | 158 (34)*** | 0.62 | 98 |
| CO | 4.37 (1.16)** | 485 (126)** | 0.80 | 21 |
| WO | 2.40 (2.64)NS | 1,300 (24)** | 0.86 | 12 |
| Dormant season | | | | |
| WP | 3.47 (0.71)*** | 92 (41)* | 0.68 | 38 |
| CO/BG | 1.87 (0.70)* | 58 (42)NS | 0.41 | 37 |
| CO | 1.13 (0.50)* | 66 (28)* | 0.64 | 18 |
| WO | 2.46 (4.43)NS | 672 (22)* | 0.63 | 13 |

throughfall of within-season changes in deposition is evident from the volume-weighted mean deposition data in Table 2.   Therefore, annual and seasonal variability in atmospheric $SO_2$ and $SO_4{}^{2-}$ concentrations also could contribute to this underestimation, particularly in late summer when these samples were collected.

The annual estimates of external washoff and internal leaching (Table 5) indicate that dry deposition to the canopy is a major source of the net throughfall deposition for $SO_4{}^{2-}$.   About 64% of the coniferous and 55% of the deciduous throughfall deposition is attributable to external washoff.   This compares to about 38% for the chestnut oak site at Oak Ridge where uptake and subsequent leaching of $SO_2$ seems to dominate the dry deposition process.

When compared to actual deposition values (Table 6) the regression values are within the experimental and analytical error for this investigation.   It also is apparent from the input-output comparisons that precipitation alone can account for only about half of the $SO_4{}^{2-}$ leaving the Mill Run watershed during 1983.   Even when the dry deposition flux as throughfall is added, only about 77% can be accounted for.   However, stemflow may account for another 10 to 15% and the remainder may be attributed to experimental and

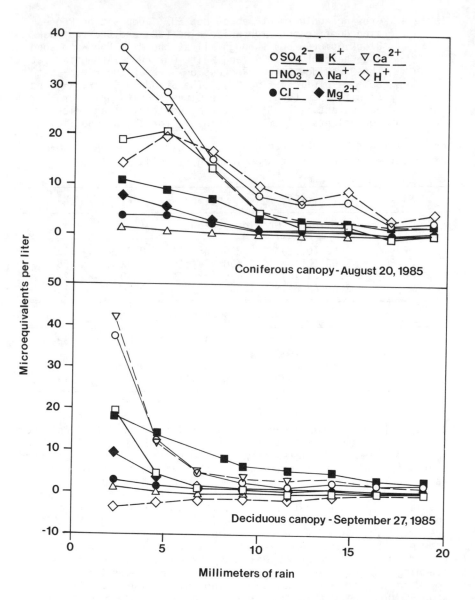

Figure 2.    Changes in net deposition under a coniferous (2a) and
deciduous (2b) canopy.  Units are microequivalents m⁻².
The event in 2a occurred on August 18, 1985, during a
2.16-cm rainfall that followed a 207-h dry period, and
was sampled on August 20, 1985.  The deciduous throughfall  event  occurred  during  a  1.95-cm  rainfall  on
September 26-27, 1985, and followed a 414-h dry period.

Table 5.   Estimates of External Washoff and Internal Leaching
Sources of $SO_4^{2-}$ in Net Throughfall. Units are Microequivalents $m^{-2}$ and Values in Parentheses are the Percent
of Total Deposition Accounted for by Each Component. The
Coniferous Throughfall Event Occurred during a 2.16-cm
rainfall on August 18, 1985, Following a 207-h Dry
Period. The Deciduous Throughfall Event Occurred during
a 1.95-cm Rainfall on September 26-27, 1985, Following a
414-h Dry Period. Sulfate Source Estimates are Based
on Uncorrected Sums of Sequential Throughfall Chemistry
(US), Corrected Sums (CS), and Regression Equations
(REG). Methods are Described in Detail in the Text.

| Method | External washoff | Internal leaching | Total |
|---|---|---|---|
| | Coniferous throughfall | | |
| US | 1,445.02 (76.2) | 452.00 (23.8) | 1,897.02 |
| CS | 1,173.82 (61.9) | 723.20 (38.1) | 1,897.02 |
| REG | 728.64 (61.6) | 453.92 (38.4) | 1,182.56 |
| ANNUAL | 28,268.20 (63.7) | 16,132.60 (36.3) | 44,400.80 |
| | Deciduous throughfall | | |
| US | 983.42 (83.0) | 201.32 (17.0) | 1,184.74 |
| CS | 862.63 (72.8) | 322.11 (27.2) | 1,184.74 |
| REG | 674.82 (68.7) | 307.81 (31.3) | 982.63 |
| ANNUAL | 14,114.10 (54.9) | 11,607.20 (45.1) | 25,721.30 |

Table 6.   Sulfate Input-Output Budget For the Mill Run Watershed
during 1983. Inputs are Based on Precipitation (P),
Deciduous Throughfall (DT), Coniferous Throughfall (CT),
and an Areally Weighted Throughfall (T) Estimate Based
on the Relative Proportion of Coniferous (0.33) and
Deciduous (0.66) Stands. Regression Equation Estimates
of the Throughfall Inputs are also Given (DTR, CTR, TR).
All Units are kg $ha^{-1}$ $yr^{-1}$.

| Site | Input | Output | Input/Output |
|---|---|---|---|
| P | 22.5 | 47.6 | 0.47 |
| DT | 35.8 | 47.6 | 0.75 |
| CT | 40.1 | 47.6 | 0.84 |
| T | 36.8 | 47.6 | 0.77 |
| DTR | 34.9 | 47.6 | 0.73 |
| CTR | 43.8 | 47.6 | 0.92 |
| TR | 37.8 | 47.6 | 0.79 |

analytical error.   Close input-output balances have been reported
at several other sites (Likens et al., 1977; Rapp, 1973; Switzer
and Nelson, 1972; Turner et al., 1980) where $SO_4^{2-}$ is not retained
in the watershed.   Clearly, studies where budgets are based only on
precipitation may grossly underestimate total inputs of some ions.


## Implications for Forested Watersheds

The input-output budgets for the Mill Run watershed (Table 6)
pose several implications for the chemical quality of water in the
hydrologic cycle.   As pointed out earlier, dry deposition to the
coniferous canopy is greater than to the deciduous canopy; therefore any negative impacts of $SO_4^{2-}$ will be realized to a greater
degree as the percent of coniferous vegetation increases in watersheds.   The importance of a mobile anion such as $SO_4^{2-}$ in acidification and leaching of cations from small watersheds is dependent
on the buffering capacity of the soils and bedrock and the hydrologic flowpaths.   When $SO_4^{2-}$ from throughfall enters the soil as an
acid or as a salt, the pH and alkalinity of the soil solution are
depressed.   The reaction to the neutral salt is due to the "salt
effect" where $H^+$ is displaced from exchange sites as a result of
the increased solution ionic strength (Reuss and Johnson, 1985).
If the residence time of the soil solution is short due to shallow
flowpaths or if it then enters a stream or lake without contacting
an easily weathered mineral, the lost alkalinity will not be
replaced and the stream may become acidified.   Even in situations
where soil $CO_2$ may be in sufficient concentration to generate
$H_2CO_3$, positive alkalinity will not be produced unless base cations
are generated through ion exchange or weathering reactions.   The
$H_2CO_3$ will then dissassociate and be lost as water and $CO_2$ when the
solution equilibrates with the atmosphere (Reuss and Johnson,
1985).   Under these conditions $SO_4^{2-}$ will become the predominant
anion, and $HCO_3^-$ alkalinity may become positive only when longer
residence-time ground water from deep flowpaths contributes a major
portion of streamflow.   An example of this scenario is shown in
Figure 3 for the Mill Run watershed.   During those months when
shallow flowpaths contribute most of the streamflow, $SO_4^{2-}$ is the
predominant anion and the stream is acidic.   However, during the
late summer and fall, deeper flowpath waters sustain streamflow,
and alkalinity becomes positive and stream pH recovers temporarily.
Acidification of natural watersheds is an issue of widespread
importance with respect to the chemical quality of water and the
hydrologic cycle, and forest canopy interactions play an important
role.   Forest canopies have a major effect on the the concentration
of dissolved solutes in precipitation.   The most important chemical
change is an increase in the solution ionic strength and concentration of $SO_4^{2-}$.   These two factors contribute to reduction of $HCO_3^-$
alkalinity, increases in stream acidity, and leaching of base
cations from the soil.   Because of greater net deposition to coni-
fers, watersheds with predominantly coniferous vegetation may be
predisposed to acidification to a greater degree than deciduous
forests.

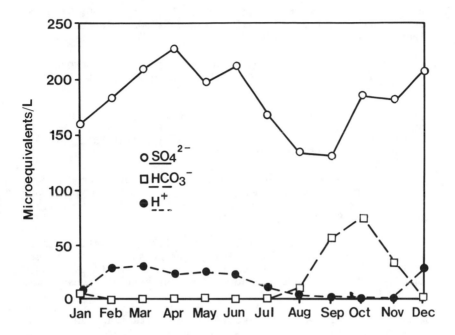

Figure 3.  Average monthly concentrations of $SO_4^{2-}$, $HCO_3^-$, and $H^+$ in Mill Run during 1983.

## DISCLAIMER

The use of trade or product names in this chapter is for iden-
tification purposes only and does not constitute endorsement by
the U.S. Geological Survey.

## REFERENCES

Bache, D. H. 1977.  Sulphur dioxide uptake and the leaching of sul-
phates from a pine forest. J. Appl. Ecol. 14:881-895.

Cole, D. W., and M. Rapp. 1981.  Elemental cycling in forest ecosystems, p. 341-409.  *In* D. E. Reichle, ed., Dynamic properties
of forest ecosystems.  Cambridge University Press, London.

Crockett, C. W. 1971.  Climate of Virginia:  Climatography of the
United States, No. 60-440.  U. S. Department of Commerce,
Washington, D.C. 25 p.

Cronan, C. S., and W. A. Reiners. 1983. Canopy processing of acidic precipitation by coniferous and hardwood forests in New England. Oecologia. 59:216-223.

Eaton, J. S., G. E. Likens, and F. H. Bormann. 1973. Throughfall and stemflow chemistry in a northern hardwood forest. J. Ecol. 61:495-508.

Fowler, D. 1980a. Removal of sulphur and nitrogen compounds from the atmosphere in rain and dry deposition, p. 22-32. In D. Drablos and A. Tollan, eds., Ecological impact of acid precipitation. Sandefjord, Norway.

_____. 1980b. Wet and dry depostion of sulphur and nitrogen compounds from the atmosphere, p. 9-27. In T. C. Hutchinson and M. Havas, eds., Effects of acid precipitation on terrestrial ecosystems. Plenum Press, New York.

Grennfelt, P., C. Bengtson, and L. Skarby. 1980. An estimation of the atmospheric input of acidifying substances to a forest eco-system, p. 29-40. In T. C. Hutchinson and M. Havas, eds., Effects of acid precipitation on terrestrial ecosystems. Plenum Press, New York.

Hayden, B. P. 1979. Atlas of Virginia precipitation. University of Virginia Press, Charlottesville. 165 p.

Johnson, D. W. 1984. Sulfur cycling in forests. Biogeochemistry. 1:29-43.

Johnson, D. W., G. S. Henderson, D. D. Huff, S. E. Lindberg, D. D. Richter, D. S. Shriner, D. E. Todd, and J. Turner. 1982. Cycling of organic and inorganic sulphur in a chestnut oak forest. Oecologia. 54:141-148.

Kielland, J. 1937. Individual activity coefficients of ions in aqueous solutions. J. Am. Chem. Soc. 59:1675-1678.

Lakhani, K. H., and H. G. Miller. 1980. Assessing the contribution of crown leaching to the element content of rainwater beneath trees, p. 161-176. In T. C. Hutchinson and M. Havas, eds., Effects of acid precipitation on terrestrial ecosystems. Plenum Press, New York.

Likens, G. E., F. H. Bormann, R. S. Pierce, J. S. Eaton, and N. M. Johnson. 1977. Biogeochemistry of a forested ecosystem. Springer-Verlag, New York.

Liljestrand, H. M., and J. J. Morgan. 1981. Spatial variations of acid precipitation in southern California. Environ. Sci. Tech. 15:333-338.

Lindberg, S. E., G. M. Lovett, D. D. Richter, and D. W. Johnson. 1986. Atmospheric deposition and canopy interactions of major ions in a forest. Science. 231:141-145.

Lovett, G. S., and S. E. Lindberg. 1984. Dry deposition and canopy exchange in a mixed oak forest as determined by analysis of throughfall. J. Appl. Ecol. 21:1013-1027.

McColl, J. G., and D. S. Bush. 1978. Precipitation and throughfall chemistry in the San Francisco Bay area. J. Environ. Qual. 7:352-357.

McFarlane, J. C., and W. L. Berry. 1974. Cation penetration through isolated leaf cuticles. Plant Physiol. 53:723-727.

Meyer, B. S., D. B. Anderson, R. H. Bohning, and D. G. Fatianne. 1983. Introduction to plant physiology. D. Van Nostrand Co., New York.

Miller, H. G., J. M. Cooper, and J. D. Miller. 1976. Effect of nitrogen supply on nutrients in litter fall and crown leaching in a stand of corsican pine. J. Appl. Ecol. 13:233-248.

Miller, H. G., and J. D. Miller. 1980. Collection and retention of atmospheric pollutants by vegetation, p. 33-40. *In* D. Drablos and A. Tollan, eds., Ecological impact of acid precipitation. Sandefjord, Norway.

Mollitor, A. V., and K. R. Berg. 1980. Effects of acid precipitation on forest soils, p. 3.1-3.80. *In* D. J. Raynall, ed., Actual and potential effects of acid precipitation on a forest ecosystem in the Adirondack Mountains. Final research report to New York State Energy Research and Development Authority. Rep. No. 80-82. NYSERDA, Albany.

Rapp, M. 1973. Le cycle biogeochemique du soufre dans une foret de *Quercus ilex* L. du sud de la France. Oecologia Plantarum. 8.325 334.

Reuss J. O., and D. W. Johnson. 1985. Effect of soil processes on the acidificaton of water by acid deposition. J. Environ. Qual. 14:26-31.

Statistical Analysis System. 1982. SAS user's guide: Statistics. SAS Institute, Raleigh, N.C. 584 p.

Schonherr, J. 1976a. Water permeability of isolated cuticular membranes: The effect of pH and cations on diffusion, hydrodynamic permeability, and size of polar pores in the cutin matrix. Planta. 128:113-126.

_____. 1976b. Water permeability of isolated cuticular membranes: The effect of cuticular waxes on diffusion of water. Planta. 131:159-164.

Schonherr, J., and R. Huber. 1977. Plant cuticles are polyelectro-
lytes with isoelectric points around three. Plant Physiol.
59:145-150.

Schonherr, J., and H. W. Schmidt. 1979. Water permeability of
plant cuticles. Planta. 144:391-400.

Switzer, G. L. and L. E. Nelson. 1972. Nutrient accumulation and
cycling in loblolly pine (*Pinus taeda* L.) plantation ecosystems: The first twenty years. Soil Sci. Soc. Am. Proc.
36:143-147.

Tukey, H. B., Jr. 1970. The leaching of substances from plants.
Ann. Rev. Plant Physiol. 21:305-324.

_____. 1980. Some effects of rain and mist on plants, with
implications for acid precipitation, p. 141-150. *In* T. C.
Hutchinson and M. Havas, eds., Effects of acid precipitation
on terrestrial ecosystems. Plenum Press, New York.

Turner, J., D. W. Johnson, and M. J. Lambert. 1980. Sulphur
cycling in a douglas-fir forest and its modification by
nitrogen applications. Oecologia Plantarum. 15:27-35.

Turner, J., and M. J. Lambert. 1980. Sulfur nutrition of forests,
p. 321-332. *In* D. S. Shriner, C. R. Richmond, and S. E.
Lindberg, eds., Atmospheric sulfur deposition: Environmental
impact and health effects. Ann Arbor Science, Ann Arbor.

# CHAPTER 2

# RIME ICE COMPOSITION AT THE ELK MOUNTAIN OBSERVATORY-ATMOSPHERIC PROCESSES, ACID AND NUTRIENT DEPOSTION RATES

J. Snider and G. Vali, University of Wyoming, Laramie, Wyoming

## ABSTRACT

The impaction of cloud water on mountain surfaces produces rime ice which is a significant source of water and nutrients, as well as acidity, in the subalpine and alpine ecosystems of the Rocky Mountains. Our observations of rime and aerosol composition with measurements of $SO_2$ at the Elk Mountain Observatory in southeastern Wyoming suggest a variety of processes which dictate cloud water chemistry and deposition rates. The nucleation scavenging of submicron acid sulfate aerosol particles was found to be an important source of sulfate and perhaps hydrogen ion and ammonium in the cloud water. Aqueous oxidation of $SO_2$ by $H_2O_2$ and by $O_3$ also can produce significant quantities of acid sulfate. A cloud model was used to investigate the effects of soluble gas dissolution on droplet composition; nitric acid and other cloud ingested components appear to be efficiently scavenged by this mechanism. The deposition of hydrogen ion and the strong acid anions by surface riming was found to be comparable to bulk wintertime precipitation inputs.

## INTRODUCTION

Mountain-top cloud droplet deposition is a vector of acid and nutrient deposition which is thought to be of considerable ecological importance. Immersion of mountain tops in wintertime clouds results in the formation of rime ice deposits via the impaction of supercooled cloud droplets on trees and other surface obstacles. These deposits have been shown to be significant both hydrologically and chemically at a mountain-top observatory in northern Colorado (Borys et al., 1983). Similar rime producing

*Chemical Quality of Water and the Hydrologic Cycle*, Robert C. Averett and Diane M. McKnight (Eds.) © 1987 Lewis Publishers, Inc., Chelsea, Michigan. Printed in the United States of America.

clouds have been studied at the University of Wyoming Elk Mountain Observatory in southeastern Wyoming (Politovich and Vali, 1983), although little work has been done at this site to quantify the rime composition or the rime deposition rates.

The Elk Mountain Observatory (EMO) is located in a region of Wyoming which is conducive to the formation of rime ice producing mountain-top clouds (Marwitz and Dawson, 1984). Other mountain ranges in Wyoming are exposed to similar clouds at frequencies comparable to Elk Mountain (Auer, 1969); however, investigations of the general importance of this phenomenon are hampered by a lack of power and accessibility during the winter months. The research facility on Elk Mountain is an ideal site to study the atmospheric processes which effect cloud and precipitation chemistry, as well as the factors (both atmospheric and terrestrial) which control the magnitudes of the various acid and nutrient deposition pathways.

It is our intention here to report measurements of the composition of cloud water collected at Elk Mountain and to attempt to identify scavenging mechanisms responsible for the observed chemistry. The utility of this approach stems from its ability to specify scavenging processes within these clouds and to therefore predict how the composition of the cloud water and the rime deposits might be altered by increases in pollutant emissions upwind of these sensitive mountain-top environments.

## METHODS

### Site Description

The Elk Mountain Observatory is located at 3.29 km MSL (all heights are reported relative to mean sea level) between the summit of Elk Mountain, to the east, and the Schaefer Ridge 0.5 km to the west. The Schaefer ridge shelters the observatory from strong and persistant westerly winds which are usually in excess of 20 m/s. Spruce and fir are the dominant trees at this elevation; however, the stands are sparce or nonexistent on the windward slopes.

Much of the work reported here was conducted at EMO during "cap" cloud events--so defined because the horizontal and vertical scales of the cloud are comparable to the dimensions of the mountain. Air parcel trajectories and in-cloud residence times are, to a first approximation, defined by the cloud extent and wind speed. Also, the observed cloud properties (i.e., cloud liquid water content, droplet size spectra, and cloud water composition) are typically stationary for several hours during cap cloud episodes. These characteristics make it possible to study time dependent processes such as contaminant scavenging within the cloud.

## Sample Collection

Rime samples were collected with a mechanical sampler consisting of a series of vertical, stainless steel bands (0.05 cm in width) rotated about a common vertical axis at 1400 rpm. These bands impact supercooled cloud droplets at tangential velocities ranging between 10 and 20 m/s. Rime is accumulated at a rate of 0.2 g/min for a liquid water content of 0.1 $g/m^3$. Three to six gram samples are collected for analysis; hence, the required collection periods range between 15 and 30 min in inverse proportion to the cloud liquid water content. The theoretical collection efficiency for the innermost collection band, spun at a rate of 10 m/s, is 90% for droplets greater than 2 µm in radius. For a typical Elk Mountain droplet size distribution, greater than 90% of the liquid water is at radii larger than this 2 µm cutoff (Politovich and Vali, 1983). Furthermore, the collection efficiency increases with droplet size, hence all but the smallest cloud particles are collected.

The frozen deposit is removed by electrically heating the support, causing the sample to fall off while still frozen. The samples are maintained frozen until immediately prior to chemical analysis. Laboratory tests indicate that neither the collection ribbon nor the sample storage procedures introduce detectable (i.e., >1 µM) contamination to the sample.

## Sulfur Dioxide Measurement

A flame photometric detector (FPD) was used to monitor the gas phase concentration of sulfur dioxide. The monitor was calibrated with known concentrations of $SO_2$ from a permeation oven calibrator. The fuel used in the FPD was hydrogen spiked with 75 ppbv $SF_6$ (gas concentrations are reported as parts per billion by volume). The analyzer has a $SO_2$ detection limit of 0.1 ppbv and a response time of 30 s (95% full scale) to a step input of 0.5 ppbv $SO_2$. Sulfur-free air also was generated by the calibrator, which removes both particulate and gas phase sulfur compounds from the sample air. The diffusion denuder consisted of a 0.3 m section of 5-mm-ID aluminum tubing coated with a slurry of lead peroxide in glycerol. The denuder selectively removes sulfur dioxide from the sample stream. Both the FPD and the calibrator were connected to a 73-mm-ID aluminum sample manifold. The flow rate through the manifold was 1.73 $m^3$/min. The manifold inlet was located on the observatory sample platform and adjacent to the rime collector. This platform is elevated and located upwind of the observatory to minimize the contamination of the rime samples and the sample air supply.

## Accessory Measurements

Standard meteorological instruments were located on the sample platform and on the Schaefer Ridge. Data from these sensors (i.e., anemometers, thermistors, etc.) and the sulfur monitor were digitally recorded by the observatory data system. The cloud liquid water content was measured using a Rotorod (Metronics, Inc., Palo Alto, Ca.). This technique is described by Rogers et al. (1983). Liquid water content measurements made with the bulk cloud water collector were found to be in good agreement with the Rotorod values. The bulk collector was therefore used for liquid water measurement when the Rotorod was not employed.

## Rime Analysis

Rime samples were analyzed for sulfate, nitrate, and chloride by ion chromatography. A peak corresponding to flouride was seen in many of the cloud water chromatograms; however, flouride was not quantified because of interferences from a negative water peak and the organic acid anions. Ammonium was analyzed in a continuous flow analysis system using the phenolhypochlorite method. The sample pH was determined with an electrode calibrated in low ionic strength sulfuric acid solutions, and the base cations were analyzed by flame atomic absorption. The precision of these methods was determined by submitting synthetic cloud water samples for analysis. The average error associated with the determination of ammonium and the base cations was 6.5 and 5.0% respectively. The error associated with the IC procedure was 7.5% for sulfate and nitrate, and 10% for chloride.

## RESULTS AND DISCUSSION

## Rime Chemistry

Rime composition is effected both by the amount of material scavenged by the cloud droplets and by dilution, as governed by the liquid water content of the cloud. The effect of dilution may be examined by multiplying the measured aqueous concentrations (expressed in micromoles per liter of melted rime or μM) by the measured liquid water content of the cloud. The resulting quantity, expressed in $nmol/m^3$, represents the amount of material dissolved in cloud water per unit volume of air. This concentration unit is useful because it can be directly compared with particulate and gas concentration measurements, and because the product of this concentration with a cloud droplet deposition velocity represents a deposition rate.

A summary of the rime composition data from both EMO and the Storm Peak Laboratory (SPL) located near Steamboat Springs, Colo., is presented in Table 1. The SPL cloud water samples were collected by R. Borys using both a passive, wind driven collector and a Rotorod. The Rotorod was used exclusively at SPL for the liquid water content measurement and for the collection of cloud water for pH analysis. Several of the components (i.e., $SO_4^=$, $NO_3^-$, $H^+$, the base cations) and the cloud water content exhibited significant differences between the two sites; the SPL concentrations were consistently higher. The differences in chemical composition can be attributed to a number of factors including the pollutant sources upwind of SPL (i.e., the town and resort of Steamboat Springs as well as the coal fired generating station near Craig, Colo.). Differences in cloud microstructure, as indicted by the significantly higher liquid water content at SPL, may also enhance the scavenging efficiency of these clouds and therefore the observed component concentrations. It should be noted that the passive collection device used at SPL is considerably less efficient in comparison to the mechanical collector used at EMO; hence, the actual compositional differences between the two sites are probably larger than indicated in Table 1.

The observed rime composition at Elk Mountain is thought to be brought about by three scavenging mechanisms. These mechanisms are discussed below using both observational data and theory.

Table 1. A Statistical Comparison of the Cloud Water Data Collected at the Elk Mountain Observatory (EMO) and at the Storm Peak Laboratory (SPL). The Rime Ice Composition Units are Nanomoles per Cubic Meter of Air ($nmol/m^3$), and the Units of Cloud Liquid Water Content (LWC) are Grams per Cubic Meter of Air ($g/m^3$).

[n = number of samples; x = mean value, $\sigma$ = standard deviation]

| | $H^+$ | $Ca^{++}$ | $Mg^{++}$ | $Na^+$ | $NH_4^+$ | $SO_4^=$ | $NO_3^-$ | $Cl^-$ | LWC |
|---|---|---|---|---|---|---|---|---|---|
| | | | Rime ice composition EMO 1985-1986 | | | | | | |
| n | 38 | 15 | 15 | 15 | 4 | 29 | 29 | 7 | 40 |
| x | 4.7 | 0.88 | 0.17 | 0.63 | 14.3 | 2.6 | 3.2 | 0.6 | 0.12 |
| $\sigma$ | 3.0 | 0.72 | 0.19 | 0.69 | 8.7 | 2.7 | 2.6 | 0.6 | 0.08 |
| | | | Rime ice composition SPL 1981-1985 | | | | | | |
| n | 141 | 8 | 8 | 12 | 12 | 20 | 20 | 20 | 141 |
| x | 20.2 | 6.0 | 1.5 | 0.98 | 8.3 | 7.9 | 4.7 | 2.1 | 0.20 |
| $\sigma$ | 24.7 | 5.3 | 1.3 | 0.90 | 8.3 | 6.8 | 3.6 | 2.9 | 0.12 |
| $T^1$ | 3.84 | 3.51 | 3.06 | 1.10 | 1.16 | 3.71 | 1.66 | 1.31 | 3.95 |
| $P^2$ | >0.99 | >0.99 | >0.99 | >0.84 | >0.85 | >0.99 | >0.95 | >0.90 | >0.99 |

[1]Computed value of the two-sample t statistic.
[2]Probability of a difference between the SPL and the EMO mean values.

Nucleation Scavenging

Particulates capable of initiating droplet formation in atmospheric clouds are known as cloud condensation nuclei or CCN (Pruppacher and Klett, 1980). Cloud condensation nuclei are typically larger than 0.01 μm in size and composed of soluble compounds which deliquesce into aqueous solution droplets at relative humidities between 50 and 100%. The largest of these solution droplets grow via vapor diffusion in the supersaturated environment near cloud base to form larger cloud droplets (vapor supersaturation is defined as the relative humidity in excess of 100%). The transition from solution droplet to cloud droplet is defined as CCN activation; this process also is known as nucleation scavenging. Two different observations have been made at EMO which can be used to calculate the quantity and composition of the aerosol scavenged in this manner. Bigg (1979) inferred the composition of the majority of the fine aerosol (i.e., particulate smaller than 1 μm in diameter) at the observatory to be ammonium sulfate or partially neutralized sulfuric acid. CCN measurements also have been made in the boundary layer upwind of Elk Mountain (Politovich and Vali, 1983). Using these measurements, the Köhler theory (Pruppacher and Klett, 1980), and assuming that particle sizes are distributed according to the power law $dN/d\log D = -A/D^3$ (where A is an integration constant defined by the cumulative concentration of aerosol activated at 0.1% supersaturation), we calculate that between 18.3 and 2.4 nmol/m$^3$ of sulfate is incorporated into the cloud water via nucleation scavenging (see Figures 1A and 1B). Considering that particles up to 3.2 μm in diameter were assumed to be composed entirely of soluble sulfate, the comparison between the observed cloud water composition and the calculation is about as good as can be expected. The calculation demonstrates that nucleation scavenging is an important source of sulfate in the Elk Mountain clouds. We also note that the calculation is rather insensitive to the cloud updraft velocity because the aerosol mass distribution (see Figure 1B) is dominated by large particles which activate to form cloud droplets at low water vapor supersaturations.

Gas Scavenging

The scavenging of gas phase constituents depends on several factors, the most important being the Henry's law solubility of the contaminant. Both nitric acid and ammonia are sufficiently soluble to be completely partitioned into the cloud water phase. However, the extent of gas partitioning can be limited by the finite amount of time required for interphase transport. Parcel transport times from the base of the Elk Mountain cap cloud to the observatory are long (500-1000 sec) in comparison to the time constant for phase equilibrium (50-200 sec). Hence, we expect these gases to be completely partitioned into the aqueous phase during rime sample collection and we expect the measured concentrations to reflect the concentrations of the gases in the subcloud air.

Wintertime gas phase measurements of $HNO_3$ at Niwot Ridge, Colo., are consistent with the assumption that the $NO_3^-$ observed at EMO is due to the scavenging of $HNO_3$ within the cloud (Parrish et al., 1986). It is difficult to make similar comparisons between observed ammonium concentrations and gas phase $NH_3$ measurements. Cloud water dissolved ammonium can originate from both nucleation scavenging and via diffusion from the gas phase. Furthermore, there are no published $NH_3$ measurements, which we are aware of, at locations representative of the Elk Mountain site.

Although the pH of nearly all the EMO rime samples were observed to be below the bicarbonate/$CO_2$ endpoint, several of the samples contained apparent cation excesses, defined as the difference between the sum of the hydrogen ion and the base cation concentrations, and the strong acid anion concentrations. Previous cloud and precipitation studies have shown that organic acids, especially formic acid because of its low $pK_a$ and large Henry's law constant, can be a source of carboxylate alkalinity in samples collected in both polluted and pristine environments (Keene et al., 1983; Jacob et al., 1986). The model simulation shown in Figure 2A and 2B lends some credence to our suspicion that unanalyzed organic acid anions are the cause of the observed cation excess. The range of formic acid levels used as input for this simulation are consistent with gas phase measurements made during the winter months in an isolated part of Arizona (Dawson et al., 1980).

## $SO_2$ Oxidation

There is an increasing amount of observational evidence which suggests that $SO_2$ is rapidly oxidized to sulfuric acid within liquid water clouds. The reaction of dissolved SIV (sulfur in the +4 oxidation state) with (1) ozone and (2) hydrogen peroxide are thought to be the two dominate mechanisms in the Elk Mountian cloud (Snider and Vali, 1987). The levels of $SO_2$ observed within the Elk Mountain cloud are usually below 0.5 ppbv; the measurements shown in Figure 3 are representative of the collected data set and have been used to calculate the reaction rates seen in Table 2. Significant acid sulfate yields are anticipated in clouds with parcel residence times on the order of 1 h. We are conducting experiments to test this prediction by artificially elevating the levels of $SO_2$ within the Elk Mountain cloud.

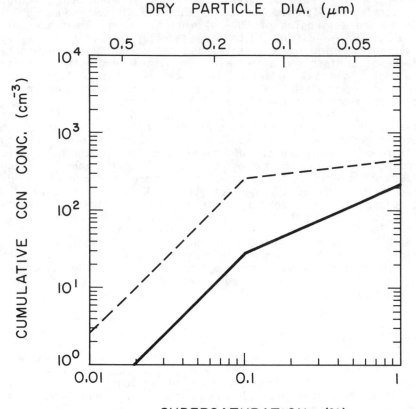

Figure 1A.    CCN concentration as a function of the supersaturation required for activation. The measurements made upwind of Elk Mountain are plotted according to the relation, $N = C \times S^k$, where N is the cumulative number of CCN activated at supersaturation S. Values for the upper curve were taken to be: $C = 451$ $cm^{-3}$ and $k = 0.24$. The lower curve is described by $C = 219$ $cm^{-3}$ and $k = 0.90$. These distributions represent range of the observations ($\pm 1\sigma$). The distributions at supersaturations less than 0.1% were generated by assuming an aerosol size distribution given by $dN/dlogD = -A/D^3$.

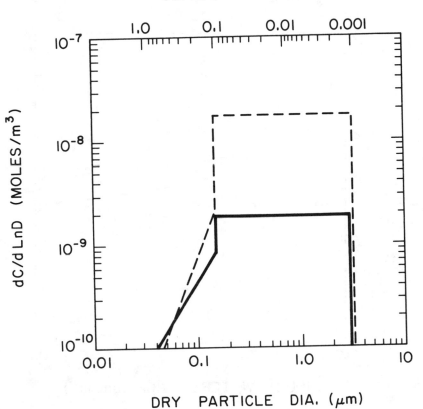

Figure 1B.   Concentration of scavenged CCN aerosol (moles per cubic meter of air per logarithmic size interval) corresponding to the measurements given in Figure 1A.

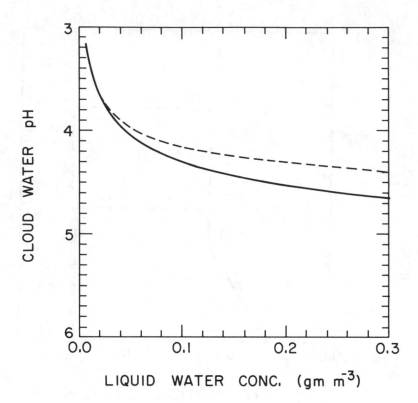

Figure 2A.  Equilibrium cloud water pH as a function of LWC calcu-
            lated using formulas presented in Brimblecombe and
            Dawson (1984) and the electroneutrality constraint.
            Gas inputs for the standard model (solid line) were
            taken to be:   0.1 ppbv $HCO_2H$, 0.05 ppbv $NH_3$, 0.3 ppbv
            $SO_2$, 0.1 ppbv $HNO_3$, and 340 ppmv $CO_2$.   The nucleation
            scavenged ammonium bisulfate concentration is assumed
            to be 2.6 $nmol/m^3$. The pecked line is for 0.5 ppbv
            $HCO_2H$.

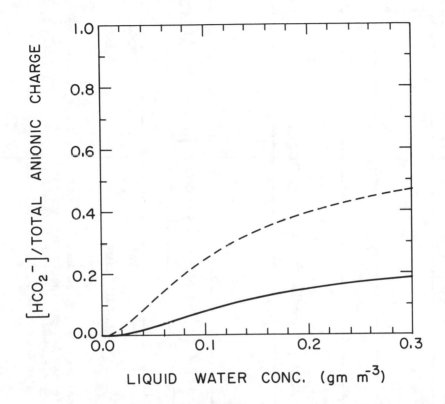

Figure 2B. Ratio of dissolved formate to the total anionic charge as a function of LWC. Other details as in Figure 2A.

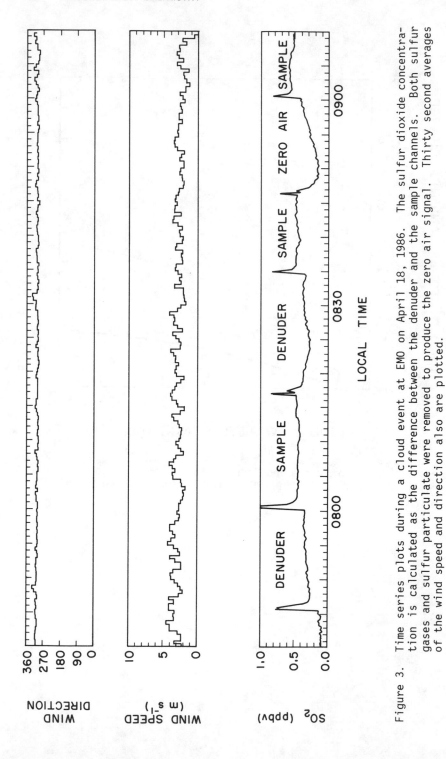

Figure 3.  Time series plots during a cloud event at EM0 on April 18, 1986.  The sulfur dioxide concentration is calculated as the difference between the denuder and the sample channels.  Both sulfur gases and sulfur particulate were removed to produce the zero air signal.  Thirty second averages of the wind speed and direction also are plotted.

Table 2.  Aqueous Concentrations and Effective Gas Phase Reaction
Rates of Ozone, Hydrogen Peroxide and $SO_2$.  Reaction
Rates Calculated Assuming First Order Kinetics.  Henry's
Law Equilibrium Assumed for all Species.

| Reagent | Aqueous conc. ($\mu$M) | Reaction rate (nmol/m$^3$/h) |
|---|---|---|
| $H_2O_2$[1] (+SIV) | 12.0 | 0.95 |
| $O_3$[2] (+SIV) | $9.5 \times 10^{-4}$ | 0.79 |
| SIV[3] $(+H_2O_2)$[4] | 1.1 | 0.95 |
| SIV $(+O_3)_5$ | 1.1 | 0.79 |

Conditions:
LWC = 0.1 g/m$^3$,     T = -10°C,     P = 700 mb,     pH = 4.5
$O_3$ = 50 ppbv,     $H_2O_2$ = 0.05 ppbv,     $SO_2$ = 0.3 ppbv

[1]$H_2O_2$ solubility data from Lind and Kok, 1986.
[2]$O_3$ solubility data from Chameides, 1984.
[3]SIV solubility data from Maahs, 1982
[4]Rate constant from Kunen et al., 1983; Calvert et al., 1985.
[5]Rate constant from Maahs, 1983.

## Chemical Deposition

Estimates of the nutrient and acid influx due to droplet
impaction can be inferred from the composition data presented in
Table 1 and models of cloud droplet interception.  Lovett et al.
(1982) report a cloud droplet deposition velocity of 0.2 m/s for
conditions quite similar to Elk Mountain (i.e., canopy structure,
modal droplet diameter, and wind velocity).  Instantaneous deposition rates were converted to integrated chemical inputs for the
months of November through May using the cloud immersion frequency
data collected by Auer (1969).  The chemical inputs due to precipitation and surface riming are compared in Table 3.  The snow
composition data used to make this comparison also was collected on
Elk Mountain; a winter season integrated precipitation amount of
875 mm was assumed (Rechard and Smith, 1972).  It appears from
these calculations that the deposition of H+ and the strong acid
anions via cloud interception is comparable to the input due to
bulk precipitation.  Moreover, the influx of base cations is an
order of magnitude larger via the latter mechanism.  This
difference is consistent with the fact that a particulate greater
than a few microns in size is efficiently scavenged by snowfall but
not by cloud droplets.
    The conclusions which can be made from this comparison are
tentative due to the paucity of the snow composition data (only
15 distinct samples were analyzed) and due to our reliance on model
generated deposition velocities.  We are developing methods to
directly measure rime deposition rates at the observatory.  Future

Table 3.   Winter Season (November-May) Deposition by Surface Riming
and Bulk Precipitation at the Elk Mountain Observatory.
A Cloud Immersion Frequency of one Day in Three (Auer,
1969), a Water Equivalent Snow Depth of 875 mm (Rechard
and Smith, 1972), and a Cloud Droplet Deposition Velocity
of 0.2 m/s (Lovett et al., 1982) are assumed.   The Snow
Composition Data was Collected at EMO during the Winters
of 1980 and 1981.

| Ion | Surface riming bulk | | | Precipitation | |
| | Aqueous conc. ($\mu$M) | Airborne conc. ($nmol/m^3$) | Deposition (mol/ha) | Aqueous conc. ($\mu$M) | Deposition (mol/ha) |
|---|---|---|---|---|---|
| $H^+$ | 31.3 | 4.7 | 48.9 | 8.0 | 70.0 |
| $Ca^{++}$ | 5.9 | 0.88 | 9.2 | 4.4 | 38.5 |
| $Mg^{++}$ | 1.1 | 0.17 | 1.8 | 0.8 | 7.0 |
| $Na^+$ | 4.2 | 0.63 | 6.6 | 2.1 | 18.4 |
| $SO_4^=$ | 17.3 | 2.6 | 27.0 | 6.9 | 60.4 |
| $NO_3^-$ | 21.3 | 3.2 | 33.3 | 4.8 | 42.0 |
| $Cl^-$ | 4.0 | 0.6 | 6.2 | 1.6 | 14.0 |

studies will be conducted to quantitate both the bulk deposition
and the rime ice inputs with sufficient temporal and spatial
resolution to elucidate the controlling chemical, meteorological,
and terrestrial factors.   This work will then be used to assess
deposition rates in remote alpine areas where comparable studies
may be difficult or impossible.


CONCLUSION

It is generally thought that acid and nutrient deposition
originates from a combination of dry and wet processes. The results
presented here and elsewhere (Borys et al., 1983) demonstrate that
alpine and subalpine environments in the Rocky Mountains are
exposed to an additional source of acidity and nutrients which
significantly augments the integrated wintertime deposit.   We also
have specified some of the physical and chemical processes which
occur within the cloud to produce the observed rime ice chemistry
and which regulate the influxes.   Both particulate and gas phase
precursor concentrations were found to be important determinants of
the cloud water and the rime ice deposit composition. Experiments
are in progress to better quantitate the reaction kinetics of $SO_2$
within the Elk Mountain cloud.   It is hoped that the effects of $SO_2$
emission increases on acid deposition at this and other mountain
top sites can be anticipated from this work.

ACKNOWLEDGMENTS

We are indebted to Randolf Borys who provided us with the Storm Peak Laboratory rime data.

DISCLAIMER

The use of trade or product names in this chapter is for iden-tification purposes only and does not constitute endorsement by the U.S. Geological Survey.

REFERENCES

Auer, A. 1969. Wyoming orographic cloud census, 1968-1969. Report to Wyoming Natural Resources Board, State Engineer's Office, Contract No. 6264.

Bigg, E. K. 1979. Private communication.

Borys, R. D., P. J. Demott, E. E. Hindman, and D. Feng. 1983. The significance of snow crystal and mountain surface riming to the removal of atmospheric trace constituents from cold clouds, p. 181-189. *In* H. R. Pruppacher, R. G. Semonin, and W. G. N. Slinn, eds., Precipitaion scavenging, dry deposition, and resuspension. Elsevier, New York.

Brimblecombe, P., and G. A. Dawson. 1984. Wet removal of highly soluble gases. J. Atmos. Chem. 2:95-107.

Calvert, J. G., A. Lazrus, G. L. Kok, B. G. Heikes, J. G. Walega, J. Lind, and C. A. Cantrell. 1985. Chemical mechanisms of acid generation in the troposphere. Nature. 317:27-35.

Chameides, W. L. 1984. The photochemistry of a remote marine stratiform cloud. J. Geophys. Res. 89:4739-4756.

Dawson, G. A., J. C. Farmer, and J. L. Moyers. 1980. Formic and acetic acids in the atmosphere of Southwest U.S.A. Geophys. Res. Let. 7:725-728.

Jacob, D. J., J. W. Munger, J. M. Waldman, and M. R. Hoffmann. 1986. The $H_2SO_4$-$HNO_3$-$NH_3$ system at high humidities and in fogs 1. Spatial and temporal patterns in the San Joaquin Valley of California. J. Geophys. Res. 91:1073-1088.

Keene, W. C., J. N. Galloway, and J. D. Holden. 1983. Measurement of weak organic acidity in precipitation from remote regions of the world. J. Geophys. Res. 88:5122-5130.

Kunen, S. M., A. L. Lazrus, G. L. Kok, and B. G. Heikes. 1983. Aqueous oxidation of $SO_2$ by hydrogen peroxide. J. Geophys. Res. 88:3671-3674.

Lovett, G. M., W. A. Reiners, and R. K. Olson. 1982. Cloud droplet deposition subalpine balsam fir forests: Hydrological and chemical inputs. Science. 218:1303-1304.

Lind, J. A., and G. L. Kok. 1986. Henry's law determination for aqueous solutions of hydrogen peroxide, methylhydrogen peroxide, and peroxyacetic acid. J. Geophys. Res. 91:7889-7895.

Maahs, H. G. 1982. Sulfur dioxide/water equilibria between $0°$ and $50°$. An examination of data at low concentrations, p. 187-195. *In* D. R. Schryer, ed., Heterogeneous atmospheric chemistry. Am. Geophys. Union. Washington, D.C.

Maahs, H. G. 1983. Kinetics and mechanism of the oxidation of S(IV) by ozone in aqueous solution with particular reference to $SO_2$ conversion in nonurban tropospheric clouds. J. Geophys. Res. 88:10721-10732.

Marwitz, J. D., and P. J. Dawson. 1984. Low-level air flow in southern Wyoming during wintertime. Mon. Wea. Rev. 112:1246-1262.

Parrish, D. D., R. B. Norton, M. J. Bollinger, S. C. Liu, P. C. Murphy, and D. L. Albritton. 1986. Measurements of $HNO_3$ and $NO_3^-$ particulates at a rural site in the Colorado mountains. J. Geophys. Res. 91:5379-5393.

Pruppacher, H. R., and J. D. Klett. 1980. Microphysics of clouds and precipitation. Reidel, Dordrecht, Holland.

Politovich, M. K., and G. Vali. 1983. Observations of liquid water in orographic clouds over Elk Mountain. J. Atmos. Sci. 40:1300-1312.

Rechard, P. A., and V. E. Smith. 1972. Physical and hydrometeorlogical characteristics of the Snowy Range Observatory. University of Wyoming Water Resources Series No. 30.

Rogers, D. C., D. Baumgardner, and G. Vali. 1983. Determination of supercooled liquid water content by measuring rime rate. J. Climate and Appl. Meteor. 22:153-162.

Snider, J. R., and G. Vali. 1987. Measurements of $SO_2$ oxidation rates in the Elk Mountain orographic cloud. Submitted for publication.

# CHAPTER 3

# THE INFLUENCE OF AN URBAN ENVIRONMENT IN THE CHEMICAL COMPOSITION OF PRECIPITATION

Leroy J. Schroder, Myron H. Brooks, and John R. Garbarino,
U.S. Geological Survey, Denver, Colorado

Timothy C. Willoughby, Goodson and Associates, Denver, Colorado

## ABSTRACT

An urban environment influences the chemical composition of precipitation through the process of washout. Concentrations of calcium, magnesium, sodium, chloride, sulfate, and nitrate follow the trend of maximum concentration in the initial precipitation sample; then these analyte concentrations vary inversely in relation to precipitation intensity. Large fluctuations in ammonium concentration obscure any specific trends. Maximum concentrations of copper, lead, strontium, and zinc often occur later in a storm, and may be related to traffic volume. Cluster analysis of the chemical composition of urban precipitation indicates that there is a linkage among magnesium, chloride, sodium, and calcium. This linkage is probably related to the scavenging of soil-derived material and road dust deposited by precipitation. The linkage between nitrate and sulfate is shown, but is not closely related to hydrogen ion.

## INTRODUCTION

Precipitation is an important component of the hydrologic cycle and provides a pathway for both natural and anthropogenic materials into the Earth's surface and ground waters. In order to understand the differences in the chemical composition of precipitation, it is necessary to sample precipitation in areas anticipated to be receiving both natural and anthropogenic materials.

The chemical composition of atmospheric aerosols and airborne particulate matter can have an effect on the chemical composition of precipitation. The most commonly reported relations are between

*Chemical Quality of Water and the Hydrologic Cycle*, Robert C. Averett and Diane M. McKnight (Eds.) © 1987 Lewis Publishers, Inc., Chelsea, Michigan. Printed in the United States of America.

precipitation pH (acidity) and sulfate and nitrate concentrations (Cogbill and Likens, 1974; Robinson and Ghane, 1982); trace-metal concentrations and particulate matter in precipitation (Ondov et al., 1983); and changing meteorological conditions influencing acidity, sulfate, and nitrate concentrations (Pellatt et al., 1984; Robinson and Ghane, 1982).

The two major processes of introducing contaminants to precipitation are rainout and washout. Rainout is the processes occurring within clouds, such as condensation, nucleation, or gas dissolution. Washout is the process occurring below the cloudbase, and it may be considered as scavenging the airborne particulates and gases. Both rainout and washout probably occur continuously during single storms because most storms have convective air-current components that add large masses of near-surface air to overlying clouds (Schroder and Hedley, 1986).

The combination of rainout, washout, and convective movement of air-borne contaminants can cause changing contaminant concentrations in precipitation during the course of a storm. The combination also can cause large contaminant-concentration variations in precipitation from apparently similar storms.

Gascoyne and Patrick (1981) indicate that an initial precipitation sample has the maximum total ionic content and that the ionic content declines to a minimum, but finite, value in subsequent samples. Gatz and Dingle (1981) have shown that an inverse relation occurs between precipitation contaminants and rainfall intensity. In urban settings, the contaminant concentrations near roadways and highways have been associated with large particles that settle out within a few dekameters of the highway (Ondov et al., 1983).

Investigations were conducted to determine the effects of an urban environment on: (1) The relations determined by Gascoyne and Patrick (1981) and Gatz and Dingle (1981); (2) the relation between hydrogen ion concentration and sulfate and nitrate concentrations; and (3) the trace-metal concentrations in precipitation.


## EXPERIMENTAL DESIGN

The sampling site discussed here is about 75 m south of a major street in the northwest portion of the Denver, Colo., metropolitan area and is situated on the boundary between a residential and commercial area. The local wind speed and direction are monitored on a tower at the site by the Colorado Health Department. Although prevailing wind direction at a sampling site is of interest, wind direction during a storm also needs to be monitored. The local wind direction can indicate the potential source of the airborne particulates, especially in an urban area. The prevailing wind direction at the site is from the west, but the wind is from the south during many of the storms. The wind direction during storms is from the west 38% of the time and from the south 30% of the time.

Sequential sampling of storms provides a large amount of information; however, sequential sampling of many storms is manpower intensive and impractical. Instead of sequential sampling, usually single storms are sampled. In this study, manually operated wet-precipitation samplers were used to collect sequential samples during individual storms. An automatic wet-precipitation sampler was used to collect composite samples for individual storms. These samplers have been described by Volchock and Graveson (1976) and Schroder et al. (1985) and have a collection efficiency of 97 ± 4%. All precipitation samples were collected in polyethylene sacks (Good and Schroder, 1984) that had been rinsed three times with deionized water.

After the determination of specific conductance and pH, all samples were filtered and split into three subsamples as soon as possible after collection. The first subsample, to be analyzed for cations and trace elements using inductively coupled plasma spectrometry (Garbarino and Taylor, 1980), was acidified with nitric acid. The second subsample, to be analyzed for anions by ion exchange chromatography, was chilled to 4°C until analyzed. The third subsample of each composite sample was stored for subsequent analysis. The third subsample of each sequential sample was frozen in high-density polyethylene bottles.

## RESULTS AND DISCUSSION

### Major Ions

Maximum concentrations of calcium, magnesium, sodium, chloride, sulfate, and nitrate generally occur in the initial precipitation sample; the concentrations decrease in subsequent samples as the storm continues (Table 1). Concentrations of these constituents vary inversely in relation to the precipitation intensity until near the end of a storm. Usually, concentrations of these major ions remain nearly constant during the end of a storm, unless the final part of the storm occurs at a peak-traffic period. Sulfate and nitrate concentrations in urban precipitation usually increase during periods of increased traffic flow.

Ammonium concentrations do not seem to respond directly to either the rainout or washout processes but are often high (compared to the entire event mean) in the initial part of the storm. Then, ammonium concentrations fluctuate above and below the entire event mean during most of the remaining part of the storm. However, ammonium concentrations often remain nearly constant near the end of the storm. These large fluctuations in the ammonium concentrations occur during most storms of 2 or more hours and obscure any identification of specific trends.

Specific conductance, which may be used as an estimate of ionic strength of the samples, follows the trends of the measured major ions. Precipitation pH (acidity) does not always follow the patterns of the major ions. The maximum pH/minimum acidity normally occurs in the initial part of the storm in the western United States. Sulfur and nitrogen oxides and their reaction products are returned to the Earth's surface along with the soil-based atmospheric particulates. The soil particulates seem to neutralize the acidity that would be caused by large concentrations

Table 1.  Sequential Samples of Major Ions during a Typical Storm in an Urban Area.

[Analyte concentrations in mg/L; specific conductance in µS/cm at 25°C; pH in units]

| Time sampled (military) | Percentage of total snowfall per hour | Chemical analyte | | | | | | | | Specific conductance |
|---|---|---|---|---|---|---|---|---|---|---|
| | | Ca | Mg | Na | Cl | $SO_4$ | $NO_3$ | $NH_4$ | pH | |
| 2015-2300 | 0.33 | 1.34 | 0.12 | 0.80 | 0.85 | 6.2 | 6.4 | 4.3 | 6.4 | 44 |
| 2300-0140 | .98 | .20 | .01 | .18 | .32 | 2.0 | 2.3 | 2.3 | 6.0 | 14 |
| 0140-0600 | 1.62 | .10 | .05 | .09 | .15 | 1.2 | 1.0 | 2.9 | 5.3 | 5.0 |
| 0600-0925 | .40 | .42 | .10 | .50 | .50 | 2.5 | 1.2 | 3.0 | 4.5 | 21 |
| 0925-1125 | .55 | .54 | .02 | .25 | .43 | 2.1 | 2.0 | .7 | 4.6 | 18 |
| 1125-1330 | 1.90 | .33 | .03 | .13 | .10 | 1.1 | 1.2 | .3 | 5.0 | 9.0 |
| 1330-1425 | 2.84 | .25 | .01 | .22 | .35 | 1.0 | 1.0 | 1.7 | 5.2 | 7.4 |
| 1425-1555 | 2.53 | .15 | .05 | .15 | .15 | .7 | .8 | .6 | 5.5 | 5.8 |
| 1555-1820 | 4.83 | .14 | .01 | .20 | .05 | .4 | .5 | 1.0 | 5.1 | 2.8 |
| 1820-2120 | 4.11 | .10 | .11 | .18 | .05 | .4 | .5 | 1.4 | 5.4 | 1.9 |
| 2120-0400 | 1.33 | .65 | .07 | .24 | .24 | 1.3 | 1.1 | .7 | 5.4 | 7.4 |
| 0400-0600 | 5.04 | *--- | --- | --- | --- | --- | --- | --- | --- | --- |
| 0600-0700 | 11.8 | .20 | .01 | .20 | .25 | .5 | .5 | .3 | 5.0 | 4.0 |
| 0700-0800 | 7.92 | .15 | .01 | .10 | .10 | .3 | .5 | .9 | 5.1 | 10.5 |
| 0800-0900 | 5.43 | .10 | .01 | .15 | .10 | .2 | .6 | .8 | 5.2 | 3.2 |
| 0900-1110 | 3.74 | .10 | <.01 | .10 | .05 | .2 | .4 | 1.7 | 5.2 | 2.5 |
| 1110-1245 | .48 | .15 | .02 | .15 | .18 | .2 | .8 | .6 | 5.3 | 5.4 |

*Sample lost.

of sulfur and nitrogen oxides and their reaction products. Precipitation acidity normally increases as the storm progresses and
the maximum acidity in a metropolitan area occurs during periods
of maximum traffic volume.

Summary statistics for 46 individual storms sampled at the
urban site during the summer of 1985 are given in Table 2. Single-
storm data for the urban site were used to calculate the weekly
constituent concentrations; the week was defined as Tuesday to
Tuesday, to conform with the National Trends Network (NTN)
protocol.

Cluster analysis was the technique chosen to compare data
collected at the urban site with data collected at three NTN sites
located in Colorado. The three sites chosen for comparison with
the urban site were: (1) Site CO19, located in Rocky Mountain
National Park, Colo.; (2) site CO21, located west of Colorado
Springs, Colo.; and (3) site CO22, located east of Fort Collins,
Colo.

Cluster analysis determines the similarity between pairs of
constituents, and then groups the constituents into clusters. The
analysis begins with a fixed number of clusters, and each subsequent analysis reduces the number of clusters by one. The
analysis terminates with all constituents combined into one large
cluster. This large cluster is represented in a linkage tree, or
dendrogram. Constituent pairs that represent similar groups are
placed near each other in the dendrogram. Dissimilar constituents
are connected in the dendrogram simply to complete the linkage
tree.

Table 2.    Major Ions, Specific Conductance, and pH for Urban
           Precipitation Samples Collected during the Summer 1985.

[Analyte concentrations in mg/L; specific conductance in µS/cm at
25°C; pH in units]

| Chemical | Range | | Mean | Number of |
|---|---|---|---|---|
| constituent | Minimum | Maximum | | samples |
| Ca | 0.10 | 4.1 | 0.84 | 38 |
| Mg | <.01 | .48 | .09 | 38 |
| Na | .14 | 1.42 | .56 | 38 |
| $NH_4$ | .10 | 4.64 | .96 | 29 |
| Cl | .08 | .99 | .37 | 32 |
| $SO_4$ | .59 | 9.3 | 2.79 | 32 |
| $NO_3$ | .01 | 10.2 | 3.16 | 32 |
| $NO_3:SO_4$ ratio | .49 | 1.35 | .96 | 32 |
| Specific conductance | 8.1 | 89 | 32 | 41 |
| pH | 3.4 | 6.5 | 4.7 | 46 |

Results of a cluster analysis made by Gorham et al. (1984) for major ions in precipitation from the eastern United States during 1980-81 are used as an example in Figure 1.  These authors concluded that the linkage (relation) of hydrogen ion, sulfate, nitrate, ammonium, and calcium concentrations were caused by air pollution, and that sulfate plus nitrate concentrations caused the acid rain measured in the eastern United States.  Their conclusion was based on the probable sources of the major ions found in precipitation as given in Table 3.  Results of a cluster analysis of the major ion concentrations in precipitation at the urban site are shown in Figure 2.  Compared to precipitation in the eastern United States (Fig. 1), there is less relation between sulfate and nitrate, and even less relation between these two constituents and hydrogen ion at the urban site.  The relation among magnesium, chloride, sodium, and calcium concentrations at the urban site is probably related to the scavenging of soil-derived material and road dust deposited by precipitation.  Results of cluster analysis of the three NTN precipitation sampling sites (Figs. 3-5) indicate similar linkage patterns.  Sulfate and nitrate are related closely at all three sites, but are related less closely to the other groups compared to precipitation in the eastern United States.  The cluster analysis for each NTN site indicates a relation between calcium and ammonium that was not indicated at the urban site.  The relation between calcium and ammonium at the NTN sites is probably caused by a combination of soil-based particles and agricultural use of ammonium.  Ammonium is usually produced by fossil fuel combustion, and this source probably causes the slight relation between calcium and ammonium in precipitation at the urban site. The most noteworthy difference between precipitation at the urban sampling site and precipitation at the NTN sites, probably is the relation of hydrogen ion to other major ions.  Hydrogen ion is only slightly related to other major ions in precipitation at the urban site, while it is closely related to the ions normally attributed to soil-derived materials in precipitation at the three NTN sites. The effect of sampling-period differences between the urban site and the NTN sites cannot be determined by this type of statistical analysis.

## Trace Metals

Trace metals in precipitation usually can be separated into three groups:

1. Beryllium, cobalt, lithium, and vanadium concentrations always remain less than the analytical detection limits during a storm;

2. Barium, cadmium, iron, and manganese maximum concentrations usually occur during the initial part of a storm;

3. Copper, lead, strontium, and zinc maximum concentrations usually occur later in a storm.

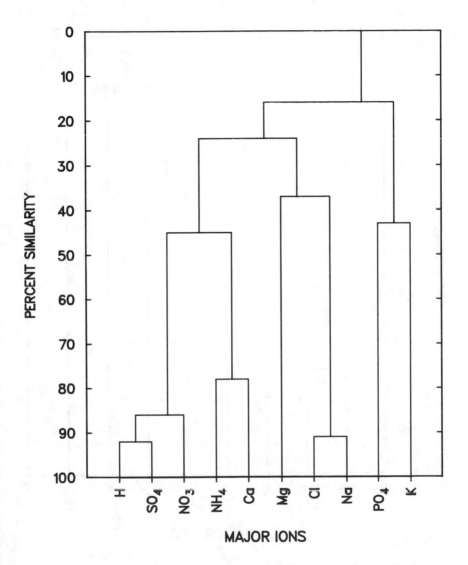

Figure 1.  Cluster analysis of major ions in precipitation in the eastern United States during 1980-81 (Gorham et al., 1984).

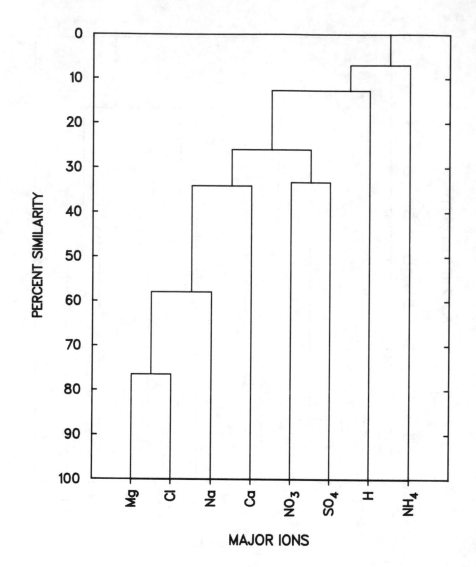

Figure 2.  Cluster analysis of major ions in precipitation collected at the sampling site located in the Denver,
Colo., metropolitan area, summer 1985.

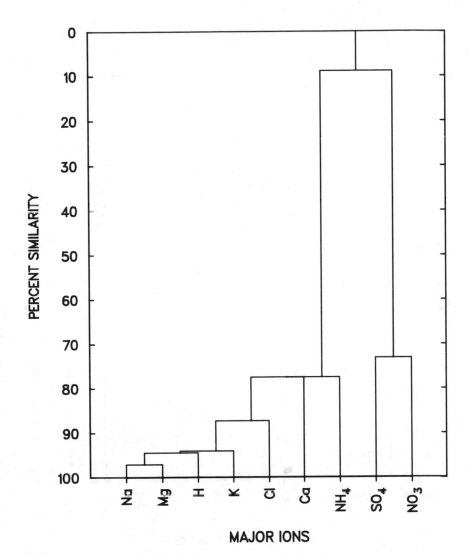

Figure 3.   Cluster analysis of major ions in precipitation collected at National Trends Network sampling site C019,
summer 1985.

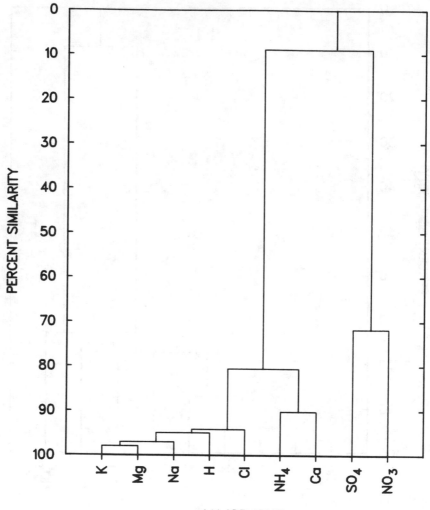

Figure 4.   Cluster analysis of major ions in precipitation col-
            lected at National Trends Network sampling site CO21,
            summer 1985.

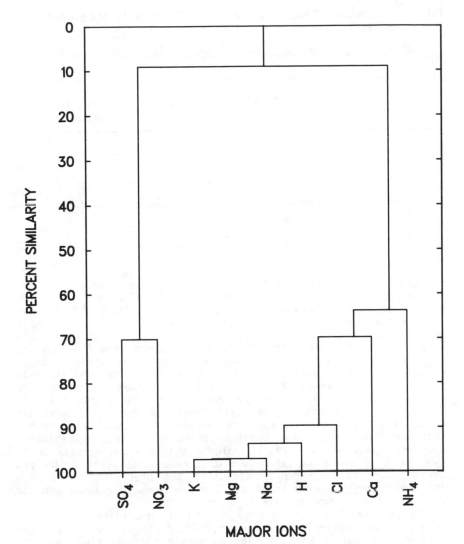

Figure 5.   Cluster analysis of major ions in precipitation collected at National Trends Network sampling site CO22,
summer 1985.

Table 3.    Sources of Major Ions in Precipitation in the United
States.

[1, primary source; 2, secondary source; 3, possible source]

| Major ion | Source | | | |
|-----------|--------------|------|------------|----------------------|
| | Air pollution | Soil | Fertilizer | Fossil-fuel combustion |
| H | 1 | | | 2 |
| Ca | 3 | 1 | 2 | |
| Mg | | 1 | 2 | |
| Na | | 1 | | 2 |
| Cl | | 2 | 3 | 1 |
| $SO_4$ | 1 | 3 | | 2 |
| $NO_3$ | 1 | | | 2 |
| $NH_4$ | | 3 | 2 | 1 |

Concentrations of iron and lead normally are inversely related
to precipitation intensity; however, this relation is not as
pronounced as that between major ions and precipitation intensity.
Trace-metal concentrations in an initial precipitation sample
usually are larger than trace-metal concentrations in the second
sample (Table 4). This trend follows the trend of major ions and
conforms to the relation described by Gascoyne and Patrick (1981).
Assuming that the source of lead in urban precipitation primarily
is from the combustion of leaded gasoline, a linear relation among
lead, nitrate, sulfate, and hydrogen concentrations should exist.
This relation is not statistically significant at alpha equals
0.05. The input of particulates derived from emissions probably
increases the trace-metal concentrations in urban precipitation.
However, Ondov et al. (1983) indicate that a sampling site would
need to be within a few dekameters of a roadway for traffic-derived
particulates to be directly affecting the precipitation sample.
The urban sampling site is about 75 m from a major roadway and the
effect of the roadway appears to be minimal.

Trace-metal concentrations in rural precipitation (Laird
et al., 1986; Jefferies and Snyder, 1981; Lindberg et al., 1982)
vary considerably. For example, manganese concentration range in
rural precipitation is greater than three orders of magnitude
(Fig. 6). As discussed before, the trace-metal concentrations are
largest during the initial part of a storm, and small-volume
precipitation storms commonly have larger trace-metal concentrations than do long-duration, large-volume storms. Mean trace-metal
concentrations at the urban study site are elevated compared to the
mean trace-metal concentrations in rural precipitation. However,
as shown in Figure 6, the mean trace-metal concentrations at the
urban site are often an order of magnitude lower than the maximum
concentrations at rural sites.

Table 4.   Trace Metals in Sequential Samples Collected during a
Typical Storm in an Urban Area.

| Time sampled (military) | Trace metals (μg/L) | | | | | | | |
|---|---|---|---|---|---|---|---|---|
| | Ba | Cd | Cu | Fe | Mn | Pb | Sr | Zn |
| 2015-2300 | 5.3 | 8.7 | 4.1 | 14 | 15 | 7.3 | 6.9 | 19.5 |
| 2300-0140 | 1.1 | .4 | 1.7 | 5.6 | 3.6 | 4.4 | 1.7 | 11.4 |
| 0140-0600 | .6 | <.3 | 7.6 | 3.1 | 1.0 | <2.5 | <1.1 | 14.0 |
| 0600-0925 | 2.3 | <.3 | 3.9 | 8.0 | 4.5 | 8.5 | 2.1 | 25.1 |
| 0925-1125 | 3.3 | 5.3 | 2.1 | 6.4 | 6.6 | 6.7 | 2.6 | 33.8 |
| 1125-1330 | 1.6 | <.3 | <2.0 | 3.7 | 2.8 | 5.6 | 1.6 | 11.4 |
| 1330-1425 | 1.2 | <.3 | <2.0 | 4.5 | 3.0 | <2.5 | 1.2 | 5.5 |
| 1425-1555 | .9 | <.3 | <2.0 | 1.7 | 1.2 | <2.5 | <1.1 | 2.6 |
| 1555-1820 | <.5 | <.3 | <2.0 | 1.6 | .3 | <2.5 | <1.1 | 1.0 |
| 1820-2120 | <.5 | <.3 | <2.0 | <.8 | <.3 | <2.5 | <1.1 | 4.8 |
| 2120-0400 | 1.2 | <.3 | <2.0 | 6.4 | 2.7 | 3.6 | 7.8 | 3.4 |
| 0400-0600 | *---- | ---- | ---- | ---- | ---- | ---- | ---- | ---- |
| 0600-0700 | <.5 | <.3 | 4.3 | 2.7 | .5 | <2.5 | 1.2 | 12.0 |
| 0700-0800 | <.5 | <.3 | <2.0 | 1.5 | <.3 | <2.5 | <1.1 | 3.8 |
| 0800-0900 | <.5 | <.3 | <2.0 | 1.2 | <.3 | <2.5 | <1.1 | 4.9 |
| 0900-1110 | <.5 | <.3 | 6.7 | <.8 | <.3 | <2.5 | <1.1 | 3.3 |
| 1110-1245 | .6 | <.3 | 3.0 | 3.5 | .6 | 4.5 | <1.1 | 25.0 |

*Sample lost.

## SUMMARY

Contaminants in urban precipitation can vary during a storm by as much as one order of magnitude. During the initial part of a storm, most of the measured inorganic constituents are at maximum concentration. These contaminant concentrations decrease rapidly, usually to some finite concentration, and vary in relation to the precipitation intensity until near the end of a storm. If the final part of the storm occurs at a peak-traffic period, hydrogen ion, nitrate, and sulfate concentrations usually increase.

Acidity, as measured by pH, and ammonium do not follow the general trend of maximum concentration during the initial part of a storm, but decrease to some finite value. Soil-derived particulates seem to neutralize the acidity that would be caused by sulfur and nitrogen oxides. The fluctuation of ammonium concentrations during a storm are sufficiently large that they obscure any specific trends.

Cluster analysis of the major ions in urban precipitation indicates a relation between sulfate and nitrate concentrations. This relation, however, is not connected clearly to the hydrogen ion concentrations. The relation of magnesium, chloride, sodium, and calcium concentrations probably is due to the scavenging of soil-derived material and road dust by precipitation. Cluster

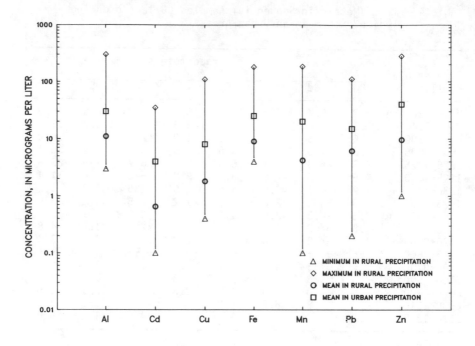

Figure 6.  Comparison between trace-metal concentrations in urban
precipitation and trace-metal concentrations in rural
precipitation.

analysis of data from three rural sites indicate that ammonium is
closely related to calcium and soil components. Ammonium is virtually independent of the other major ions in urban precipitation.
    Mean trace-metal concentrations in urban precipitation tend to
be higher than mean trace-metal concentrations in rural precipitation. Small-volume precipitation storms often have higher trace-
metal concentrations than do large-volume storms that occur over a
period of several hours.

REFERENCES

Cogbill, C. V., and G. E. Likens. 1974. Acid precipitation in the
    northeastern United States. Water Resources Res. 10:1133-1137.

Garbarino, J. R., and H. E. Taylor. 1980. A Babington-type nebu-
    lizer for use in the analysis of natural water samples by in-
    ductively coupled plasma spectrometry. Appl. Spectros.
    34:585-590.

Gascoyne, M., and C. K. Patrick. 1981. Variation in rainwater chemistry, and its relation to synoptic conditions, at a site in northwest England. J. Environ. Studies. 17:209-214.

Gatz, D. F., and A. N. Dingle. 1981. Trace substances in rainwater concentration variations during convection rains and their interpretation. Tellus. 23:14-27.

Good, A. B., and L. J. Schroder. 1984. Evaluation of metal ion absorptive characteristics of three types of plastic sample bags used for precipitation sampling. J. Environ. Sci. Health. A19:631-640.

Gorham, E., F. B. Martin, and J. T. Litzau. 1984. Acid rain: Ionic correlations in the eastern United States, 1980-1981. Science. 225:407-409.

Jefferies, D. S., and W. R. Snyder. 1981. Atmospheric deposition of heavy metals in central Ontario. Water Air Soil Pollut. 15:127-152.

Laird, L. B., H. E. Taylor, and Vance Kennedy. 1986. Snow chemistry of the Cascade-Sierra Nevada mountains. Environ. Sci. Tech. 20:275-290.

Lindberg, S. E., R. C. Harris, and R. R. Turner. 1982. Atmospheric deposition of metals to forest vegetation. Science. 215:1609-1611.

Ondov, J. M., W. H. Zoller, and G. E. Gordon. 1983. Trace element emissions on aerosols from motor vehicles. Environ. Sci. Technol. 16:318-328.

Pellatt, G. L., R. Bustin, and R. C. Harris. 1984. Sequential sampling and variability of acid precipitation in Hampton, Virginia. Water Air Soil Poll. 21:33-49.

Robinson, J. W., and H. Ghane. 1982. Occurrence of acid rain in Baton Rouge, Louisiana, Summer, 1981. J. Environ. Sci. Health. 217:129-136.

Schroder, L. J., and A. G. Hedley. 1986. Variation in precipitation quality during a 40-hour snowstorm in an urban environment--Denver, Colorado. Intern. J. Environmental Studies. 28:131-138.

Schroder, L. J., R. A. Linthurst, J. E. Ellson, and S. F. Vozzo. 1985. Comparison of daily and weekly precipitation sampling effeciencies using automatic collectors. Water Air Soil Poll. 24:177-187.

Volchock, H. L., and R. T. Graveson. 1976. Wet/dry fallout collection, p. 256-264. *In* Proceedings of the Second Federal Conference on the Great Lakes, Great Lakes Basin Commission.

CHAPTER 4

ANALYTICAL METHODOLOGY FOR THE MEASUREMENT OF THE
CHEMICAL COMPOSITION OF SNOW CORES FROM THE
CASCADE/SIERRA NEVADA MOUNTAIN RANGES

Howard E. Taylor, U.S. Geological Survey, Denver, Colorado

ABSTRACT

Collection techniques, sample preparation, and analytical
procedures are described for determining the composition of
chemical constituents in snow cores. Specific conductance, pH,
dissolved organic carbon, calcium, magnesium, sodium, potassium,
chloride, sulfate, nitrate, fluoride, phosphate, ammonium, iron,
aluminum, managanese, copper, cadmium, and lead were determined in
carefully collected and prepared samples from snow cores from the
Cascade/Sierra Nevada Mountains. Analytical techniques, including
inductively coupled plasma emission; atomic absorption and
ultraviolet-visible-infrared absorption spectrometry; ion exchange
chromatography; and electrometric methods, were adapted for maximum
sensitivity in low ionic strength samples (specific conductance <5
microsiemens per centimeter at 25°C). Extensive use of quality
assurance techniques verified the validity of solute concentration
measurements.

INTRODUCTION

The measurement of chemical constituents in snowpack has been
used to assess the geographic variations and trends in atmospheric
deposition in Washington, Oregon, and California (Laird et al.,
1986). Snowpack sampling was utilized for several reasons
including: The ability to integrate all precipitation events
occurring during a single winter season, as well as to accumulate
dry fallout during non-storm periods; the collection of samples
containing a minimal amount of windblown dust and other terrestrial
materials due to the protective snow cover; and the ability to
sample and compare results from a large geographic area.

*Chemical Quality of Water and the Hydrologic Cycle*, Robert C. Averett and Diane M. McKnight (Eds.) © 1987 Lewis Publishers, Inc.,
Chelsea, Michigan. Printed in the United States of America.

This report describes techniques that were used for collecting snow core samples, preparing samples for laboratory analysis, laboratory procedures for measuring the concentration of selected chemical constituents in the snowmelt, and an evaluation of the quality and characteristics of analytical results. The scope of this report addresses only the soluble constituents in the melted snow cores. Insoluble particulate matter was carefully separated, and its characterization will be the subject of future reports. The constituents and properties measured include hydrogen ion, calcium, magnesium, sodium, potassium, chloride, sulfate, nitrate, fluoride, phosphate, ammonium, iron, aluminum, manganese, copper, cadmium, lead, dissolved organic carbon, and specific conductance.

## METHODS

### Sampling

To avoid the loss of chemical constituents from the snowpack by leaching from partial snowmelt or rainfall, most core samples were collected at sites above 1,600 m in elevation, and sample collection was performed in late winter before the spring melt. The snow cores were collected with a stainless steel sampler of the Federal Snow Sampler design (cutter head opening 3.77 cm). The lower 7 to 8 cm of the sample was routinely discarded to eliminate the possibility of contamination from the underlying soil. All samples were hermetically sealed in doubled polypropylene bags immediately after collection, labeled, and maintained frozen by mechanical refrigeration during transport to the laboratory.

### Sample Preparation

To insure uniform handling of all samples, the melting, filtration and bottling process followed a strict protocol. The double-bagged samples were placed in a clean polyethylene open bucket. The bucket was placed between two infrared heat lamps (approximately 0.5 m above the top of the bags) and the sample was allowed to begin thawing. After approximately 90% of the snow had melted (7-8 hr), the samples in their respective buckets were placed inside a large walk-in refrigerator with the temperature regulated to 4°C, and left overnight. The next morning the last 10% of the snow had melted while the temperature of the water was maintained at 4°C. By this procedure, the overall temperature of the sample never exceeded 4°C, and hence, any bacterial deterioration of the nutrient constituents, as well as adsorption of trace metals on the surfaces of the polypropylene bag, were minimized.

Immediately after thawing the samples were filtered, preserved, and bottled. After agitating the bag to suspend any settled material, pressure filtration was performed in two steps using a Cole Parmer Masterflex peristaltic pump and 2.5-mm I.D. silicone tubing. In the first step, approximately 100 mL of the sample was passed through a 0.45-μm pore diameter, 47-mm diameter silver membrane filter housed in a stainless steel in-line holder. The sample was collected in a glass screw-cap bottle and reserved for dissolved organic carbon analysis. Approximately 15 to 20 mL of sample was used to rinse the filter and holder prior to collecting the sample.

Subsequently, as much as 2 L of sample (depending on the total volume of the melted snow core) was filtered through a 90-mm, 0.4-μm pore diameter polyethylene membrane filter. The filtration assembly as described by Shockey and Taylor (1984) consisted of an apparatus to permit the pumping of the sample through the membrane filter. Figure 1 shows a diagram depicting the assembly of the two-head peristaltic pump with tubing routed in such a way as to effect a push-pull force across the filter membrane. The purpose of this configuration is to expedite the filtering process when filtering large volumes of water which contain sufficient sediment to plug the pores of the filter. The filter holder was constructed of a polycarbonate material which was thoroughly washed and leached in deionized water overnight between samples. This insured a minimum of cross contamination.

Approximately 20 mL of sample was used to rinse the pump tubing and the filter apparatus. After rinsing, 1 L of the filtered sample was collected in a polyethylene bottle (previously rinsed with nitric acid) and reserved for metals analysis. The sample was preserved by the addition of 1 mL of double-distilled, high-purity concentrated nitric acid for each liter of sample. Up to 1 L additional filtrate was collected and saved for nutrient determinations in a polyethylene bottle that had been rinsed with deionized water. Finally, a 20-mL portion of filtrate was transferred to a plastic beaker for the immediate determination of pH and specific conductance.

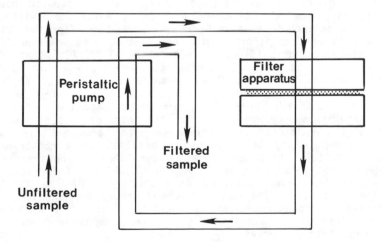

Figure 1.   Apparatus for filtering snow melt samples.

As each 1-L bottle of nutrient sample was collected, it was immediately preserved by a freezing technique. This technique is illustrated in Figure 2. By a diffusion process, ions migrate to the walls of the container (Fig. 2) where the cations and anions become irreversibly adsorbed to active sites, thus depleting the concentration of both in the bulk solution. A quick-freezing technique was employed to inhibit this process. The sample bottle was rapidly immersed in a dewar of liquid nitrogen. This quickly formed a thin layer of ice on all internal surfaces of the bottle. By the principle of zone refining, ions were preferentially excluded from this ice shell, thereby creating a diffusion barrier to active sites on the container walls. After this ice-layer forming process, the bottles were transferred to a conventional deep freezer where the samples were completely frozen and were stored until all samples had been filtered and were ready for analysis.


ANALYTICAL METHODOLOGY

A variety of methods was used for the analysis of specific constituents in the samples obtained from the preparation of the melted snow cores. The techniques and procedures are summarized by groups of constituents determined by a specific analytical technique. In all cases, stock solutions of calibration standards were prepared from ultra-high-purity metals or from metal salts dissolved in distilled deionized water and stored in ultra-high purity, nitric-acid-rinsed (for metals) Teflon bottles.


Inductively Coupled Plasma Emission Spectrometry

Techniques for the simultaneous determination of selected trace metals in natural water samples by inductively coupled plasma emission spectrometry have been reported by Garbarino and Taylor (1979). These techniques have been adapted and modified for use in the analysis of precipitation and snow melt samples. All analyses were performed on a modified, Jarrell-Ash Model 975 Plasma AtomComp emission spectrometer. Determinations were made by direct nebulization, using a crossflow pneumatic nebulizer, of filtered and acidified (nitric acid) samples into a 27.1-mHz argon plasma operating at 1.25 kW. Both atomic and ionic optical emission from this excitation source was focused on the entrance slit of a 0.75-m focal-length, multichannel spectrometer that provided for background correction by the use of an oscillating refractor-plate spectrum shifter. Each determination consisted of the average of two replicate, 24 s integrations, each of which was corrected by subtracting background measured on either side of the spectral emission line. The instrument was calibrated by running known concentration standards and a reagent blank. Trace metal concentrations of the samples were computed from these frequently-updated calibration curves.

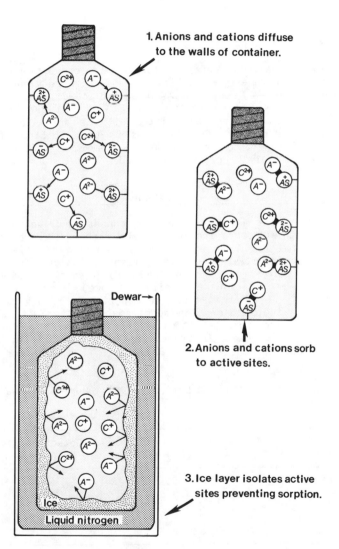

Figure 2.    Quick-freeze preservations of filtered snow melt.

Table 1 lists the elements which were determined by this technique, the wavelengths of the specific emission spectral lines which were used, and the detection limits which were obtained under these operating conditions. In each case, these specific spectral lines were chosen to minimize spectral interferences normally encountered in natural water matrices while optimizing the intensity of the spectral line to provide maximum sensitivity for the analysis.

## Atomic Absorption Spectrometry

Potassium analyses were performed by conventional atomic absorption spectrometry utilizing an air-acetylene flame-atomization device as reported by Skougstad et al. (1979, p. 229). All determinations were performed on a Perkin Elmer Model 5000 Atomic Absorption spectrometer fitted with an automated sampler. Absorption measurements were made using the 766.5 nm resonance spectral line using a slightly reducing flame. By this procedure, a detection limit of 10 μg/L was obtained.

## Electrothermal Atomization Atomic Absorption Spectrometry

The trace metals aluminum, cadmium, copper, and lead were found to be present at concentration levels (sub-μg/L) substantially below those that could be determined by conventional flame atomic absorption or inductively coupled plasma emission spectrometry. Therefore, samples were analyzed for these elements by electrothermal-atomization (graphite furnace) atomic absorption spectrometry. All determinations were performed on a Varian Model 975 automated atomic absorption spectrometer fitted with a Varian automated graphite furnace atomizer equipped with an automatic sample injection system. Instrument operation was similar to that reported by Shrader and others (1983).

Table 1. Elements Determined by Inductively Coupled Plasma Emission Spectrometry.

| Element | Wavelength (nm) | Detection limit (μg/L) |
|---------|-----------------|------------------------|
| Calcium | 396.8 | 2 |
| Iron | 259.9 | 0.1 |
| Magnesium | 279.5 | 3 |
| Manganese | 257.6 | 0.1 |
| Sodium | 588.9 | 10 |

Improved sensitivity and precision for cadmium, copper and lead were obtained by using five successive 50-µL aliquots of sample dispensed into the graphite furnace tube with solvent evaporation at 110°C occurring between each aliquot deposition. This procedure resulted in the analyte residue being concentrated reproducibly in the center of the graphite tube at the time of atomization rather than being dispersed over the entire bottom of the tube as would have been the case if a single 250-µL aliquot was used. Aluminum was determined by dispersing 10 successive 15-µL aliquots onto a graphite platform mounted in the graphite tube, evaporating solvent between each aliquot. Use of the graphite platform has been described by Kaiser et al. (1981).

Table 2 lists the spectral wavelengths that were used for the absorbance measurements, along with the furnace atomization temperatures, and the achievable detection limits for each of the four trace metals. All analytical determinations were based on the average of two replicate absorption measurements with background correction.

## Ultraviolet-Visible and Infrared Absorption Spectrometry

Ammonium, nitrate, and phosphate ions were determined spectrophotometrically by specially optimized, low-concentration-level absorption spectrophotometry. The absorbance measurements were made on a Technicon Model AA2 automated analysis system with appropriate plumbing and configurations for automated analysis. All analyses were performed on chemically unpreserved samples that were maintained frozen until just prior to analysis, to avoid contamination from the use of antibacterial preservatives and to inhibit adsorption on the wall surfaces of the sample containers.

Ammonium ion determination has been described by Skougstad et al. (1979, p. 389) and consists of reacting the sample solution with sodium salycilate, sodium nitroprusside, and sodium hypochlorite in an alkaline medium and measuring the absorbance of the developed color in a 50-mm cell at 660 nm. By this procedure, a detection limit of 3 µg/L, as nitrogen, was achieved.

Table 2. Operation Conditions and Detection Limits for Graphite-Furnace Atomic Absorption Spectrometry.

| Element | Wavelength (nm) | Atomization temp. (°C) | Detection limit (µg/L) |
|---------|-----------------|------------------------|------------------------|
| Cadmium | 228.8 | 1800 | 0.002 |
| Copper | 324.8 | 2300 | 0.005 |
| Lead | 217 | 2000 | 0.02 |
| Aluminum | 309.3 | 2700 | 0.03 |

The determination of nitrate ion, as described by Skougstad et al. (1979, p. 437), involved the chemical reduction of all nitrate ions to nitrite ions by passing the sample through a granular copper-cadmium alloy packed column. After reduction the nitrite was reacted with sulfanilamide dihydrochloride to form a colored species whose absorbance was measured in a 50-mm cell at 520 nm. A detection limit of 2 µg/L, as nitrogen, was easily obtained by this procedure.

Phosphate ion was determined by converting all phosphate to orthophosphate by an acid persulfate digestion followed by reaction with ammonium molybdate which, after reduction with ascorbic acid, formed a colored species as described by Skougstad et al. (1979, p. 453). Absorbance of this species was measured in a 50-mm cell at 660 nm, which produced a detection limit of 2 µg/L as phosphorus.

Dissolved organic carbon was determined by an infrared absorption spectrometric technique (Wershaw et al., 1983). After acidification of the sample, all inorganic carbon (carbonate and bicarbonate) was decomposed to carbon dioxide and removed by purging the sample with an inert gas. All dissolved organic carbon was oxidized with potassium persulfate at 120°C in a closed container and the resulting carbon dioxide oxidation product was measured with a Dohrman model DC-80 carbon analyzer. This procedure permitted the collective determination of all dissolved organic carbon species to a detection limit of 0.1 mg/L.

## Ion Exchange Chromatography

Chloride, fluoride, and sulfate ions were determined by ion exchange chromatography. Techniques described by Hedley and Fishman (1982) were modified to obtain maximum sensitivity and lowest detection limits. A Dionex model 2/20i Ion Chromatograph fitted with a standard anion S4 separator column and a conductivity detector were used for these determinations. A 600-µL sampling loop was used to inject samples into the chromatograph where separation was achieved during elution with a standard 0.003 M sodium bicarbonate/0.0024 M sodium carbonate solution. Peak areas from the elution chromatogram were integrated and compared to the areas obtained from standard calibration solutions to calculate ion concentrations. Detection limits of 10 µg/L were achieved for all three anions.

## Electrometric Techniques

Specific conductance was measured at 25°C using a Radiometer Model CDM 3 direct-reading conductance bridge equipped with a dip-type cell with a cell constant of 0.326. Specific conductance was measured at the time of sample preparation immediately after the sample filtration step.

An Orion Model 601A Ionanalyzer with a custom-designed, low-ionic-strength glass electrode was used to measure pH potentio-metrically. The pH meter was calibrated using special low-ionic-strength solutions prepared by diluting a sulfuric acid stock solution sufficiently to obtain theoretical pH values of 1.97, 3.65, 6.66, and 7.52. These solutions were used to calibrate the pH meter periodically and to serve as quality control samples to insure satisfactory performance. All pH determinations were made at the same time as the specific conductance measurements.

## RESULTS AND DISCUSSION

### Quality Assurance

Because of their extremely dilute nature, great care was exercised in sampling, handling, and analysis. All preparation steps were carried out in a dust-free laboratory, isolated from all other activities. Personnel were required to handle all samples and apparatus with polyethylene gloves. All equipment was carefully washed and rinsed with deionized water between each preparation.

Blanks were carried through all steps of the preparation and analysis, and appropriate corrections were made to the data to accommodate any blank values found. In substantially all cases blanks were measured at levels below the detection capability of the analytical technique employed. Considerable effort was made to identify and eliminate all sources of blank contamination during pilot studies that were performed before actual sample handling and analysis were initiated.

To minimize the possibility of systematic errors arising from day-to-day variations in instrumentation, standardization, and any other non-random sources, samples were retained until all preparation was completed. Analyses were then performed as simultaneously as possible for a given parameter. For all parameters, samples were analyzed in a random order, with reference standards and blanks interspersed at a 10% frequency. In addition, duplicate samples, chosen at random and comprising 25% of the total sample set, were randomly interspersed with the other samples during the analysis campaign. Field duplicates were also randomly collected at 20% of the sampling sites to establish the relative precision of field procedures and techniques.

The standard reference materials consisted of U.S. Geological Survey Standard Reference Water Samples (SRWS) Nos. 69, 73, 74, 75, 80, and N7 and National Bureau of Standards SRM 1643a. To approximate the concentration levels anticipated in the snow melt samples, SRWS Nos. 69 and 75 were diluted 1/100, and SRWS Nos. 74 (for chloride and sulfate only), 80, and N7 were diluted 1/10. Diluted SRWS's were not used for major cation quality assurance because the low concentration levels found were not anticipated during the experimental design and prior to actual commencement of analytical work. However, this does not present a problem in data interpretation as all major cations were determined by inductively coupled plasma emission spectrometry which exhibits a linear dynamic concentration response of more than five orders of magnitude. Table 3 lists each of the parameters along with the mean values of the analysis of the standard reference material and the confidence

Table 3.  Summary of Standard Reference Material Accuracy and
          Precision.

| Parameter | Units | SRM[1] | SRM mean | S.E.[2] | Experimental mean | Experimental S.E. |
|-----------|-------|--------|----------|---------|-------------------|-------------------|
| Calcium | mg/L | 74 | 7.48 | 0.15 | 7.46 | 0.01 |
| Magnesium | mg/L | 74 | 1.94 | 0.04 | 1.94 | 0.01 |
| Sodium | mg/L | 74 | 2.77 | 0.08 | 2.81 | 0.01 |
| Potassium | mg/L | 80 | 0.437 | 0.008 | 0.426 | 0.003 |
| Chloride | mg/L | 74 | 0.140 | 0.03 | 0.141 | 0.007 |
| Sulfate | mg/L | 74 | 1.43 | 0.07 | 1.43 | 0.01 |
| Nitrate(N) | mg/L | 74 | 2.43 | 0.11 | 2.48 | 0 |
| Phosphate(P) | mg/L | N7 | 0.029 | 0.002 | 0.030 | 0.002 |
| Iron | µg/L | 1643a | 88 | 4 | 85 | 2 |
| Manganese | µg/L | 1643a | 31 | 2 | 31 | 1 |
| Copper | µg/L | 73 | 0.626 | 0.023 | 0.613 | 0.049 |
| Cadmium | µg/L | 75 | 0.079 | 0.004 | 0.084 | 0.006 |
| Lead | µg/L | 69 | 0.023 | 0.006 | 0.027 | 0.009 |
| Aluminum | µg/L | 69 | 3.10 | 0.32 | 3.37 | 0.08 |

[1]SRM = Standard reference material.
[2]S.E. = Standard error.

interval at the 95% confidence level of the determinations.  Also
listed  is  the  distribution  of  values  obtained  from  multiple-
laboratory  round-robin  analysis  (Fishman, M.J., U.S. Geological
Survey, written commun., 1979, 1980, and 1982; Janzer, V.J., U.S.
Geological  Survey,  written  commun.,  1983)  of  the  U.S. Geological
Survey Standards.
    As  can  be  seen  from  Table 3,  a  comparison  of  the  mean  concen-
tration values and the standard errors of the means show that all
parameters  were  determined  well  within  the  experimental  error  of
the analytical procedures.
    Because  each  of  the  samples  in  this  study  was  collected  in  a
systematic  fashion  (i.e.,  the  sample  strategy  was  based  upon
geographic considerations) rather than randomly selected, it cannot
be  anticipated  that  the  distribution  of  concentrations  of  any
specific  constituent  or  of  a  subsample  of  the  distribution  is
Gaussian in nature, precluding conventional statistical computation
(i.e., comparison of means, standard deviations, etc.).  The paired
sampling technique described by Koch and Link (1971) was employed
to establish a normal distribution of test data which allows basic
statistical  calculations  to  be  performed.   The  paired  values  of  the
duplicates  for  each  analytical  parameter  were  arranged  with
corresponding samples adjacent to each other in the order in which
they were determined.  The algebraic differences between the paired
values were used to establish a data set which was random in nature
and  hence  followed  a  normal  sample  distribution.   These  characteristics of the data set were verified by plotting the frequency
distribution of these differences of paired values on normal probability paper.  The  linear  nature  of  the  resulting  plot  confirmed

the hypothesis that the distribution was, in fact, Gaussian and hence represented random behavior. This treatment was applied independently in a similar fashion to both field and laboratory duplicate samples. A typical example of this is shown in Figure 3 for copper.

A hypothesis was formulated that if the analytical procedure for a given parameter was producing data that was systematically accurate over the entire data set, then the differences in values between replicated samples should differ, only to the extent of the experimental error of the methodology and in a random fashion. Therefore, the mean of the algebraic differences of the paired determinations should be equivalent to zero. Student's t-test can be used to test the validity of this hypothesis (Dixon and Massey, 1969). Computation of data sets for each parameter showed that at the 99% confidence level, the mean value of the differences of the paired replicates was not signficicantly different from zero for either the field or laboratory duplicates. Therefore, the original hypothesis was not disproven, and it is assumed with a high degree of confidence that all paired analytical replicates for each parameter data set represent the same statistical distribution. This, in turn, suggests that the analytical results within a given analysis set (samples and standards) are fully within accuracy and precision control established by the nature of the methodology and no systematic errors are biasing the data. This conclusion was found to be valid for both the field and laboratory duplicates.

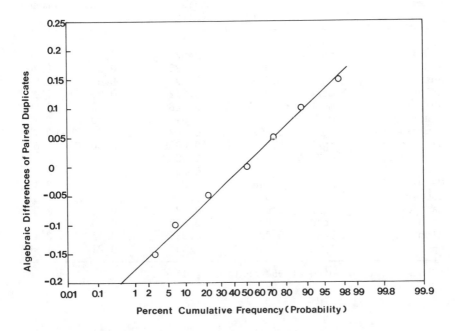

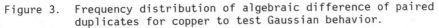

Figure 3.    Frequency distribution of algebraic difference of paired duplicates for copper to test Gaussian behavior.

Table 4 summarizes the result of performing these statistical tests on each parameter. As seen in Table 4, each experimentally determined Student's t (Mandel and Nanni, 1978) is less than the theoretical value at the 95% confidence level (verifying hypothesis).

From this set of paired data for each parameter, overall (field plus laboratory) and laboratory analysis precision was estimated. Table 5 lists a calculated laboratory precision at the average concentration of the paired replicate subset for each of the parameters studied. This precision was calculated based upon estimating the average standard deviation from the average range of the paired replicates (Crow et al., 1960, p. 13). From the average standard deviation, the average standard error of the mean was computed. From this the 95% confidence intervals were established which are listed in Table 5.

Table 5 lists calculated overall precisions (based on field duplicates) at typical concentration levels for each of the parameters. Precisions and the 95% confidence intervals were calculated as specified above and represent the overall experimental error.

Table 4.    Student's t-test of Means of Paired Replicate
            Determinations.

| Parameter | Units | Mean (diff) | Standard deviation (diff) | t(exp) | t(theo)[1] |
|-----------|-------|-------------|---------------------------|--------|-----------|
| Calcium | mg/L | 0.0047 | 0.045 | 0.31 | 2.26 |
| Magnesium | mg/L | -0.0047 | 0.024 | 0.34 | 3.20 |
| Sodium | mg/L | 0.0012 | 0.023 | 0.16 | 2.20 |
| Potassium | mg/L | 0.029 | 0.17 | 0.98 | 2.01 |
| Chloride | mg/L | 0.017 | 0.032 | 0.90 | 3.1 |
| Sulfate | mg/L | -0.0057 | 0.028 | 0.52 | 2.36 |
| Fluoride | mg/L | 0.0025 | 0.012 | 0.39 | 2.77 |
| Ammonium(N) | mg/L | 0.0015 | 0.007 | 0.46 | 2.77 |
| Nitrate(N) | mg/L | -0.00089 | 0.0088 | 0.30 | 2.62 |
| Phosphate(P) | mg/L | 0.00025 | 0.0017 | 0.29 | 2.77 |
| Iron | µg/L | 0 | 0.25 | 0 | 2.77 |
| Manganese | µg/L | 0.046 | 0.27 | 0.62 | 2.16 |
| Copper | µg/L | 0.011 | 0.086 | 1.1 | 2.01 |
| Cadmium | µg/L | 0.0025 | 0.029 | 0.56 | 2.01 |
| Lead | µg/L | -0.0078 | 0.080 | 0.66 | 2.01 |
| Aluminum | µg/L | -0.004 | 0.035 | 0.51 | 2.1 |
| pH | | 0.00073 | 0.037 | 0.21 | 1.96 |

[1]Student's t computed at 95% confidence level.

Table 5.  Measurement Precision of Constituents in Snow Melt.

| Parameter | Unit | Concentration[1] | | Precision[2] |
|---|---|---|---|---|
| Calcium | mg/L | 0.115 | ± | 0.02 |
| Magnesium | mg/L | 0.081 | ± | 0.03 |
| Sodium | mg/L | 0.293 | ± | 0.01 |
| Potassium | mg/L | 0.124 | ± | 0.02 |
| Chloride | mg/L | 0.258 | ± | 0.05 |
| Sulfate | mg/L | 0.017 | ± | 0.02 |
| Fluoride | mg/L | 0.019 | ± | 0.01 |
| Ammonium | mg/L | 0.0415 | ± | 0.008 |
| Nitrate | mg/L | 0.0257 | ± | 0.002 |
| Phosphate | mg/L | 0.0069 | ± | 0.002 |
| Iron | µg/L | 2.68 | ± | 0.02 |
| Manganese | µg/L | 1.47 | ± | 0.05 |
| Aluminum | µg/L | 1.97 | ± | 0.08 |
| Copper | µg/L | 0.95 | ± | 0.2 |
| Cadmium | µg/L | 0.807 | ± | 0.004 |
| Lead | µg/L | 0.400 | ± | 0.02 |
| pH | -- | 5.56 | ± | 0.01 |
| Specific conductance | µS/cm | 2.90 | ± | 0.08 |

[1]Average concentration of paired replicate subsets.
[2]95% confidence internal standard error of mean.


## Data Summary

Although a detailed tabulation of all data for 102 snow cores will be published elsewhere, a data summary is presented in Table 6. Maximum, minimum, mean, and median concentrations for all measured constituents are reported. Laird et al. (1986) and Taylor and Laird (1985) have published geographic distribution profiles of several of the constituents based on data collected in this study and have offered suggestions as to the significance and meaning of these distributions.


## CONCLUSIONS

This report demonstrates the capability and procedures to accurately and precisely measure the concentration of dissolved chemical constituents in snow melt. As shown in Table 6, definitive analytical data were obtained for all reported parameters. The sampling techniques described allow a statistically representative sample of the snow pack to be collected. With careful laboratory preparation, sample integrity can be maintained until analyses are performed.

Table 6.   Statistical Data on Snow Core Samples.

| Constituent | Maximum | Minimum | Mean | Median |
|---|---|---|---|---|
| Specific conductance[1] | 6.10 | 1.78 | 2.8 | 2.52 |
| pH | 5.88 | 5.11 | -- | 5.58 |
| Hydrogen Ion[2] | 0.008 | 0.0005 | 0.003 | 0.003 |
| Calcium[3] | 0.16 | 0.005 | 0.042 | 0.036 |
| Magnesium[3] | 0.051 | 0.003 | 0.009 | 0.003 |
| Sodium[3] | 0.26 | 0.01 | 0.07 | 0.06 |
| Potassium[3] | 0.30 | 0.01 | 0.019 | 0.01 |
| Sulfate[3] | 0.32 | 0.04 | 0.14 | 0.13 |
| Chloride[3] | 1 | 0.01 | 0.22 | 0.16 |
| Fluoride[3] | 0.10 | 0.01 | 0.03 | 0.03 |
| Nitrate (N)[3] | 0.12 | 0.002 | 0.025 | 0.022 |
| Ammonium (N)[3] | 0.18 | 0.003 | 0.08 | 0.010 |
| Phosphate (P)[3] | 0.022 | 0.002 | 0.005 | 0.004 |
| Aluminum[4] | 100 | 0.18 | 3.2 | 1.7 |
| Cadmium[4] | 0.96 | 0.024 | 0.12 | 0.073 |
| Copper[4] | 3.8 | 0.100 | 0.44 | 0.28 |
| Iron[4] | 8.5 | 0.1 | 0.92 | 0.1 |
| Lead[4] | 1.1 | 0.05 | 0.33 | 0.28 |
| Manganese[4] | 7.8 | 0.01 | 1 | 0.7 |
| Dissolved organic carbon[4] | 4.1 | 0.20 | 0.60 | 0.50 |

[1] In microsiemens per centimeter.
[2] Milliequivalent.
[3] In milligrams per liter.
[4] In micrograms per liter.

## DISCLAIMER

The use of trade or product names in this chapter is for iden-
tification purposes only and does not constitute endorsement by the
U.S. Geological Survey.

## REFERENCES

Crow, E. L.,   F. A. Davis,   and   M. W. Maxfield. 1960.   Statistics
    manual.   Dover, N.Y., p. 13.

Dixon, W. J., and F. J. Massey, Jr. 1969.   Introduction to statis-
    tical analysis.   McGraw-Hill, N.Y., p. 98.

Fishman, M. J.   1979, 1980, 1982.   Private communication.

Garbarino, J. R.,   and   H. E. Taylor. 1979.   An  Inductive-coupled
    plasma atomic-emission spectrometric method for routine water
    quality testing.   Appl. Spectrosc. V. 33, No. 3., p. 220-226.

Hedley, A. G., and M. J. Fishman. 1982. Automation of an ion chromatograph for precipitation analysis with computerized data reduction. U.S. Geol. Survey Water-Resources Inv. 81-78., p. 1-33.

Janzer, V. J. 1983. Private communication.

Kaiser, M. L., S. R. Koirtyohann, E. J. Hinterberger, and H. E. Taylor. 1981. Reduction of matrix interferences in furnace atomic absorption with the L'vov platform. Spectrochim. Acta 36B, 8:773-783.

Koch, G. S., Jr., and R. F. Link. 1971. Statistical analysis of geological data, V. 1. Dover, N.Y., p. 154.

Laird, L. B., H. E. Taylor, and V. C. Kennedy. 1986. Snow chemistry of the Cascade-Sierra Nevada Mountains. Environ. Sci. Tech. 20:275-290.

Mandel, J., and L. F. Nanni. 1978. Measurement evaluation, p. 220. In S. I. Inhorn, ed., Quality assurance practices for health laboratories. Amer. Pub. Health Assoc.

Shockey, M. W., and H. E. Taylor. 1984. A method for pretreatment of snow samples to ensure sample integrity for analytical testing. 26th Rocky Mountain Conf., Denver, p. 19.

Shrader, D. E., L. M. Voth, and L. A. Covick. 1983. The Determination of toxic metals in waters and wastes by furnace atomic absorption. Varian Inst. at Work, No. AA-31, p. 1-8.

Skougstad, M. W., M. J. Fishman, L. C. Friedman, D. E. Erdmann, and S. S. Duncan. 1979. Methods for the determination of inorganic substances in water and fluvial sediments. U.S. Geol. Survey Techniques of Water-Resources Inv., 5, Chap. A1, p. 626.

Taylor, H. E., and L. B., Laird. 1985. The measurement and distribution of heavy metals from snow cores obtained from the Cascade-Sierra Nevada Mountains in the Western United States. Proc. Internat. Conf. Heavy Metals in the Environ. Athens, Greece, p. 162-164.

Wershaw, R. L., M. J. Fishman, R. R. Grabbe, and L. E. Lowe. 1983. Methods for the determination of organic substances in water and fluvial sediments. U.S. Geol. Survey Techniques of Water-Resources Inv., Book 5, Chap. A3, p. 22-24.

SECTION II

CHEMISTRY OF GROUND WATER

# CHAPTER 5

# HYDROGEOLOGICAL AND GEOCHEMICAL ASPECTS OF GROUND AND SURFACE WATER POLLUTION ASSOCIATED WITH LEAD AND ZINC MINES IN THE TRI-STATE MINING DISTRICT

Douglas C. Kent, Zuhair Al-Shaieb, David W. Vaden, and Peter W. Bayley, Oklahoma State University, Stillwater, Oklahoma

## ABSTRACT

Waters discharged from flooded mines in the Picher, Oklahoma, field become acidic and transport significant concentrations of metals into surface waters. The oxidation process of the sulfide mineralogy responsible for the generation of acidity begins in waters within the mines and is completed soon after discharge. Conceptual chemical models have been developed and verified using data from selected sampling sites. Four distinct zones have been recognized and include water within the mines, water at discharge sites, and surface waters located proximally and distally downstream from discharge sites. Differences in water chemistry exist between the Picher, Oklahoma, and the Joplin, Missouri, fields within the mineralized district. The most important factor controlling this disparity is the rate of ground-water recharge to the Mississippian aquifer, which serves as the ore host. Differences in recharge can be attributed to differences among the geologic units overlying the mineralized zone.

## INTRODUCTION

The Tri-State mining district occupies portions of northeast Oklahoma, southeast Kansas, and southwest Missouri (Fig. 1). It is the site of zinc and lead sulfide mineralization of the Mississippi Valley type. McKnight and Fischer (1970) report that peak production occurred in 1925, when 387,000 tons of zinc and 101,000 tons of lead were produced. The Joplin, Mo., field is located 20 to 30 mi east of the Picher, Okla., field (Fig. 1), and the two areas are very similar geologically. In each area the ore and gangue deposits are developed in a carbonate host of early to

*Chemical Quality of Water and the Hydrologic Cycle*, Robert C. Averett and Diane M. McKnight (Eds.) © 1987 Lewis Publishers, Inc., Chelsea, Michigan. Printed in the United States of America.

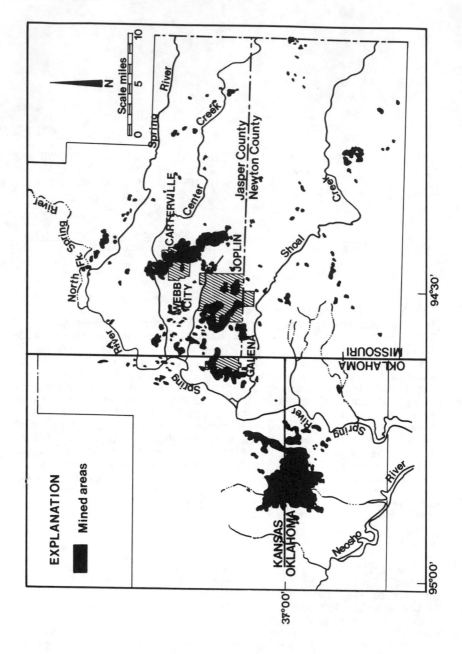

Figure 1.  A map of the Tri-State mining district of Oklahoma,
Kansas, and Missouri which illustrates the distribution
of mine workings.  Two areas are important in this
study:  the Picher, Okla., field and the Joplin, Mo.,
field.

middle Mississippian age (the Boone Formation), which also serves as a shallow, regional aquifer. Both areas are underlain by early to middle Paleozoic limestones, dolomites, and shales. The ore body itself was developed in response to ascending magmatic fluids in a host which was prepared for mineralization by earlier brecciation (McKnight and Fischer, 1970).

Ground waters discharged from flooded zinc and lead mines in the Picher, Okla., and Joplin, Mo., areas may become acidic and transport metals (particularly iron and zinc) into area surface waters. The following four zones have been identified based on the spatial partitioning of ground and surface water geochemistries: ground water within the mines (Zone 1); mine waters surfacing through seeps (Zone 2); surfaced mine waters immediately downstream of the seeps (Zone 3); and surface waters located downstream from Zone 3 (Zone 4).

Geochemical zonation is particularly well developed in the Tar Creek area near Picher, Okla. (Fig. 1). Tar Creek drains the surface of the Picher mining field and receives input from four known sites of mine water discharge. The spatial relationships of geochemical zonation in the Picher field are shown using a three-dimensional perspective in Figure 2A.

The concept of geochemical zonation was applied to the mining fields developed near Joplin, Mo. (Fig. 2B). Ground and surface water geochemistries displayed in the Joplin area are known to show less impact from mine-related contaminants than their correlatives in the Picher area (Feder et al., 1969; and this chapter). Geochemical zonation is believed to exist in the Joplin area, but is much less well developed than in the Picher area.

Hydrologic studies conducted during the course of this investigation indicate that rapid recharge and dewatering of the mines occurs in the Joplin area. The Joplin area lacks the cover of an overlying Pennsylvanian shale, which is present in the Picher area. Principal recharge to the aquifer and ore host (the Mississippian Boone Formation) in the Picher area is limited to the region east of the contact with overlying shales.

## GEOCHEMISTRY OF ACID GENERATION

The generation of acidity ($H^+$) occurs in two stages. The initial stage occurs within the flooded mines (Zone 1) as a result of the oxidation of sulfide to sulfate (Eq. 1). One mole of sulfuric acid is generated for each mole of iron sulfide which is oxidized. This reaction proceeds in the presence of dissolved oxygen, which is provided by surface waters recharging the shallow aquifer. Barnes and Romberger (1968) report that a dissolved oxygen concentration of as low as $10^{-60}$ atmospheres is sufficient to drive the reaction.

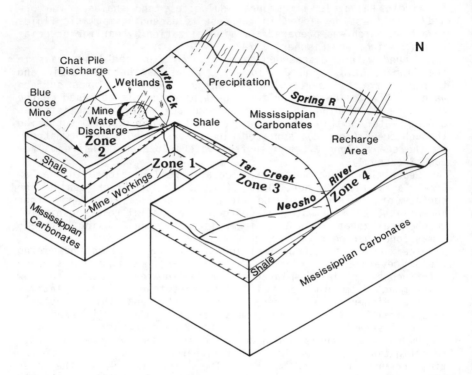

Figure 2A.    Conceptualized model of the Picher, Okla., area showing
              spatial relations exhibited by four recognized geo-
              chemical zones. Ground waters within the mines
              (Zone 1) surface through seeps (Zone 2) and are
              discharged into Tar Creek, which displays Zone 3
              characteristics. Waters of Zone 3 flow into the
              Neosho River, which functions as Zone 4 waters.

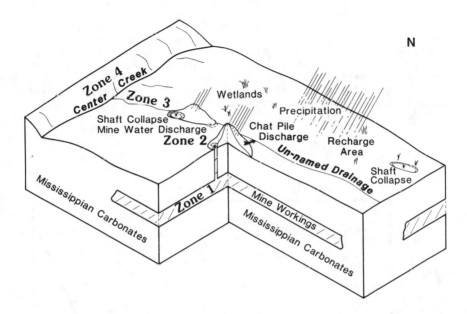

Figure 2B.  Conceptualized model of the Joplin, Mo., area showing
spatial relations exhibited by four recognized geochemical zones.  Ground waters within the mines
(Zone 1) surface through seeps (Zone 2) and are
discharged into an unnamed drainage, which displays
Zone 3 characteristics.  Waters of Zone 3 flow into
the Center Creek, which functions as Zone 4 waters.

Acidity generated within the mines is largely neutralized via oxidation of ferrous iron to ferric iron (Eq. 2) and during the migration of the contaminated ground waters through the carbonate aquifer (Eq. 3). Sites of mine-water discharge (Zone 2) reflect the neutralization described in Equation 3. Samples of these discharging waters exhibit a moderate pH (Fig. 3) and are buffered with bicarbonate concentrations ranging between 200 and 500 mg/L (Fig. 4). $CO_2$ gas is evolved from surfacing mine waters at sites of discharge, occurring as a product of acid neutralization when the saturation point of carbonic acid in solution has been exceeded. In addition, discharging mine waters contain large concentrations of iron, zinc, and sulfate. Analyses by atomic absorption spectrometry indicate that concentrations of iron and zinc range between 100 and 300 mg/L each. Sulfate, as analyzed colorimetrically, ranges between 800 and 2,500 mg/L, commonly exceeding 1,500 mg/L.

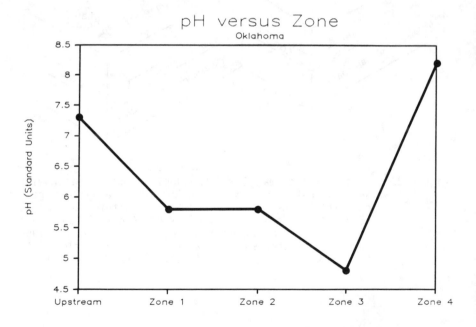

Figure 3.   Average values of pH from the Oklahoma study area plotted with respect to the geochemical zone represented.

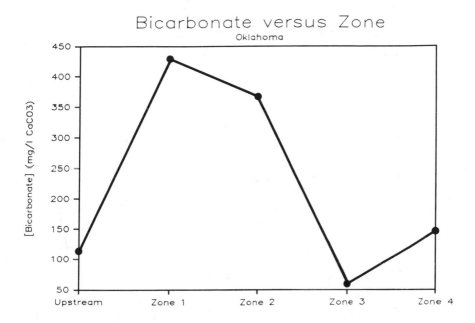

Figure 4.    Average bicarbonate concentrations from the Oklahoma
study area plotted with respect to the geochemical zone
represented.

The second stage of acid generation occurs in surfaced mine
waters near points of discharge (Zone 3), in response to the
precipitation of ferric hydroxide (Eq. 4).  For every two moles of
ferric hydroxide precipitated, three moles of sulfuric acid are
produced.  This stage of acid generation is sufficient to overwhelm
the high bicarbonate concentrations observed at Zone 2 sites and
results in the gradual depression of pH in Zone 3.  The pH of
Zone 3 waters observed in Oklahoma averages 4.8 (Fig. 3).

Field observations indicate that the pH of Zone 3 waters
decreases as they flow downstream, away from points of discharge
(Fig. 5).  The precipitation of ferric hydroxide is initiated soon
after discharge, and an orange floc is deposited as a sediment in
the surface drainage channels.  The acid generated is neutralized
for about 2 mi downstream of the discharge sites as a result of the
reaction with bicarbonate produced in Zone 2 (Fig. 5).  This
buffering agent is reduced downstream (Fig. 6) as the precipitation
of ferric hydroxide and the generation of sulfuric acid continues.
After 2 to 3 mi bicarbonate concentrations are very low or
nonexistent, and the pH is significantly depressed (Figs. 5 and 6).

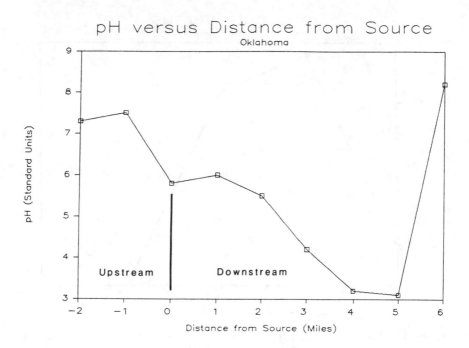

Figure 5.    Average values of pH from the Oklahoma study area
plotted with respect to approximate distances (in miles)
up and downstream from a site of mine water discharge.
Note that in this diagram distance = 0 is the discharge
site.    Upstream distances are indicated by negative
values, and downstream distances are indicated by positive values.    The data represented by this plot were
obtained along Tar Creek, in the Oklahoma study area.

The average value of pH at locations 4 to 5 mi downstream of
discharge sites along Tar Creek is about 3.2 (Fig. 5).    The pH
remains depressed until the waters flow into larger bodies of
relatively well-buffered surface waters (Zone 4).    Bicarbonate
concentrations in Zone 4 waters average 146 mg/L (Fig. 4).    The
acidic waters of distal Zone 3 flow into Zone 4 surface waters,
where they are neutralized (Fig. 3).    Equation 5 illustrates the
neutralization of Zone 3 acids reacting with the bicarbonate of
Zone 4 waters.
The reactions of the recognized geochemical zones are summarized below.    Important publications concerning the generation of
acids as a result of the oxidation of sulfide minerals have
provided the basis for the equations presented below (Barnes and
Romberger, 1968; Garrels and Thompson, 1960; Hawley and Shikaze,
1971; Krauskopf, 1979; and Stumm and Morgan, 1970).

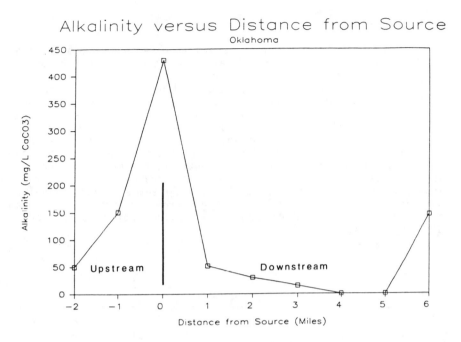

Figure 6. Average values of bicarbonate concentrations from the Oklahoma study area plotted with respect to approximate distances (in miles) up and downstream from a site of mine-water discharge. Note that in this diagram distance = 0 is the discharge site. Upstream distances are indicated by negative values, and downstream distances are indicated by positive values. The data represented by this plot were obtained along Tar Creek, in the Oklahoma study area.

Zone 1

$$2FeS_2 + 2H_2O + 7O_2 \longrightarrow 2FeSO_4 + 2H_2SO_4 \tag{1}$$

$$4FeSO_4 + 2H_2SO_4 + O_2 \longrightarrow Fe_2(SO_4)_3 + 2H_2O \tag{2}$$

Zone 2

$$CaCO_3 + H^+ \longrightarrow Ca^{2+} + HCO_{3-} \tag{3}$$

Zone 3

$$Fe_2(SO_4)_3 + 6H_2O \longrightarrow 2Fe(OH)_3 + 3H_2SO_4 \tag{4}$$

Zone 4

$$HCO_{3-} + H^+ \longrightarrow H_2CO_3 \tag{5}$$

REGIONAL VARIABILITY IN WATER QUALITY

Instances of acidic surface waters associated with discharging mine waters were found to be largely restricted to the portion of the district centered around Picher, Okla. Feder et al. (1969) report that surface waters near the Joplin, Mo., portion of the district may exhibit values of pH as low as 4.6. Values of pH measured in the Joplin area during the course of this study were decidedly neutral. The average pH of Zone 3 waters in Missouri was 6.8 (Fig. 7) and 4.8 in Oklahoma (Fig. 3). The concentrations of iron and zinc in waters from Oklahoma discharge sites average 217 and 180 mg/L, respectively. In contrast, the concentrations of iron and zinc in waters from Missouri discharge sites were much lower, averaging 2.3 and 1.7 mg/L for iron and zinc respectively. Bicarbonate concentrations in waters from Oklahoma discharges averaged 367 mg/L (Fig. 4) and 172 mg/L from waters of Missouri discharge sites (Fig. 8).

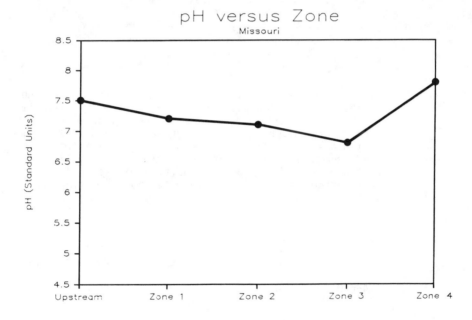

Figure 7.  Average values of pH from the Missouri study area plotted with respect to the geochemical zone represented.

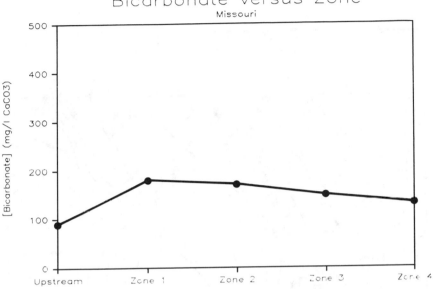

Figure 8.    Average values of bicarbonate concentrations from the
            Missouri study area plotted with respect to the
            geochemical zone represented.

A significant geologic dissimilarity exhibited in the two
areas is the presence or absence of an overlying Pennsylvanian
shale cover. Such a shale is present in the Picher, Okla., area
but is absent in the Joplin, Mo., area because of erosion (Fig. 1).
The presence or absence of this shale cover affects the rate of
recharge to the Mississippian aquifer. Differences in the rate of
recharge is believed to be the most significant factor controlling
the geochemical variability between the Picher and Joplin study
areas.

## HYDROGEOLOGY OF THE TRI-STATE DISTRICT

Figure 9 is a potentiometric surface map of the Oklahoma-Kansas
portion of the Tri-State District. The map illustrates an
unconfined, gaining stream condition in areas east of the contact
with the Pennsylvanian shale cover. Mounding of the surface
indicates that recharge to the Mississippian aquifer is occurring
in this unconfined area. Minor recharge to the aquifer may also
occur in shale covered areas through mine shafts, wells, and
collapse features. The map (Fig. 9) illustrates a confined, but
locally gaining stream condition in the area overlain by shale.

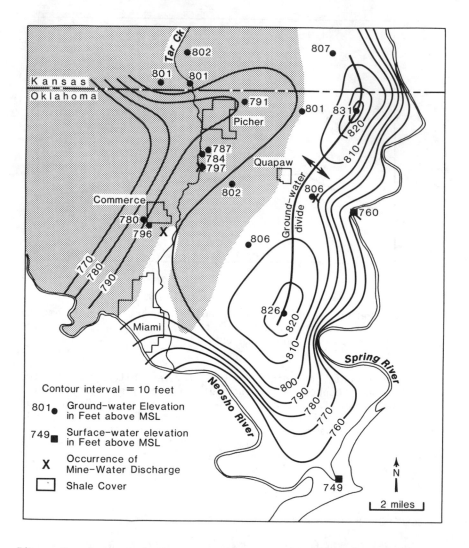

Figure 9.   A map of the potentiometric surface developed in the Oklahoma and Kansas portion of the Tri-State district. The map indicates that principal recharge to the shallow Mississippian aquifer occurs east of the contact with overlying Pennsylvanian shales.

Ground water from within the mines is allowed to discharge in areas where the potentiometric surface intersects the ground surface. The intersection of these two surfaces occurs along portions of Tar Creek in places where the overlying, low-permeability shales have been removed by downcutting in the channel.

The potentiometric surface map of the Joplin, Mo., area (Fig. 10) illustrates an unconfined, gaining stream condition. The Joplin area lacks an overlying shale cover and recharge to the Mississippian aquifer in this area is indicated. Observations in the field indicate that discharges in the unconfined intervals east of overlying shale cover are fewer in number and of shorter duration than the discharges occurring in shale-covered areas. Sites in nonshale-covered areas commonly discharge surfacing mine waters for only one or two days following a precipitation event. Two such sites have been identified in Missouri and one in Oklahoma (Figs. 9 and 10). By contrast, mine waters are continuously discharged from at least four sites in the shale-covered area in Oklahoma (Fig. 9).

These observations indicate that the overlying Pennsylvanian shale cover, where present, retards recharge. The retardation serves to prolong the duration of mine-water discharges in shale covered areas. Dewatering of the mines following a precipitation event is more rapid in areas lacking shale cover. Rapid dewatering of the mines in nonshale-covered areas shortens the duration time of mine-water discharges, resulting in a high volume release within a short period of time. Short term high volume discharges of ground and surface waters buffered with bicarbonate in the nonshale-covered areas have a greater potential for diluting the concentrations of mine-related contaminants occurring at the discharge sites. Lower concentrations of mine-related contaminants in waters discharged from areas east of the shale contact (including the Joplin area) are believed to be the result of this dilution associated with the rapid recharge and dewatering of the mines, following a precipitation event. Since dewatering of the mines is less rapid in shale-covered portions of the district, the residence time of ground waters within the mines is increased and geochemical reactions occurring within the mines (Eqs. 1 and 2) are allowed to proceed more nearly to completion. The result is a continuous discharge of mine waters with high concentrations of mine-related contaminants in areas confined by overlying Pennsylvanian shales.

## CONCLUSIONS

Four geochemical zones for ground waters discharged from mines to area surface waters in the Picher, Okla., and Joplin, Mo., areas have been established. They are: Ground water within the mines (Zone 1); surfacing mine waters (Zone 2); surfaced mine waters proximal to discharge sites (Zone 3); and surfaced waters distal to discharge sites (Zone 4).

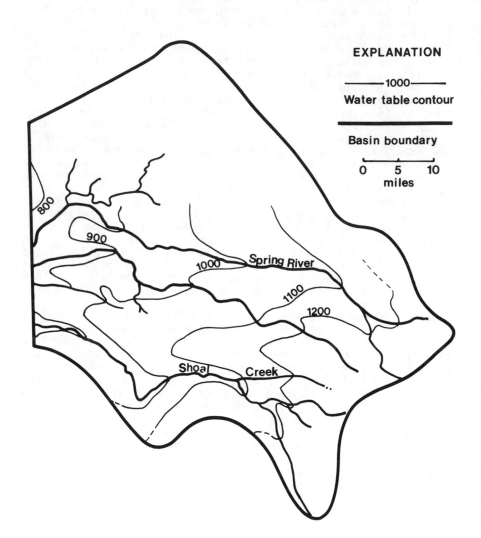

Figure 10.   A map of the potentiometric surface developed in the
             Missouri portion of the Tri-State district.   The map
             indicates an unconfined, gaining stream condition to
             exist in this area.   The Joplin area lacks shale cover,
             and recharge to the shallow Mississippian occurs.

Acid generation occurs in two stages: In ground waters occurring within the mines as a result of the oxidation of sulfide to sulfate and within surface waters as a result of the precipitation of ferric hydroxide ($Fe(OH)_3$). Acid generated within the mines is largely neutralized during the migration of mine-related waters through the carbonate aquifer (the Mississippian Boone Formation). Acid is also consumed during the oxidation of ferrous iron to ferric iron, occurring within the mines. Surfacing mine waters exhibit a nearly neutral pH and a high bicarbonate concentration. Quantities of acid generated as a result of the precipitation of $Fe(OH)_3$ are sufficient to overcome the high bicarbonate concentrations present at discharge sites. Once the bicarbonate buffer is destroyed, the pH decreases sharply. The pH remains low in impacted surface waters until confluence with larger, well-buffered surface waters. Acid is then neutralized as a result of reacting with bicarbonate buffers developed in the larger bodies of surface water. Zinc and lead sulfide bearing Mississippian carbonates are aerially extensive aquifers and receive principal recharge in areas lacking an overlying shale cover. A minor source of recharge may exist in shale-covered areas as surface waters enter through mine shafts, wells, and collapse features.

The Pennsylvanian shale cover acts as a confining unit. In places where the shale is locally absent, such as along portions of Tar Creek, mine waters are allowed to surface as seeps and springs. The shale cover retards the dewatering of the mines following a precipitation event, prolonging the duration of the discharge. Negative impacts to surface-water quality are much more significant in areas where the shale cover is present (Oklahoma). The discharges of prolonged duration time occurring in shale-covered areas exhibit higher contaminant concentrations than discharges occurring in nonshale-covered areas. This effect is probably related to the increased potential for dilution of mine waters associated with the short-term, high-volume discharges observed in the nonshale-covered areas. Persistent discharges in shale-covered areas suggest a longer residence time for waters within the mines, so that oxidation reactions occurring in the mines may proceed more nearly to completion than those occurring in nonshale covered areas.

## REFERENCES

Barnes, H. L., and S. B. Romberger. 1968. Chemical aspects of acid mine drainage. J. Water Pollut. Contr. Fed. 40:371-384.

Feder, G. L., J. Skelton, H. G. Jeffery, and E. J. Harvey. 1969. Water resources of the Joplin area, Missouri. U.S. Geol. Survey, Water Resources Division, Report 24.

Garrels, R. M., and M. E. Thompson. 1960. Oxidation of pyrite by iron sulfate solutions. Am. J. Sci. 258:57-67.

Hawley,  J.  R.,  and  K.  H.  Shikaze.  1971.   The  problem  of  acid  mine
    drainage  in  Ontario.   Canadian  Mining  J.  92:82-93.

Krauskopf,  K.  B.  1979.   Introduction  to  geochemistry.   McGraw-Hill
    Book  Co.  New  York.

McKnight,  E.  T.,  and  R.  P.  Fischer.  1970.   Geology  and  ore  deposits
    of  the  Picher  field  Oklahoma  and  Kansas.  U.S.Geol.  Survey
    Prof.  Paper  588.

Stumm,  W.,  and  J.  J.  Morgan.  1970.   Aquatic  chemistry:   an  intro-
    duction  emphasizing  chemical  equilibria  in  natural  waters.
    Wiley-Interscience.   New  York.

CHAPTER 6

GEOCHEMISTRY OF GROUND WATER IN THE SHALLOW ALLUVIAL
AQUIFER, CARSON DESERT, WESTERN NEVADA

Michael S. Lico, Alan H. Welch, and Jennifer L. Hughes,
U.S. Geological Survey, Carson City, Nevada

ABSTRACT

The ground-water chemistry of the shallow alluvial aquifer
(<10 m deep) in the Carson Desert, 90 km east of Reno, Nevada, was
studied by using a mass-balance model that was constructed from
water-quality data, mineralogic data, and thermodynamic calculations.  The purpose of this study is to describe the geochemical
and hydrologic processes responsible for the observed water
composition.  A small part of the aquifer was chosen as the study
area because it adequately represents the range in observed water
chemistry in the aquifer throughout the Carson Desert.
The shallow alluvial aquifer is composed of deltaic and fluvial
unconsolidated sand, silt, and clay of Pleistocene to Holocene age.
The sediments consist of quartz, feldspars, volcanic-lithic
fragments, micas, chlorite, illite, calcite, smectite, and other
weathering products from volcanic and granitic rocks.  The ground-
water quality in the shallow alluvial aquifer ranges from a
slightly saline, sodium-calcium-bicarbonate-sulfate type (about
1,100 mg/L dissolved solids) to a very saline type (about 11,000
mg/L dissolved solids) dominated by sodium, sulfate, and bicarbonate.  The water, in places, contains relatively high concentrations of dissolved organic carbon (>80 mgC/L) which is predominantly humic substances as indicated by the yellow color.  The pH
ranges from near neutral to 9.2, whereas the $E_h$ (electrochemical
potential) ranges from strongly oxidizing to slightly reducing.
The mass-balance model indicates that the major-element
chemistry is controlled by precipitation-dissolution reactions,
adsorption-desorption processes, and evapotranspiration.  Ground
water that moves along a flowpath dissolves minerals (the most
significant are feldspars, chlorite, ferric oxyhydroxide, and
calcite).  Silica, ferrous monosulfide, calcium montmorillonite, and calcite are precipitating in parts of the aquifer.

*Chemical Quality of Water and the Hydrologic Cycle*, Robert C. Averett and Diane M. McKnight (Eds.) © 1987 Lewis Publishers, Inc.,
Chelsea, Michigan. Printed in the United States of America.

Exchange reactions on clay-mineral surfaces involve the adsorption of $Ca^{2+}$ and the desorption of $Na^+$. In the deeper part of the aquifer, below about 3.5 m, organic matter of sedimentary origin is being oxidized to bicarbonate, sulfate is being reduced to sulfide, and ferrous monosulfide is being precipitated.

## INTRODUCTION

The Carson Desert in western Nevada (Fig. 1) is an arid, hydrologically closed basin. Water entering the basin discharges into one of two terminal sinks, the Carson Sink in the north or the complex of Carson Lake-Fourmile Flat-Eightmile Flat sinks to the south.

In 1905, the Newlands Reclamation Project began delivering water to the Lahontan Valley, near Fallon, for irrigation purposes. Since that time, the ground-water levels have risen as much as 18 m in some parts of the valley. Most of the rural community that surrounds Fallon uses ground water from the shallow alluvial aquifer. Thus, understanding the geochemistry of this water system, including changes that occur due to irrigation and the related rise in the water table, is important to the health and economic life of the community.

In this chapter, the major geochemical and hydrologic processes responsible for the observed chemistry of the ground water in the shallow alluvial aquifer are identified. A small ($0.12$ km$^2$) part of the shallow alluvial aquifer was selected as the study area because it adequately represents the range in water composition found in the aquifer in the Carson Desert. The study area, termed Dodge Ranch, is one of many ground-water "cells" created by a network of canals and drains in the Fallon area. Detailed chemical analysis of the ground water and sediment, mineralogic characterization of the sediment, and thermodynamic calculations of the ground waters along the flow path were used to determine the geochemistry of this area.

## METHODS OF STUDY

Wells used for sampling were drilled with a 17.8-cm hollow-stem auger without drilling fluids. These wells were cased with 5-cm i.d. polyvinyl chloride (PVC) pipe and premanufactured machine-slit PVC well screens (0.15-mm slots). Washed Monterey sand was added to the annulus to cover the screened interval, then bentonite pellets were added to create a seal above the screen of at least 0.3-m vertical thickness. The remaining annular space was filled with drill cuttings. A surface seal was emplaced with bentonite pellets. Piezometers were installed in 7.6-cm diameter augered holes and consist of 5-cm i.d. PVC pipe with premanufactured PVC screens (0.15-mm slots) at the bottom. Annular space was filled with drill cuttings and a surface seal was emplaced with bentonite pellets.

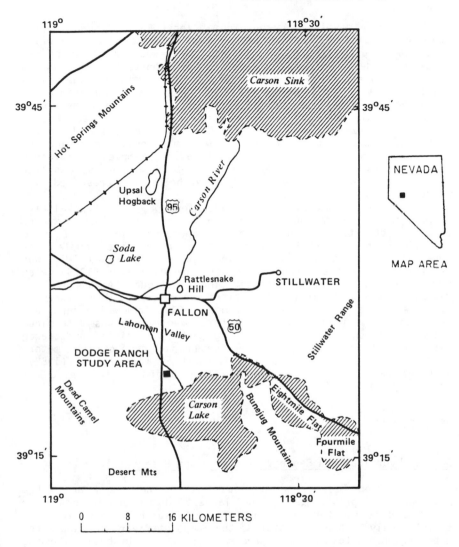

Figure 1.  Location of study area and other geographic features.

During drilling operations, cores were taken through the hollow-stem auger with a stainless-steel split-spoon drive sampler fitted with a polycarbonate liner.  Bulk sediment samples also were obtained as the auger brought cuttings to the land surface.  Lithologic logs of the drilled sediments are presented in Lico et al. (1986).  X-ray diffraction analysis, grain-size distribution, thin-section analysis, and porosity and hydraulic conductivity tests were performed on the sediment samples.  Details of the analytical procedures can be found in Lico et al. (1986).

Wells were developed by pumping at least 10 well-bore volumes (the volume of water within the casing) of water from the well. Prior to sampling ground water, each well was pumped until pH, specific conductance, and temperature were constant. The sampled water was filtered through a 0.45-μm pore-size membrane housed in a polycarbonate filter holder. Filtered water was used to rinse all prewashed sample bottles and was collected directly into the rinsed 500-mL polyethylene bottle. Samples were preserved according to Brown et al. (1970), with the exception of samples for dissolved iron which were filtered through a 0.1-μm pore-size membrane and acidified with ultra-pure hydrochloric acid. Samples for organic carbon analysis were filtered through a 0.45-μm pore-size silver filter housed in a stainless-steel filter holder and preserved with mercuric chloride.

Field determinations were made for pH, $E_h$, alkalinity (expressed as bicarbonate and carbonate), dissolved oxygen, specific conductance, and sulfide. Temperature, pH, and $E_h$ were determined in a flow-through cell. The pH of the samples was measured using a calibrated combination electrode. $E_h$ was determined using the method of Thorstenson and Fisher (1979) which employs a platinum-calomel combination electrode. Alkalinity was determined using the incremental titration method of M. C. Yurewicz (U.S. Geological Survey, written comm., 1981), from which carbonate and bicarbonate were calculated. Dissolved oxygen was determined using a polarographic oxygen-sensing probe. Specific conductance was measured using a temperature-compensated conductivity meter. Sulfide was determined using the method of Lico et al. (1982). Aluminum was extracted in the field according to the method of Barnes (1975).

HYDROGEOLOGY

## General Hydrology and Geology of the Carson Desert

The Carson Desert, which lies within the Basin and Range physiographic province, is the largest intermountain basin in Nevada and is the terminal sink for the Carson River and, in years of relatively high precipitation, the Humboldt River. Ground water generally flows toward one of the major sinks in the basin (Morgan, 1982; Olmsted et al., 1984; Glancy, 1986). North of Rattlesnake Hill (Fig. 1) ground water flows north-northeast toward Carson Sink. South of this latitude, including the area of study, ground water generally flows south-southeast toward Carson Lake, Fourmile Flat, and Eightmile Flat.

The Carson Desert is surrounded by mountains composed of a wide variety of igneous, sedimentary, and metamorphic rocks ranging in age from Triassic to Quaternary. The Carson River probably has been the primary source of sediment supplied to the study site in the southern Carson Desert since the Pleistocene age. The Carson River derives most of its sediment from granitic terrane in the Sierra Nevada and from acidic to basic volcanic rocks that are present throughout much of the drainage basin. The basin itself is underlain, in descending order, by (1) Holocene post-Lake Lahontan interbedded fluvial and eolian sediments, (2) Pleistocene sediments of Lake Lahontan, (3) sedimentary and volcanic rocks of Quaternary and Tertiary age, and (4) pre-Tertiary igneous and sedimentary rocks. Morrison (1964) and Willden and Speed (1974) provide more detail on the stratigraphy and geology of the Carson Desert area.

## Hydrogeology of the Dodge Ranch Site

The alluvial sediments forming the shallow aquifer at the study site are deposits of the Carson River. Soil types in the study area range from a sandy loam to clay (Dollarhide, 1975). Sediments beneath the soil range from coarse, poorly sorted sand to silty clay (Lico et al., 1986) and locally range in thickness from about 6 to 8.2 m (Fig. 2). The alluvium is generally poorly sorted fine to very fine sand with grain-size distributions skewed towards the finer grain sizes. Clay also is present both as small lenses and pods in the sandy sediment or as laterally extensive, thick layers. Coarse-grained sand is present locally, but typically it is very poorly sorted and contains much fine-grained sediment. These deposits represent the Fallon Formation of Holocene age which typically has a heterogeneous lithology characteristic of flood-plain sediments (Morrison, 1964). Hydraulic conductivities in the shallow aquifer range from $8.35 \times 10^{-7}$ to $5.47 \times 10^{-3}$ m/d. This wide range results from the varying amounts of fine sediment (silt and clay) in the sands.

A thick clay sequence (>10 m) underlies the alluvium and probably is a lacustrine clay unit of the Sehoo Formation of Quaternary age. It is almost impermeable in comparison to the overlying sands and silts of the shallow aquifer. The low permeability of the clay and the small potentiometric head difference between the shallow water-table aquifer and the deeper aquifer beneath the clay indicates that there is little or no vertical flow between the aquifers.

Well locations were selected to allow sample collection along a ground-water flow path. Sampling water along a flow path permits the use of mass-balance calculations to model geochemical reactions. The shallow flow system is relatively simple with the flow directions largely controlled by drainage ditches that border the field to the west and south. These drainage ditches act as line sinks for the shallow ground water (Fig. 3). Ground-water flow in the vicinity of well sites 15 through 17 is approximately from east to west as indicated by over 20 synoptic head measurements, taken over a period of about 10 months (Lico et al., 1986). The water table is shallow, ranging from 0.1 to 2.8 m below land surface. The water levels are controlled primarily by irrigation schedules and vary annually by as much as 0.5 m in the vicinity of the sampled wells. The aquifer is recharged by flood irrigation in the study area approximately 10 times per year. During each application about 740 $m^3$ of water is applied to the field from an irrigation ditch on the east side. Drainage occurs

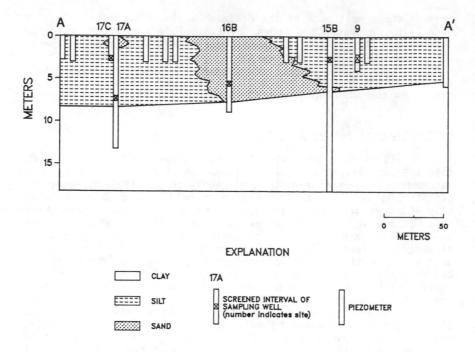

Figure 2.    Geologic section along line A-A' showing lithology of
sediments at Dodge Ranch study area.   Location of A-A'
is shown in Figure 3.

on the west and south sides of the field.   The drains merge at the
southwest corner of the field and flow to Carson Lake.   The crop
grown on the field is pasture grass with typical root depths of
between 1 and 2 m.   The pasture grass is grazed by cattle
throughout the year and ammonium sulfate fertilizer is applied to
the field once a year at a rate of 50 g/m$^2$.

## Mineralogy of Sediment at Dodge Ranch Site

The minerals comprising the aquifer sediments are the sources
and sinks for dissolved constituents in ground water.   As such,
they affect the aqueous geochemistry of the ground water.   Textural
relations between grains and mineral-surface features are indicative of whether a mineral is precipitating or dissolving.   Thus,
mineralogic and petrologic examinations are useful in evaluating
the internal consistency of reaction-path models.

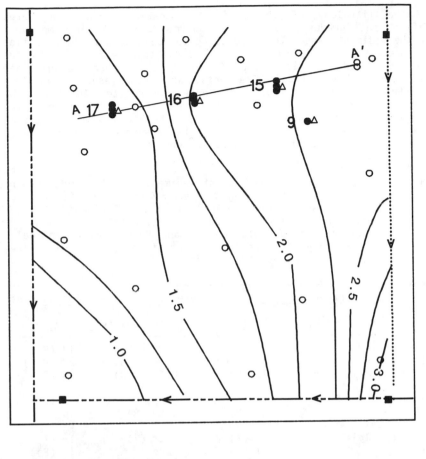

## EXPLANATION

0                    100

METERS

17 ● SAMPLING WELL

   ○ PIEZOMETER

   ■ STAFF GAUGE

   △ SOIL—MOISTURE SAMPLER

-- — IRRIGATION DRAIN (SINK), ARROWS SHOW FLOW DIRECTION.

......... IRRIGATION DITCH (SOURCE), ARROWS SHOW FLOW DIRECTION.

——— CONTOUR OF WATER TABLE(IN METERS ABOVE 1200 METERS)

Figure 3.   Water-level   contours   of   potentiometric   surface   on
            August 31, 1984 at the Dodge Ranch study area (T.17N
            R.29E, SW1/4NW1/4 of sec. 6). (Water-level contours are
            in meters above 1,200 meters).

The lithology of the sediments underlying the study area reflect the igneous nature of the source rocks. Major components of the sediment were determined using optical and X-ray diffraction techniques. Microscopic examination of thin-sections indicates that the sediment is composed mostly of quartz, plagioclase feldspar, volcanic-lithic fragments, and orthoclase. Other less abundant minerals include hornblende, muscovite, biotite, and opaque minerals. Table 1 shows the mineralogic composition of the sediments at the Dodge Ranch site and whether the minerals composing the sediment are primary (detrital), secondary (precipitated in situ), or show evidence of dissolution.

Table 1.   Mineralogic Composition of Sediment from the Dodge Ranch Study Area. Percentages of each Mineral were Calculated from X-ray Diffraction Analysis of the Sediment. Primary, Detrital (P), Secondary, Precipitated (S), and Evidence for Dissolution (D) were Determined by Optical Microscopy. (Plagioclase is sum of anorthite plus albite, np = not present, tr = trace.)

| Mineral | Percentage of sediment core in indicated mineral class | | |
| | Well--- Zone--- | 15B Shallow | 16B Intermediate | 17A Deep |
|---|---|---|---|---|
| Quartz | | 24 (P,S) | 32 (P,S) | 13 (P,S) |
| Plagioclase | | 51 (P,D) | 45 (P,D) | 23 (P,D) |
| Orthoclase | | 18 (P) | 21 (P,D) | 9 (P,S) |
| Calcite | | 1 (S) | np | 51 (S)[1] |
| Chlorite | | tr (P) | tr (P) | np |
| Heulandite/ Clinoptilolite | | tr (?) | tr (?) | tr (?) |
| Illite | | 1 (P) | np | 1 (P) |
| Calcium- Montmorillonite | | 3 (S) | 1 (S) | 1 (S) |
| Other | | 2 | 1 | 2 |
| Total | | 100 | 100 | 100 |

[1]Calcite present as coatings on grains which was precipitated before irrigation began in the area. This calcite may have been a caliche zone before the rise in water table.

Many of the feldspar grains, particularly plagioclase, show evidence of dissolution. However, quartz grains appear fresh with no evidence of dissolution; instead, secondary silica coats some quartz grains. Calcite is present in the deeper part of well 17A (7.3-7.6 m). The calcite is clearly secondary at this depth interval, occurring as concentric rinds on grains of feldspar, quartz, and volcanic-lithic fragments, and possibly is an old caliche layer formed before irrigation began in the area. X-ray diffraction patterns of the sediment fractions larger than clay size (2 µm) confirms the presence of quartz, orthoclase, albite, anorthite, hornblende, and muscovite. Other minerals identified were microcline, gypsum, hypersthene, actinolite, augite, biotite, tremolite, and hematite. Dolomite and calcite were identified in some samples. Traces of the zeolites heulandite and clinoptilolite were found in some samples. The clay-sized sediment fraction (<2 µm) of samples from the Dodge Ranch site also were analyzed by X-ray diffraction techniques. Smectite (most likely montmoril- lonite), illite, chlorite, mixed-layer illite-montmorillonite, and mixed-layer montmorillonite-chlorite were identified using the techniques of Carroll (1970). Halloysite, dickite, and attapulgite were tentatively identified in a few samples. For a complete list of minerals found by X-ray diffraction see Lico et al. (1986).


AQUEOUS GEOCHEMISTRY

The shallow alluvial aquifer at the Dodge Ranch study area can be divided into four zones on the basis of the major-element composition of the ground water and position within the flow system. These zones are: (1) A shallow zone; (2) an intermediate zone; (3) a deep zone; and (4) an evapotranspiration zone (Fig. 4). These zones were defined on the basis of five synoptic samplings from wells 9, 15A-C, 16A-C, and 17A, C, and D and soil-moisture samplers located at these same well sites. The ground-water chemistry is variable and ranges from a slightly saline (about 1,100 mg/L dissolved solids) sodium-calcium-bicarbonate-sulfate type water to a very saline (about 11,000 mg/L dissolved solids) sodium sulfate-bicarbonate type (Fig. 5). The dissolved solids content of the water increases as the water moves along the east- to-west flow path. Analyses representative of the shallow, intermediate, deep, and evapotranspiration zones are shown in Table 2. Geochemical reactions for the evapotranspiration zone are not modeled in this report because of the difficulty in quantifying the inputs into this zone. However, water from both flood irrigation and the shallow zone probably contribute to the water quality of the evapotranspiration zone.

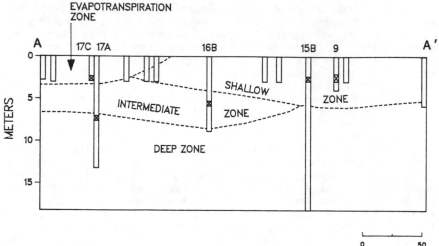

Figure 4.   Cross-section along line A-A' showing ground-water zones
            in the shallow alluvial aquifer at Dodge Ranch study
            area.  Location of A-A' is shown in Figure 3.


## Geochemical Evolution of the Ground Water

Mass balance calculations, as described by Plummer and Back
(1980), were used to tentatively identify the reactions controlling
the observed water chemistry at the Dodge Ranch study area.  This
approach consists of deducing plausible reactions on the basis of
differences in water chemistry at discrete points along a flow
path.  The net change in dissolved species between points represents the evolutionary history of a ground water system from a
chemical standpoint.  The mass-balance calculations solve a series
of simultaneous equations representing chemical reactions.  Results
consist of a series of stoichiometric coefficients for the
specified reactions that are consistent with the net changes in
dissolved species between the two points. Coefficients for the
geochemical reactions were calculated by using the computer program
BALANCE (Parkhurst et al., 1982).   For the purpose of this
discussion, the analytical data for irrigation water and four wells
(Table 2) have been selected to represent various parts of the
hydrologic system.   Three reaction steps were modeled; recharge
water to shallow water, shallow water to intermediate water, and
intermediate water to deep water.

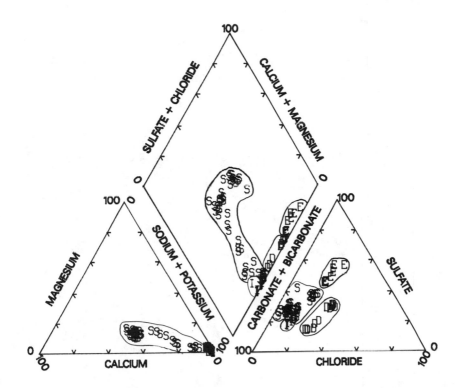

Figure 5.    Chemical characteristics of ground water at the Dodge
Ranch study area showing changes in water composition as
water proceeds along flow path. (Symbols in diagram
are:   S = shallow ground water, I = intermediate ground
water, D = deep ground water, and E = evapotranspiration
ground water.)

The approach employed here involves the assumption that the
water samples are along a flow path and represent the chemical
evolution of water resulting from geochemical processes.   Because
mixing and dispersion are not included in the mass balance calculations, this assumption is basically one of piston flow.   The phases
included in the plausible reactions have been selected on the basis
of mineralogic information, hydrologic data, isotopic composition
of solid and aqueous phases, and thermodynamic calculations.   The
phases chosen for the model are pure end-member compositions and
these are shown in Table 3.   The model is not a unique solution for
the data set but does represent a set of reactions which are consistent with presently available aqueous and solid phase data and
thermodynamic calculations.    The mass balance approach cannot
distinguish between congruent and incongruent dissolution, thus
these processes may both be occurring in this ground-water system.

Table 2.  Chemical Composition of Ground Water from the Dodge Ranch Study Area[1].

| Constituent | Irrigation water Recharge 4-10-85 | Well 15B Shallow 2-26-85 | Well 16B Intermediate 2-26-85 | Well 17A Deep 2-25-85 | Well 17C Evapotranspiration 2-25-85 |
|---|---|---|---|---|---|
| Temperature (°C) | 20.0 | 14.0 | 12.0 | 11.5 | 11.0 |
| pH (pH units) | 8.17 | 7.18 | 9.20 | 8.28 | 7.94 |
| $E_h$ (millivolts) | -- | 340 | 193 | 143 | 420 |
| Calcium (mg/L) | 30 | 120 | 1.3 | 1.4 | 48 |
| Magnesium (mg/L) | 7.2 | 23 | .8 | 21 | 75 |
| Sodium (mg/L) | 41 | 160 | 510 | 2,800 | 3,400 |
| Potassium (mg/L) | 4.9 | 3.8 | 9.5 | 55 | 36 |
| Bicarbonate (mg/L) | 140 | 400 | 660 | 3,100 | 2,600 |
| Carbonate (mg/L) | <1 | <1 | 140 | 48 | <1 |
| Sulfate (mg/L) | 46 | 260 | 200 | 1,800 | 3,500 |
| Chloride (mg/L) | 18 | 86 | 87 | 1,300 | 1,100 |
| Fluoride (mg/L) | .3 | .7 | 3.7 | 4.8 | 1.7 |
| Silica (mg/L) | 14 | 36 | 34 | 24 | 21 |
| Dissolved organic carbon (mg/L) | -- | 10.3 | 35.7 | 81.4 | 26.3 |
| Dissolved oxygen (mg/L) | 8.5 | .7 | <.2 | <.2 | -- |
| Log $pCO_2$ (atm) | -3.07 | -1.69 | -3.42 | -1.94 | -1.70 |
| Sulfide (µg/L) | -- | <3 | 12 | 15 | -- |
| Dissolved solids (mg/L)[2] | 302 | 1,090 | 1,660 | 9,220 | 10,800 |
| Manganese (µg/L) | 13 | 1,100 | 80 | 90 | 1,500 |
| Iron (µg/L) | 36 | 70 | 67 | 11 | 56 |

Table 2.  Chemical Composition of Ground Water from the Dodge Ranch Study Area--Continued.

| Constituent | Sample name---<br>Zone----------<br>Date----------- | Irrigation water<br>Recharge<br>4-10-85 | Well 15B<br>Shallow<br>2-26-85 | Well 16B<br>Intermediate<br>2-26-85 | Well 17A<br>Deep<br>2-25-85 | Well 17C<br>Evapotranspiration<br>2-25-85 |
|---|---|---|---|---|---|---|
| Arsenic (µg/L) | | 13 | 26 | 2,200 | 1,800 | 440 |
| Molybdenum (µg/L) | | 9 | 9 | 71 | 410 | 590 |
| Boron (µg/L) | | 260 | 910 | 10,000 | 31,000 | 20,000 |
| Aluminum (µg/L) | | -- | .4 | 2.9 | 1.7 | 7.6 |
| | | | | | | |
| Uranium (µg/L) | | -- | 20 | 130 | 740 | -- |
| δSulfur-34 (o/oo) | | 4.6 | -- | 8.1 | 5.0 | 6.2 |
| δDeuterium (o/oo) | | -75.8 | -87.6 | -98.5 | -107.6 | -91.4 |
| δOxygen-18 (o/oo) | | -8.2 | -10.4 | -12.6 | -13.8 | -11.3 |
| δCarbon-13 (o/oo) | | -4.3 | -14.9 | -13.1 | -7.2 | -- |

[1]Analyses were performed by U.S. Geological Survey, Arvada, Colo.; Menlo Park, Calif.; and Denver, Colo., by the methods of (Fishman and Friedman, 1985).  Sulfur-34 isotopic analysis done by Global Geochemistry Corporation, Canoga Park, Calif.
[2]Sum of analyzed dissolved constituents.

Table 3.  Saturation Indices for Selected Minerals as Calculated by WATEQ2 (Ball et al., 1980).

| Mineral | Irrigation water Recharge | Well 15B Shallow | Well 16B Intermediate | Well 17A Deep | Well 17C Evapotranspiration |
|---|---|---|---|---|---|
| | | | Saturation index[1] | | |
| Quartz ($SiO_2$) | .45 | 1.0 | .93 | .84 | .80 |
| Chalcedony ($SiO_2$) | -.057 | .48 | .39 | .30 | .26 |
| Silica glass ($SiO_2$) | -.57 | .079 | -.12 | -.21 | -- |
| Chlorite ($Mg_5Al_2Si_3O_{10}(OH)_8$) | -2.1 | -12 | .92 | -1.9 | -1.6 |
| Anorthite ($CaAl_2Si_2O_8$) | -5.8 | -7.8 | -5.6 | -7.1 | -4.9 |
| Halite ($NaCl$) | -7.7 | -6.4 | -6.0 | -4.2 | -4.2 |
| Albite ($NaAlSi_3O_8$) | -2.9 | -2.2 | .18 | .031 | .37 |
| Orthoclase ($KAlSi_3O_8$) | -1.5 | -1.4 | .96 | .83 | .90 |
| Illite ($K_{.6}Mg_{.25}Al_{2.3}Si_{3.5}O_{10}(OH)_2$) | -2.3 | -2.3 | -.21 | -.017 | 1.2 |
| Gypsum ($CaSO_4$) | -1.9 | -.84 | -3.0 | -2.5 | -.79 |
| Calcite ($CaCO_3$) | .26 | .044 | .25 | -.29 | .75 |
| Analcime ($NaAlSi_2O_6 \cdot H_2O$) | -- | -4.6 | -2.1 | -2.2 | -1.8 |
| Leonhardite ($CaAl_2Si_4O_{10} \cdot 2H_2O$) | -- | 4.4 | 8.2 | 4.9 | 9.1 |
| FeS PPT. ($FeS$) | -- | -- | -1.5 | -2.1 | -- |
| Calcium montmorillonite ($Ca_{.17}Al_{2.33}Si_{3.67}O_{10}(OH)_2$) | -- | 1.1 | -7.3 | .15 | 1.0 |
| Rhodochrosite ($MnCO_3$) | -1.3 | -.36 | .95 | .41 | .85 |

[1]Saturation index = Log $\dfrac{[\text{Activity product}]}{K_T}$ ; where $K_T$ = equilibrium constant at temperature T.

By convention, a positive value indicates a mineral can precipitate from solution; whereas, a negative value indicates a mineral can dissolve if present.

Oxidation-reduction reactions are important in many geochemical systems. In this model the only redox species considered are organic carbonbicarbonate, ferrous iron-ferric oxyhydroxide, and sulfide-sulfate although others also may be important. Evapotranspiration is considered only in the shallow ground-water zone (<3 m deep) because plant roots should not reach the deeper ground water zone.

The shallow zone is located at the upper end (east) of the flow system and reaches depths of up to 6.1 m (Fig. 4). This zone is characterized by dilute ground water containing from 820 to 1,450 mg/L dissolved solids. The dominant cation is sodium, with calcium being present in subordinate to subequal quantities. Bicarbonate is the dominant anion, with sulfate present in lesser amounts. The ground water has dissolved-oxygen concentrations ranging from 0.3 to 1.0 mg/L, $E_h$ values greater than 86 mv, and sulfide concentrations less than 10 µg/L. pH values generally are between 7.2 and 7.7. Dissolved organic carbon is low in this zone with concentrations near 10 mg/L. Manganese concentrations are high, ranging from 1.1 to 2.4 mg/L while iron ranges from <2 to 70 µg/L.

Water is applied to the study area by flood irrigation during the period from about mid-April to October each year. The irrigation water is a dilute $Na-Ca-HCO_3$ water containing about 300 mg/L dissolved solids (Table 2). The model results indicate that as irrigation water percolates through the unsaturated zone it dissolves ammonium sulfate fertilizer (applied to the field once a year at the rate of about 50 g/m$^2$), chlorite, sodium chloride, calcite and silica (probably volcanic glass). High $pCO_2$ values in the water indicate that the atmosphere (log $pCO_2$ about -3.5) is not the only source of $CO_2$. Oxidation of sedimentary organic matter and plant-root respiration are other possible sources. Dissolved $Ca^{2+}$ replaces $Na^+$ on clays as a result of ion-exchange processes. Exchange reactions can be considered to be at equilibrium since the kinetics of these reactions are fast compared to dissolution or precipitation reactions (Berner, 1980). The mass-balance equation for these processes (in mmols/L) is:

RECHARGE WATER (concentrated by evapotranspiration)
+ 0.37 $CO_2$(g) + 2.2 $SO_4^{2-}$ + 3.2 SODIUM-EXCHANGE
+ 0.13 CHLORITE + 0.43 $SiO_2$ + 3.9 CALCITE + 1.9 NaCl        (1)
= SHALLOW WATER + 1.6 CALCIUM-EXCHANGE
+ 0.10 Ca-MONTMORILLONITE + 0.028 ORTHOCLASE.

The geochemical processes reflected by Equation (1) are dissolution, precipitation, and exchange. These processes are superimposed on the concentration of solutes by evapotranspiration, which is approximately 145 cm/y in the Fallon area (Pennington, 1980). Chloride is largely balanced by the effects of evapotranspiration. Sodium chloride also may be added to the shallow ground water by dissolution of halite in the vadose zone or possibly from inclusions in the dissolving lithic fragments. Halite has not been identified by X-ray diffraction analysis in the shallow zone. The observed increase in sulfate in the shallow ground water as compared to the recharge water is accounted for by solution of applied ammonium sulfate fertilizer and concentration by evapotranspiration. Nitrogen from the fertilizer was not included in the model because the biologic uptake was not measured. Calculated saturation indices, as calculated by the computer program WATEQ2 (Ball et al., 1980) (Table 3) show that chlorite, NaCl, and silica glass should dissolve and calcium montmorillonite should precipitate in the shallow ground-water zone. Leonhardite, a partially dehydrated form of the zeolite laumontite, might also precipitate as suggested by stability-field diagrams (Fig. 6) and saturation indices.

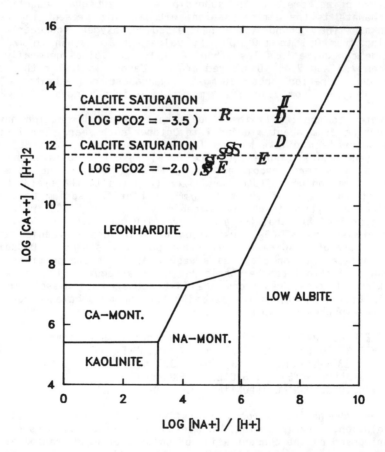

Figure 6.    Stability fields in the Ca-Na-HCl system at 25 C.    (Data from the Dodge Ranch study area are shown as R = re-charge water, S = shallow ground water, I = intermediate ground water, D = deep ground water, and E = evapotrans-piration ground water.)

The intermediate zone ranges in depth from about 3.5 to 9.2 m. Ground water in the intermediate zone is slightly more saline than that of the shallow zone with dissolved solids ranging from 1,320 to 1,870 mg/L. The dominant cation is sodium with very low calcium and magnesium concentrations. Bicarbonate is the major anion with sulfate and chloride present in subordinate amounts. The $E_h$ ranges from 92 to 330 mv, indicating a wide range in oxidation potential. Sulfide ranges from 12 to 21 µg/L. The pH is high in this zone ranging from 8.7 to 9.2. Dissolved organic carbon ranges from 18 to 36 mg/L. Manganese and iron concentrations range from 20 to 360 and from 21 to 67 µg/L, respectively.

Water that moves from the shallow zone to the intermediate zone changes composition by a combination of processes. Anorthite, gypsum, ferric oxyhydroxide, silica, and orthoclase are dissolving. Calcium montmorillonite, ferrous monosulfide, and chlorite are being precipitated in this zone. Exchange reactions that involve clay minerals remove $Ca^{2+}$ ions from solution and add $Na^+$ ions to the water. Organic carbon, mostly present as humic acids, is being oxidized to $HCO_3^-$. These processes are summarized in Equation (2):

$$\text{SHALLOW WATER} + 15.2 \text{ SODIUM-EXCHANGE} + 2.7 \text{ ANORTHITE} +$$
$$6.6 \text{ CH}_2\text{O} + 2.9 \text{ FeOOH} + 0.10 \text{ ORTHOCLASE} + 2.3 \text{ GYPSUM} +$$
$$2.9 \text{ SiO}_2 = \text{INTERMEDIATE WATER} + 2.9 \text{ FeS} + \qquad (2)$$
$$7.6 \text{ CALCIUM-EXCHANGE} + 2.2 \text{ Ca-MONTMORILLONITE} +$$
$$0.18 \text{ CHLORITE}.$$

The oxidation of organic carbon to $HCO_3^-$ probably is the result of bacterial sulfate reduction via the reaction:

$$2 \text{ CH}_2\text{O} + \text{SO}_4{}^{2-} = \text{H}_2\text{S} + 2 \text{ HCO}_3^-. \qquad (3)$$

This is further supported by the appearance of sulfide in the intermediate zone and a concomitant lowering of the E . Carbon isotope ($\delta^{13}C$) values near -13 o/oo indicate the dissolved $HCO_3^-$ is derived from organic matter.

Decreases in the concentrations of Fe and Mn are the result of precipitation of ferrous monosulfide and possibly rhodochrosite within the intermediate zone. Manganese concentrations also may decrease because of precipitation of manganous sulfide. The large increase in pH (from 7.18-9.20) from the shallow to the inter-mediate zone probably is the result of the hydrolysis of feldspar.

The deep ground-water zone is generally located below the intermediate zone and is present at depths that range from 7.6 to 9.2 m. The water in the deep zone is moderately saline with dissolved solids ranging from 5,020 to 9,230 mg/L. Sodium is the dominant cation with calcium and magnesium present only in low concentrations. Bicarbonate is the most plentiful anion with chloride and sulfate present in almost equal but lesser amounts. $E_h$ values are lower in this zone, ranging from 108 to 143 mv. Sulfide ranges from 15 to 100 µg/L. Dissolved oxygen ranges from below detection (<0.2 mg/L) to 0.6 mg/L. The pH ranges from about

8.3 to 8.7 in this zone. Dissolved organic carbon increases in the deep zone to concentrations of between 81 and 96 mg/L and appears to be largely composed of humic substances. Manganese and iron range from 80 to 90 and from 11 to 120 $\mu$g/L, respectively.

As water moves along the flow path from the intermediate zone to the deep zone, it becomes more saline because of further interaction with the sediment. Gypsum, ferric oxyhydroxide, orthoclase, chlorite, and sodium chloride are dissolving, adding their respective constituents to the ground water. Silica (as chalcedony), calcite, calcium montmorillonite, and ferrous monosulfide are precipitating from the water in the deep zone. Organic carbon is being oxidized via Equation (3) producing high $HCO_3^-$ concentrations. Exchange reactions are removing calcium and adding sodium to the water. These processes can be summarized by the following equation:

$$
\begin{aligned}
&\text{INTERMEDIATE WATER} + 66.2 \text{ SODIUM-EXCHANGE} + 34.9 \text{ GYPSUM} + \\
&1.2 \text{ ORTHOCLASE} + 34.2 \text{ NaCl} + 0.17 \text{ CHLORITE} + 40.6 \text{ CH}_2\text{O} + \\
&18.0 \text{ FeOOH} = \text{DEEP WATER} + 18.0 \text{ FeS} + 1.7 \text{ CALCITE} + \\
&1.8 \text{ SiO}_2 + 0.65 \text{ Ca-MONTMORILLONITE} + 33.1 \text{ CALCIUM-EXCHANGE}.
\end{aligned} \quad (4)
$$

The pH decline, from 9.18 to 8.28, is attributed to the release of protons during the sulfate-reduction process.

The above processes are based on a model which includes several assumptions and limitations. A single unique set of phases cannot be chosen to explain the observed water chemistry at the Dodge Ranch site; thus, this model is only one of several possible models that can be formulated. Some of the assumptions in this model are that (1) ground water moves by piston flow, (2) evapotranspiration only occurs in the shallow zone, (3) end-member solid phases were used, and that (4) silica is precipitating as chalcedony. Limitations include no direct physical evidence for exchange reactions, insufficient thermodynamic data for clay minerals and some zeolites, and the inherent inaccuracies of evapotranspiration values. Regardless of these limitations, this type of multifaceted approach is useful for identifying processes that may be active in water-rock systems. Water-chemistry data, solid-phase chemistry and mineralogy, stable isotopes, and hydrologic data should all be used to create an internally consistent model that describes the processes acting in ground-water systems.

## SUMMARY AND CONCLUSIONS

Geochemical and hydrologic processes that affect the water chemistry in the shallow alluvial aquifer at the Dodge Ranch study area include dissolution, precipitation, exchange reactions on clay minerals, oxidation-reduction reactions, and evapotranspiration. The aquifer is composed of complexly interbedded silty and clayey sand and is bounded on the bottom by a thick clay sequence. The sediment is composed of quartz, anorthite, albite, illite, orthoclase, montmorillonite, chlorite, mixed-layer clays, muscovite, volcanic-lithic fragments, and other less abundant minerals.

The ground water evolves from a slightly saline, sodium-calcium-bicarbonate-sulfate type through a sodium-bicarbonate type to a moderately saline sodium-bicarbonate-sulfate-chloride type. Evapotranspiration is active in the shallow zone of the aquifer. In the extreme case, evapotranspiration has produced a very saline sodium-sulfate-bicarbonate type water at the distal end of the flow path. Plant-root respiration and the oxidation of organic matter results in a high $pCO_2$ in the soil zone. Sulfate is increased in the shallow zone by the dissolution of ammonium sulfate fertilizer, and, in the intermediate and deep zones, by the dissolution of gypsum. Volcanic glass is dissolving in the shallow and intermediate zones.

Cation exchange is a major process that affects the water chemistry in this shallow ground-water system. This process is responsible, in part, for the large increase in dissolved sodium as the water evolves along the flow path. Calcium is adsorbed by clay minerals releasing sodium into solution.

In the deep zone, precipitation of silica (as chalcedony) limits the concentration of silica in the ground water to slightly under saturation with respect to amorphous silica. Calcium montmorillonite, ferrous monosulfide, and calcite are precipitating in the deeper part of the system.

Organic carbon, mostly present as humic substances, plays an important part in the evolution of the ground water at Dodge Ranch. Beginning in the shallow zone, organic carbon is being oxidized to create a high $pCO_2$ content in the shallow water. In the intermediate and deep zones, organic carbon is being oxidized by sulfate-reducing bacteria to produce bicarbonate and sulfide. The sulfide is subsequently precipitated as ferrous monosulfide.

## DISCLAIMER

The use of trade or product names in this chapter is for identification purposes only and does not constitute endorsement by the U.S. Geological Survey.

## REFERENCES

Ball, J. W., D. K. Nordstrom, and E. A. Jenne. 1980. Additional and revised thermochemical data and computer code for WATEQ2--A computerized chemical model for trace and major element speciation and mineral equilibria of natural waters. U.S. Geological Survey Water Resources Inv. Rept. 78-116, 109 p.

Barnes, R. B. 1975. The determination of specific forms of aluminum in natural water. Chem. Geol. 15:177-191.

Berner, R. A. 1980. Early diagenesis, a theoretical approach. Princeton University Press, New Jersey, 241 p.

Brown, E., M. W. Skougstad, and M. J. Fishman. 1970. Methods for collection and analysis of water samples for dissolved minerals and gases. U.S. Geological Survey, Tech. of Water-Resources Inv., Bk. 5, Chap. Al, 160 p.

Carroll, D. 1970. Clay minerals: A guide to their x-ray identification. Geol. Soc. of Am. Spec. Paper 126, 80 p.

Dollarhide, W. E. 1975. Soil survey of Fallon-Fernley area, Nevada, parts of Churchill, Lyon, Storey, and Washoe Counties. U.S. Dept. of Agriculture, Soil Conserv. Service, Washington, D.C., 112 p.

Fishman, M. J., and L. C. Friedman, eds. 1985. Methods for determination of inorganic substances in water and fluvial sediments. Techniques of Water-Resources Investigations of the United States Geological Survey, Bk. 5, Chap. Al, U.S. Geological Survey Open-File Rept., 85-495, 709 p.

Glancy, P. A. 1986. Geohydrology of the basalt and unconsolidated sedimentary aquifers in the Fallon area, Churchill County, Nevada. U.S. Geological Survey Water-Supply Paper 2263, 62 p.

Lico, M. S., Y. K. Kharaka, W. W. Carothers, and V. A. Wright. 1982. Methods for collection and analysis of geopressured geothermal and oil-field waters. U.S. Geological Survey Water-Supply Paper 2194, 21 p.

Lico, M. S., A. H. Welch, and J. L. Hughes. 1986. Lithologic and chemical data for sediments at two sites in the shallow alluvial aquifer near Fallon, Churchill County, Nevada, 1984-1985. U.S. Geological Survey Open-File Rept. 86-250, 43 p.

Morgan, D. S. 1982. Hydrogeology of the Stillwater geothermal area, Churchill County, Nevada. U.S. Geological Survey Open-File Rept. 82-345, 95 p.

Morrison, R. B. 1964. Lake Lahontan--Geology of southern Carson Desert, Nevada. U.S. Geological Survey Prof. Paper 401, 156 p.

Olmsted, F. H., A. H. Welch, A. S. Vandenburgh, and S. E. Ingebritson. 1984. Geohydrology, aqueous geochemistry, and thermal regime of the Soda Lakes and Upsal Hogback Geothermal Systems, Churchill County, Nevada. U.S. Geological Survey Water Resources Inv. Rept. 84-4054, 166 p.

Parkhurst, D. L., L. N. Plummer, and D. C. Thorstenson. 1982. BALANCE--A computer program for calculating mass transfer for geochemical reactions in ground water. U.S. Geological Survey Water Resources Inv. Rept. 82-14, 29 p.

Pennington, R. W. 1980. Evaluation of empirical methods for estimating crop water consumptive use for selected sites in Nevada. State of Nevada, Dept. of Conserv. and Natural Resources, Information Series Water Planning Rept. 3, 206 p.

Plummer, L. N., and W. Back. 1980. The mass balance approach: Application to interpreting the chemical evolution of hydrologic systems. Am. J. of Sci. 280:130-142.

Thorstenson, D. C., and D. W. Fisher. 1979. The geochemistry of the Fox Hills-Basal Hell Creek aquifer in southwestern North Dakota. Water Resources Res. 15:1479-1498.

Willden, Ronald, and R. C. Speed. 1974. Geology and mineral deposits of Churchill County, Nevada. Nevada Bur. of Mines and Geol. Bull. 83, 95 p.

Yurewicz, M. C. U.S. Geological Survey. Written communication. 1981.

# CHAPTER 7

ANAEROBIC MICROBIAL TRANSFORMATIONS OF AZAARENES
IN GROUND WATER AT HAZARDOUS-WASTE SITES

Wilfred E. Pereira and Colleen E. Rostad, U.S. Geological Survey,
Denver, Colorado

David M. Updegraff and Jon L. Bennett, Colorado School of Mines,
Golden, Colorado

ABSTRACT

    Processes affecting the behavior of organic contaminants in
ground-water environments are not well understood. Although
microbial transformation reactions of many organic contaminants in
surface waters are fairly well documented, very little information
is known about these processes in ground-water systems. Studies
are being conducted at hazardous-waste sites in Pensacola, Fla.,
and St. Louis Park, Minn., to better understand the movement and
fate of polynuclear azaheterocyclic compounds or azaarenes in
aquifers contaminated by wood-treatment chemicals. Two- and
three-ring azaarenes and their oxygenated and methylated
derivatives were identified in ground water by gas chromatography-
mass spectrometry (GC-MS). The presence of oxygenated azaarenes in
anaerobic zones of ground water suggested that these compounds
probably were microbial transformation products.
    Laboratory anaerobic degradation studies were designed to study
metabolic pathways of quinoline, isoquinoline, 2-methylquinoline,
and 4-methylquinoline. Microbial metabolic transformation products
of quinoline, isoquinoline, and 4-methylquinoline identified in
laboratory anaerobic cultures were identical to those detected in
contaminated ground water at the two hazardous-waste sites.
Microbial N-, C-, and O-methylation reactions are reported for the
first time. Evidence for microbial degradation of two-ring
azaarenes in anaerobic zones of ground water is presented.
Oxygenated derivatives of azaarenes are more water soluble, mobile,
and biorefractory than parent azaarenes; hence, they are more
persistent in contaminated aquifers.

*Chemical Quality of Water and the Hydrologic Cycle*, Robert C. Averett and Diane M. McKnight (Eds.) © 1987 Lewis Publishers, Inc.,
Chelsea, Michigan. Printed in the United States of America.

INTRODUCTION

Ground water constitutes approximately 4% of all water in the hydrologic cycle. Although most ground-water aquifers in the United States are uncontaminated by hazardous chemicals, numerous reports have occurred of ground-water contamination problems in localized industrial areas arising from improper disposal practices. Because of the persistence of these chemicals in subsurface environments, contamination of ground water is likely to be the major water quality issue for the next 2 to 3 decades.

The U.S. Geological Survey is conducting studies of the fate and movement of wood-treatment wastes in contaminated aquifers (Hult et al., 1981; Ehrlich et al., 1982; Pereira et al., 1983; Troutman et al., 1984; Pereira and Rostad, 1986; Rostad et al., 1984; Mattraw et al., 1985; Pereira et al., 1985; Goerlitz et al., 1985). There are approximately 400 wood-preserving plants in our Nation (U.S. Environmental Protection Agency, 1981). Many of these plants use creosote as the primary wood-preserving chemical. Creosote is a distillate of coal tar and consists of different classes of organic compounds, such as phenols and polycyclic aromatic compounds (Pereira et al., 1983; Pereira and Rostad, 1986). One class of polycyclic aromatic compounds present in creosote, coal tar, fossil fuels, and energy related effluents are the nitrogen heterocycles or azaarenes (Pereira et al., 1983; Santodonato and Howard, 1981). Some members of the series have carcinogenic and mutagenic properties (Santodonato and Howard, 1981). Since these compounds are potentially hazardous to human health and the environment, it is essential to understand their fate and movement in contaminated aquifers, so that relevant remedial strategies may be implemented.

Investigations at two wood-treatment facilities, one near Pensacola, Fla., and the other near St. Louis Park, Minn., showed that ground water was contaminated with azaarenes derived from creosote. These compounds included quinoline, isoquinoline, and their oxygenated and alkylated derivatives. The presence of oxygenated azaarenes in aquifers contaminated by creosote had not been reported previously. Processes affecting the behavior of these compounds in contaminated aquifers are not well understood.

Grant et al. (1976) showed that a *Moraxella sp.* isolated from garden soil converted quinoline to 2-hydroxyquinoline. Bennett et al. (1985) showed that aerobic pseudomonads, isolated from creosote contaminated soil from the Pensacola site, converted quinoline to its oxygenated derivative 2(1H)quinolinone. The 2(1H)quinolinone was degraded further to unknown products. These quinoline degrading pseudomonads were unable to degrade 2(1H)quinolinone under anaerobic conditions. More recently, Shukla (1986) reported that a *Pseudomonas sp.* isolated from sewage by enrichment culture on quinoline, metabolized this substrate by a novel pathway involving 8-hydroxycoumarin. Biotransformation reactions of quinoline, thus far reported, were all aerobic reactions, requiring atmospheric oxygen. Dissolved oxygen measurements in ground water, however, indicated that a large anaerobic zone was present in the aquifer at the Pensacola site (Mattraw et al., 1985), suggesting that the oxygenated azaarenes were biotransformation products of reactions mediated by indigenous consortia of anaerobic microorganisms. These findings led us to study the anaerobic degradation of azaarenes in laboratory cultures, and to investigate these transformation reactions in contaminated ground water at two sites.


METHODS AND MATERIALS


Field Studies:    Site Description

The 18 acre plant site is situated approximately 400 m north of Pensacola Bay, Fla. During the course of operation of the plant (1902-1981), wood-treatment wastes containing creosote were discharged into two unlined surface impoundments. These wastes infiltrated the relatively thin unsaturated zone and contaminated the aquifer down to a depth of approximately 80 m, and laterally about 500 m. A north-south geologic cross section through the affected part of the aquifer, showing location of multidepth well sites J1 and J3 through J7, is shown in Figure 1. Site J1 is situated up field of the plant and is the control site. Site J3 is located near the overflow pond, while site J7 is situated closer to the bay. Regional ground-water flow generally is to the south toward Pensacola Bay. Details of the study site and geohydrology of the area have been reported elsewhere (Troutman et al., 1984; Mattraw et al., 1985).

The site near St. Louis Park, Minn., was a coal-tar distillation and wood-preserving facility that operated between 1918 and 1972. Details of this site have been reported previously (Hult, 1981).


Sample Preparation and Analysis

Azaarenes in ground water from the Pensacola site were isolated using a bonded phase extraction procedure followed by GC-MS analysis, as reported in an earlier publication (Rostad et al., 1984). Azaarenes in ground water from the St. Louis Park site were isolated and analyzed by a previously published procedure (Pereira et al., 1983).


Instrumentation

Low resolution mass spectra were obtained on a Finnigan OWA or a Finnigan MAT TSQ 46-B GC-MS system. The GC was equipped with a fused silica capillary column 30 m long by 0.26 mm i.d. with 0.25-µm bonded film of DB-5 (J. & W. Scientific). Accurate mass measurements were made on a VG 7070HS high-resolution mass spectrometer (HRMS), at a resolution of approximately 4000. The GC contained a 25 m by 0.2 cm i.d. (Hewlett Packard) cross linked column with 5% phenylmethylsilicone. Azaarenes in ground water and

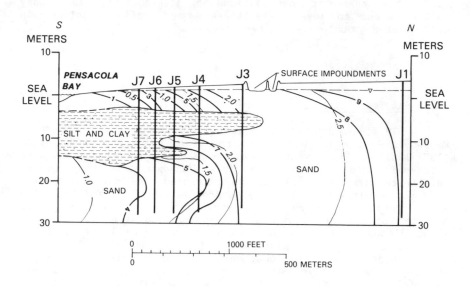

Figure 1.    Geologic cross section (north-south) of contaminated aquifer with indication of potentiometric surface and location of multidepth well sites J1 and J3 through J7.

from anaerobic digestor studies were confirmed by comparison of their mass spectra with standard reference compounds analyzed under identical conditions. Elemental compositions of major ions of biotransformation products of quinoline and 4-methylquinoline were confirmed by HRMS.


## Microbiological Methods

To evaluate the anaerobic microbial degradation of the azaarenes, four compounds (quinoline, 2-methylquinoline, 4-methylquinoline, and isoquinoline) were tested by both the anaerobic toxicity assay and the methane potential assay described by Owen et al. (1979).

The anaerobic toxicity assay was carried out by weighing from 2.2 to 5.8 mg of the azaarene into a 120-mL serum bottle. Then 50 mL of anaerobic medium, containing the acetate-propionate mixture described by Owen et al. (1979) as a nutrient, and 20% by volume of actively digesting sewage sludge from the Denver Metropolitan Sewage Plant, was added under an atmosphere of 70% nitrogen and 30% carbon dioxide. The cultures were incubated at 36°C for 194 days. During this period, gas was measured at frequent intervals using the glass syringe method. Toxicity was indicated by inhibition of gas production.

The methane potential assay was carried out in the medium used by Owen et al. (1979) with no added nutrients, except for 8.3 to 24.5 mg of the test azaarene. The inoculum was 20% by volume of a mixed anaerobic culture; one-third from an anaerobic digestor containing sludge from the Denver Metropolitan Sewage Plant; one-third from a second anaerobic digestor that had been treated with creosote and creosote-contaminated soil from the Pensacola site, and one-third from an upflow fixed film anaerobic reactor that had been treated with quinoline. All three inoculum cultures were grown at 36°C; all three contained active populations of methanogenic bacteria, as indicated by continuous production of methane and carbon dioxide. The cultures containing azaarenes, along with controls to which no azaarenes were added, were incubated at 36°C for 203 days. The volume of gas generated was measured at frequent intervals by the glass syringe method. Methane in the gas samples was determined by gas chromatography. Details of the microbial studies will be published elsewhere (Updegraff et al., written communication, 1986).

At the end of the incubation periods in both assays, 1-mL aliquots were withdrawn from the digestors with a glass syringe. Each aliquot was extracted in a centrifuge tube with 2 mL of diethyl ether (previously cleaned by passage through a column of basic alumina) on a vortex mixer. The ether layer was withdrawn with a pasteur pipette, transferred to a 2-mL glass vial, and evaporated under a stream of dry nitrogen to a volume of 1 mL. After addition of benzene (100 µL), the solution was evaporated under nitrogen to a volume of 100 µL. Each extract was analyzed by GC-MS.


RESULTS AND DISCUSSION


Azaarenes in Ground Water

Ground-water samples, obtained from wells of different depths, at sites J3 through J7 at the Pensacola site were analyzed. Several two- and three-ring azaarenes and their alkylated and oxygenated analogs were identified and confirmed by mass spectrometry. Similar compounds were identified in ground-water samples collected at the St. Louis Park site. Distributions of quinoline and isoquinoline and their corresponding oxygenated analogs quinolinone and 1(2H)isoquinolinone in ground-water samples from the Pensacola site are shown in Figures 2 and 3. Wells at site J3 (located near the impoundment and source of contamination) showed high concentrations of quinoline, 2(1H)quinolinone, isoquinoline, and 1(2H)isoquinolinone. Ground-water samples from wells at more distant sites, however, only contained quinolinone and 1(2H)isoquinolinone, with no detectable amounts of quinoline and isoquinoline. These results suggested that the oxygenated azaarenes were possible products of microbial transformation reactions, taking place in anaerobic zones of the contaminated aquifer. Accordingly, laboratory experiments were designed to study anaerobic transformations of quinoline, isoquinoline, 2-methylquinoline, and 4-methylquinoline.

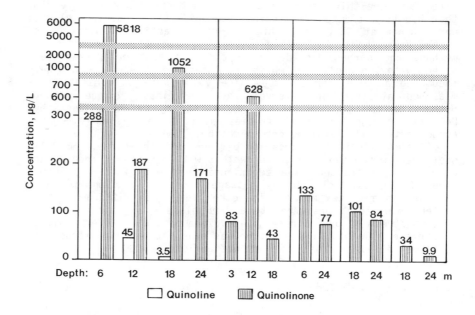

Figure 2.    Distributions of quinoline and quinolinone in ground water at the Pensacola site.


## Anaerobic Toxicity Assay

This bioassay measures the adverse effect of a compound on the rate of total gas production from an easily utilized methanogenic substrate (Owen et al., 1979). In other words, toxicity is indicated by inhibition of gas production. In this assay 2-methylquinoline, 4-methylquinoline, and isoquinoline showed little or no toxicity to the methanogenic bacteria, as indicated by lack of inhibition of gas production. Quinoline, however, inhibited gas production during the first 29 days of incubation. This inhibition of gas production was only temporary since cultures containing quinoline eventually produced nearly as much gas as the control cultures containing no azaarenes (Fig. 4). GC-MS analysis of the cultures from the anaerobic toxicity assay indicated that quinoline, 4-methylquinoline, and isoquinoline were completely degraded in the presence of the carbon sources, acetate, and propionate. The 2-Methylquinoline, however, was only partially degraded, compared to the sterilized controls.

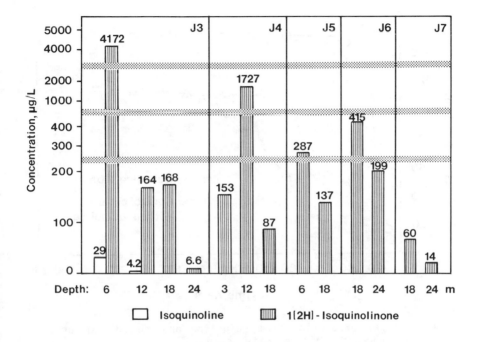

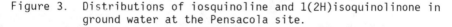

Figure 3.    Distributions of iosquinoline and 1(2H)isoquinolinone in ground water at the Pensacola site.

## Methane Potential Assay

This bioassay is a measure of substrate biodegradability and is determined by monitoring cumulative methane production from a sample that is anaerobically incubated in a chemically defined medium.  This bioassay showed that controls containing no azaarenes produced as much or more methane and carbon dioxide as any of the cultures containing quinoline, isoquinoline, 2-methylquinoline, and 4-methylquinoline, indicating that these azaarenes probably are not biodegraded to methane and carbon dioxide.  GC-MS of extracts of cultures from the methane potential assay, upon termination of the experiment, indicated the presence of various oxygenated metabolic transformation products.  Sterile controls, and controls to which no azaarenes were added, showed no evidence of these metabolites, thereby establishing their genesis from an anaerobic biological process.

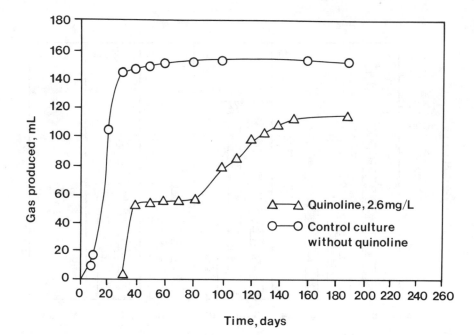

Figure 4.    Gas production from quinoline and control cultures in the anaerobic toxicity assay.

## Anaerobic Biodegradation of Quinoline (I)

Metabolic products identified in laboratory cultures by GC-MS are shown in Figure 5. Compound (III) was a major component; compounds (IV, V, and VI) were minor components. Mass spectra of these compounds have been published elsewhere (Pereira et al., 1986). Compounds identified by GC-MS in laboratory cultures were identical to those found in contaminated ground water at the two sites. Oxygenation of (I) to (II) takes place through the postulated oxaziridine intermediate. The lactim (II) tautomerizes to the lactam (III), which is the predominant tautomer. Lactam (III) undergoes enzymatic N-methylation to give (IV); C-methylation at the activated 4-position of the ring to give (V); and both N- and C-methylation to give (VI). Because the initial hydroxylation of (I) to give (II) takes place under both aerobic (Bennett et al., 1985; Shukla, 1986) and anaerobic (Pereira et al., 1986) conditions, the bacterial enzyme system involved probably splits a water molecule and incorporates the oxygen atom of water, as hydroxyl, into the 2-position of the quinoline ring, without participation of atmospheric oxygen. A postulated mechanism for this reaction is shown in Figure 6. Anaerobic bacterial enzyme systems, which hydroxylate nitrogen heterocycles utilizing the oxygen atom of water, have been reported (Dagley, 1971; Imhoff et al., 1979; Imhoff-Stuckle et al., 1983; Stadtman et al., 1972).

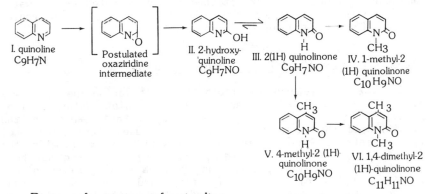

Biotransformation of quinoline

Figure 5. Anaerobic biotransformation products of quinoline in laboratory cultures.

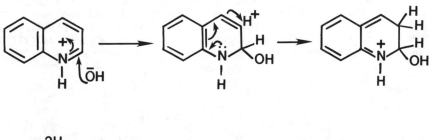

Figure 6. Postulated reaction mechanism for anaerobic enzymatic incorporation of oxygen atom of water into quinoline.

## Anaerobic Biodegradation of 4-Methylquinoline (VII)

Metabolic products identified in laboratory cultures by GC-MS (Pereira et al., 1986) are shown in Figure 7. Compound (V) was a major component and compounds (VI) and (IX) were minor components. Compounds (V) and VI) also were identified in contaminated ground water at the two sites. However, compound (IX) was identified in laboratory cultures but not detected in contaminated ground water at either site. Compound (IX) is the product of microbial enzymatic O-methylation of (VIII).

## Anaerobic Cultures Containing 2-Methylquinoline (X)

In the laboratory experiment, 2-Methylquinoline (X) was recovered unchanged without any evidence of biotransformation or degradation. The presence of a bulky methyl group in the 2-position of the quinoline ring probably causes steric hindrance, and prevents the oxygen-insertion reaction from forming the postulated oxaziridine intermediate (Fig. 8). Evidence has been presented in a previous publication (Pereira et al., 1986) to show that 2-methylquinoline is refractory and does not undergo biodegradation in contaminated ground water at the Pensacola site. Therefore, results of laboratory experiments are consistent with field observations.

## Anaerobic Biodegradation of Isoquinoline (XI)

Approximately 50% of isoquinoline was recovered unchanged from the laboratory culture. Metabolic products identified by GC-MS (Pereira et al., 1986) were one major component (XIII) and one minor component (XIV), as shown in Figure 9. Compounds (XIII) and (XIV) also were detected in ground water at both sites.

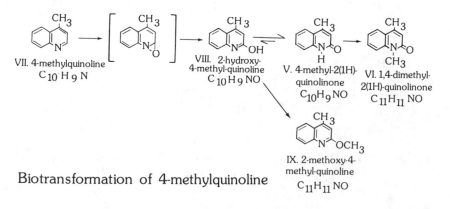

Biotransformation of 4-methylquinoline

Figure 7.    Anaerobic biotransformation products of 4-methylquino-line in laboratory cultures.

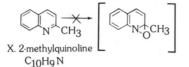

X. 2-methylquinoline
$C_{10}H_9N$

Figure 8.    Anaerobic reactions of 2-methylquinoline in laboratory cultures.

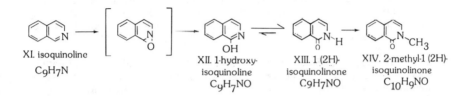

XI. isoquinoline
$C_9H_7N$

XII. 1-hydroxy-isoquinoline
$C_9H_7NO$

XIII. 1 (2H)-isoquinolinone
$C_9H_7NO$

XIV. 2-methyl-1 (2H)-isoquinolinone
$C_{10}H_9NO$

Biotransformation of isoquinoline

Figure 9.    Anaerobic biotransformation products of isoquinoline in laboratory cultures.

CONCLUSIONS

Microbial processes play a significant role in determining the chemical quality of ground water. Field studies and laboratory anaerobic degradation experiments clearly demonstrate that two-ring azaarenes are transformed by indigenous microorganisms into oxygenated and methylated derivatives in aquifers contaminated by creosote and coal tar. These oxygenated analogs have enhanced solubility and, hence, mobility in hydrogeologic environments. Biotransformation products of azaarene metabolism are refractory and may persist in anaerobic zones of the subsurface environment for relatively long periods of time. Because azaarenes are present in wastes generated by fossil fuel conversion technologies, these studies undoubtedly will have significant transfer value to aquifers contaminated from the processing of coal and oil shale.

DISCLAIMER

The use of trade or product names in this chapter is for iden-tification purposes only and does not constitute endorsement by the U.S. Geological Survey.

REFERENCES

Bennett, J. L., D. M. Updegraff, W. E. Pereira, and C. E. Rostad. 1985. Isolation and identification for four species of quinoline-degrading *Pseudomonads* from a creosote contaminated site at Pensacola, Florida. Microbios Letters 29:147-154.

Dagley, S. 1971. Microbial degradation of stable chemical structures. General features of metabolic pathways. *In* Degradation of synthetic organic molecules in the biosphere. Natural, pesticidal, and various other man-made compounds. National Academy of Sciences, Washington D.C. 1972. Proc. of a Conf., San Francisco, Calif., June 12-13, 1971.

Ehrlich, G. G., D. F. Goerlitz, E. M. Godsy, and M. F. Hult. 1982. Degradation of phenolic contaminants in groundwater by anaerobic bacteria. St. Louis Park, Minn. Groundwater 20:703-710.

Goerlitz, D. F., D. E. Troutman, E. M. Godsy, and B. J. Franks. 1985. Migration of wood-preserving chemicals in contaminated ground water in a sand aquifer at Pensacola, Florida. Environ. Sci. Technol. 19:955-961.

Grant, D. J. W., and T. R. Al-Najjar. 1976. Degradation of quinoline by a soil bacterium. Microbios 15:177-189.

Hult, M. F., and M. E. Schoenberg. 1981. Preliminary evaluation of groundwater contamination by coal-tar derivatives, St. Louis Park area, Minnesota. U.S. Geol. Survey Open-File Rept. 81-72, 53 p.

Imhoff, D., and J. R. Andreesen. 1979. Nicotinic acid hydroxylase from *Clostridium barkeri*: selenium-dependent formation of active enzyme. FEMS Microbiol. Lett. 5:155-158.

Imhoff-Stuckle, D., and N. Pfennig. 1983. Isolation and characterization of a nicotinic acid-degrading sulfate-reducing bacterium. *Desulfococcus niacini sp.* nov. Arch. Microbiol. 136:194-198.

Mattraw, H. C., Jr., and B. Franks, eds. 1985. Movement and fate of creosote waste in ground water, Pensacola, Florida. U.S. Geological Survey Toxic Waste--Ground-Water Contamination Program. U.S. Geol. Survey Water-Supply Paper 2285, 63 p.

Owen, W. F., D. C. Stuckey, J. B. Healy, Jr., L. Y. Young, and P. L. McCarthy. 1979. Bioassay for monitoring biochemical methane potential and anaerobic toxicity. Water Research 13:485-492.

Pereira, W. E., and C. E. Rostad. 1986. Investigations of organic contaminants derived from wood-treatment processes in a sand and gravel aquifer near Pensacola, Florida. Selected Papers in the Hydrologic Sciences. U.S. Geol. Survey Water-Supply Paper 2290, 154 p.

Pereira, W. E., C. E. Rostad, J. R. Garbarino, and M. F. Hult. 1983. Groundwater contamination by organic bases derived from coal-tar wastes. Environ. Toxicol. Chem. 2:283-294.

Pereira, W. E., C. E., Rostad, and M. E. Sisak. 1985. Geochemical investigations of polychlorinated dibenzo-p-dioxins in the subsurface environment at an abandoned wood-treatment facility. Environ. Toxicol. Chem. 4:629-639.

Pereira, W. E., C. E. Rostad, D. M. Updegraff, and J. L. Bennett. 1986. Fate and movement of azaarenes and their anaerobic biotransformation products in an aquifer contaminated by wood-treatment chemicals. Environ. Toxicol. Chem. (in press).

Rostad, C. E., W. E. Pereira, and S. M. Ratcliff. 1984 Bonded-phase extraction column isolation of organic compounds in groundwater at a hazardous waste site. Anal. Chem. 56:2856-2860.

Santodonato, J., and P. H. Howard. 1981. Azaarenes: Sources, distribution, environmental impact, and health effects, p. 421-438. *In* J. Saxena and F. Fisher, eds., Hazard assessment of chemicals. Vol. I. Academic Press, Inc., New York, N.Y.

Shukla, O. P. 1986. Microbial transformation of quinoline by a *Pseudomonas sp.* Appl. and Environ. Microbiol. 51:1332-1342.

Stadtman, E. R., T. C. Stadtman, I. Pastan, and L. D. S. Smith. 1972. *Clostridium barkeri sp.* n. J. Bacteriol. 110:758-760.

Troutman, D. E., E. M. Godsy, D. F. Goerlitz, and G. G. Ehrlich. 1984. Phenolic contamination in the sand and gravel aquifer from a surface impoundment of wood-treatment wastes, Pensacola, Florida. U.S. Geol. Survey Water-Resources Inv. Rept. 84-4230, 36 p.

U.S. Environmental Protection Agency. 1981. Development document for effluent limitations guidelines and standards for the timber products point source category. EPA 440/1-81/023. Washington, D.C., 498 p.

Updegraff, D. M., J. L. Bennett, W. E. Pereira, and C. E. Rostad. 1986. Oxidation of azaarenes to hydroxyazaarenes by anaerobic bacteria (unpublished results).

# CHAPTER 8

# BIOLOGICAL PROCESSES OCCURRING AT AN AVIATION GASOLINE SPILL SITE

Barbara H. Wilson, University of Oklahoma, Norman, Oklahoma

Bert Bledsoe and Don Kampbell, R. S. Kerr Environmental Research Laboratory, Ada, Oklahoma

## ABSTRACT

Petroleum-derived hydrocarbons and chlorinated solvents are the most common organic contaminants of ground waters, including those used as drinking-water supplies. Leaks and spills of storage tanks and associated piping plus inappropriate disposal methods in pits and landfills are common sources of these pollutants. The behavior of 1,1,1-trichloroethane, trichloroethylene, benzene, toluene, chlorobenzene, $m$-xylene, and $o$-xylene was studied in material from an aviation gasoline spill site. The plume of contamination has been extensively characterized hydrologically and geochemically and will be discussed in conjunction with the laboratory data. Laboratory studies were initiated to confirm field evidence of biological transformation of alkylbenzenes. The material chosen for study was obtained from an adjacent pristine area and from both oxic and anoxic zones of biological activity in the plume. Preliminary results indicate the degradation of benzene, toluene, $m$-xylene, $o$-xylene, and chlorobenzene. No degradation of 1,1,1-trichloroethane or trichloroethylene has been observed.

## INTRODUCTION

Leaks from storage tanks or associated plumbing and pipelines, plus accidental spills of petroleum products, are major sources of ground-water pollution. Petroleum products such as gasoline, fuel oils, and kerosene contain aromatic hydrocarbons such as benzene and its short-chain alkyl derivatives. The major constituents of unleaded gasoline, kerosene, and No. 2 fuel oil and their water-soluble fractions were identified by Coleman et al. (1984). The

*Chemical Quality of Water and the Hydrologic Cycle*, Robert C. Averett and Diane M. McKnight (Eds.) © 1987 Lewis Publishers, Inc., Chelsea, Michigan. Printed in the United States of America.

aromatics comprised 50% or less of the weight of the product when analyzed directly, but, the aromatics comprised greater than 93% by weight of the water-soluble fraction. Benzene and alkylbenzenes such as toluene, ethylbenzene, and the xylenes are more soluble in water than the aliphatic and higher molecular weight aromatic constituents of petroleum products, and are found dissolved in ground water to a greater extent. Because the $10^{-6}$ cancer risk level for benzene is 0.66 μg/L (Federal Register, 1984), its presence in ground waters used for drinking-water supplies is of concern. The $10^{-6}$ cancer risk level means there is a one in 1,000,000 chance of contracting cancer due to drinking water with this concentration of benzene. This is based on a 70 kg adult consuming 2 L of drinking water per day for 70 yr.

In 1969, the failure of a flange on an underground storage tank at the U.S. Coast Guard Air Station at Traverse City, Mich., resulted in the loss of 10,000 gallons of aviation gasoline into the subsurface. As a result of the spill, a shallow water-table aquifer was contaminated with alkylbenzenes dissolving from the gasoline to the ground water with total alkylbenzene concentrations of 36 to 40 mg/L near the center of the plume. The plume of contamination extended from the air station out into the East Arm of Grand Traverse Bay and impacted numerous drinking-water wells.

To confirm field evidence of aerobic and anaerobic biotransformations of the alkylbenzenes, laboratory studies were performed on aquifer material obtained from a transect extending from the heart of the plume into pristine material. The alkylbenzenes studied were benzene, toluene, o-xylene and m-xylene. Because chlorinated solvents are also common ground-water pollutants and could easily be co-occurring contaminants with alkylbenzenes at airplane refueling and servicing centers, the biological fates of 1,1,1-trichloroethane, trichloroethylene, and chlorobenzene were also studied in these materials.


## SITE DESCRIPTION AND MATERIALS

The site of the plume of contamination is the U.S. Coast Guard Air Station located at Traverse City, Mich. The plume is located in a glacial sand and gravel mix about 50 ft deep overlying at least 100 ft of impermeable glacial clay (Fig. 1). The water table varies from 12 to 18 ft below land surface; the ground water movement is approximately 5 ft per day toward the East Arm of the Grand Traverse Bay (Twenter et al., 1985). The air station is located about 4,300 ft upgrade from the bay with a downward slope approaching 2%. The narrow plume width of 200 to 400 ft is due to the relatively rapid ground water flow plus the biological weathering at the edges of the original contamination. To prevent the plume migration off Coast Guard property, an interdiction field, which functions as a hydraulic barrier to the flow of contaminated ground water, has been constructed to capture contaminated ground water for renovation by air stripping and carbon adsorption.

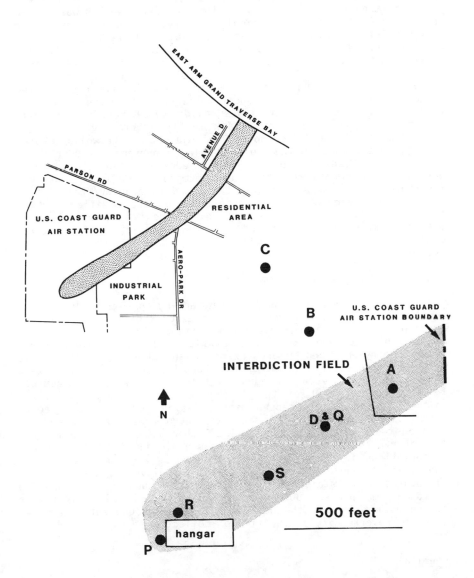

Figure 1.    Location of wells in aviation gasoline plume at Traverse
City, Mich.    Well A located in plume below the interdiction field; Well B located in the aerobic zone of
treatment; Well C located in the pristine region
surrounding the plume; Wells D, Q, R, S located in the
plume above the interdiction field.

A plan view of the study site is shown in Figure 1. In November 1985, 4-inch wells were placed at points A, B, C, and D. Wells P, Q, R, and S were placed during June 1986. Well A was placed in the plume below the interdiction field; Well B was placed in the zone of aerobic biological activity; and Well C was placed in the pristine region surrounding the plume. Wells D, Q, R, and S were placed in the plume above the interdiction field. Aquifer material for both aerobic and anaerobic fate studies was obtained from selected regions and depths in the plume. Aquifer solids were collected with an auger in a manner that did not disturb the structure of the sample. The auger was screwed into the ground and then lifted with a wench. The solids were stored in sterile canning jars then transported back to the laboratory. The core samples were collected from boreholes adjacent to the wells. Water samples were pumped to the surface using a portable bladder pump. Temperature, pH, dissolved oxygen, and redox potential of the water were measured immediately after being pumped to the surface using a flow-through cell apparatus interconnected to a pH/mV/ temperature meter. Dissolved oxygen was determined by the modified Winkler technique. Further characterization of the waters was done by measuring methane, nitrate and nitrite, ammonia, sulfate, total organic carbon, and soluble metals (Table 1).

## Construction of the Microcosms

The microcosms were constructed by adding approximately 90 g of aquifer material (wet weight) to each experimental unit (a 160-mL serum bottle). A dosing solution was prepared by adding a mixture of benzene, toluene, m-xylene, o-xylene, 1,1,1-trichloroethane, trichloroethylene, and chlorobenzene to autoclaved reverse osmosis water and then stirring overnight without headspace. The time zero concentrations of each compound in the microcosms were measured. To each bottle, 2 mL of the dosing solution was added, and the bottle was immediately capped with a Teflon-coated silicon septum and a crimp-cap seal. The sterile controls were autoclaved for 1 hr on two consecutive days at 121°C then dosed in a like manner. To preserve the integrity of the methanogenic aquifer material, all manipulations for the anaerobic fate studies were performed in an anaerobic glovebox. After removal from the glovebox, the anaerobic microcosms were stored in an anaerobic incubator jar under a nitrogen atmosphere. Both aerobic and anaerobic microcosms were stored upside down in the dark at 12°C, the ground water temperature at Traverse City at the time of collection.

## Chemicals

The alkylbenzenes used were high-purity benzene, toluene, m-xylene, and o-xylene from Chem Service, West Chester, Pa. Trichloroethylene, 1,1,1-trichloroethane, and chlorobenzene were obtained from the Aldrich Chemical Co., Milwaukee, Wis. All purities were at least 97%.

Table 1. Geochemical Characterization of the Waters at Traverse City.

| Sample | Depth to water (ft) | Total alkyl-benzenes (μg/L) | Temp. (°C) | pH | Nitrate and nitrite (mg/L as N) | Ammonia | Total organic carbon | Sulfate | Total iron | Total lead (mg/L) | Dissolved oxygen | Methane |
|---|---|---|---|---|---|---|---|---|---|---|---|---|
| In the plume running from its source to the edge of Coast Guard property | | | | | | | | | | | | |
| P3 | *17 | 6.4 | 14.6 | 7.3 | 0.6 | <0.05 | 6 | 18 | 2.0 | 0.04 | 0.4 | 0.004 |
| P2 | 22 | 0.2 | 13.3 | 7.7 | 4.1 | 0.5 | <3 | 5 | 1.2 | 0.01 | 0.5 | 0.01 |
| P1 | 30 | 0.5 | 11.3 | 7.5 | 3.2 | 0.2 | 3 | 13 | 0.83 | 0.01 | <0.1 | 0.04 |
| R3 | *13 | 1,706 | 13.1 | 6.9 | 0.4 | 0.1 | 10 | 4 | 7.3 | 0.27 | <0.1 | 9.5 |
| R2 | 22 | 165 | 12.0 | 7.6 | 2.5 | <0.05 | 3 | 8 | 3.0 | 0.04 | 3.0 | <0.001 |
| R1 | 32 | 6.1 | 11.2 | 7.5 | 1.0 | <0.05 | <3 | 12 | 1.8 | 0.01 | 0.2 | <0.001 |
| S3 | *22 | 454 | 13.3 | 7.0 | 0.7 | 0.85 | 11 | 4 | 19 | 0.07 | <0.1 | 10.9 |
| S2 | 30 | 5.5 | 11.8 | 7.4 | 0.2 | 2.6 | <3 | 7 | 4.5 | <0.005 | <0.1 | 2.7 |
| S1 | 40 | 3.0 | 12.3 | 7.7 | 2.6 | <0.05 | <3 | 14 | 0.33 | 0.02 | <0.1 | 0.001 |
| D4 | *16-18 | 11,820 | nd | 6.6 | 0.1 | 2.2 | <6 | 3 | 13 | 0.11 | <0.1 | 7.9 |
| D3 | 22-25 | 3,150 | 10.4 | 7.1 | 0.1 | 5.5 | <5 | 3 | 23 | <0.02 | <0.1 | 16.1 |
| D2 | 32-34 | 43 | 10.7 | 7.6 | <0.1 | 0.93 | <5 | 20 | 4.4 | <0.02 | <0.1 | 1.07 |
| D1 | 42-44 | 32 | 10.8 | 7.8 | 0.1 | 0.16 | <5 | 16 | 2.3 | <0.02 | 0.1 | 0.04 |
| Q5 | *17 | 273 | 15.8 | 7.1 | nd | nd | nd | nd | nd | nd | <0.1 | 9.5 |
| Q4 | 24 | 149 | 11.8 | 7.0 | 0.4 | 2.8 | 6 | 2 | 18 | 0.05 | <0.1 | 17.4 |
| Q3 | 30 | 83 | 11.7 | 7.1 | 0.2 | 1.9 | 3 | 2 | 14 | 0.02 | <0.1 | 15.9 |
| Q1 | 35 | 5.3 | 12.1 | 7.4 | 0.2 | 1.1 | 3 | 4 | 8.8 | 0.01 | <0.1 | 8.3 |
| Q0 | 45 | 2.8 | 12.1 | 7.7 | 1.2 | <0.05 | <3 | 15 | 2.4 | 0.01 | <0.1 | 0.06 |
| A4 | *15-17 | 394 | nd | 6.6 | 1.5 | 0.27 | <5 | 9 | nd | 0.30 | 4.5 | 0.78 |
| A3 | 22-24 | 2,250 | 12.4 | 6.5 | <0.1 | 0.42 | 10 | 2 | 9.2 | <0.02 | <0.1 | 10.6 |
| A2 | 39-41 | 90 | 11.6 | 6.6 | <0.1 | 0.47 | <5 | 6 | 3.6 | 0.03 | <0.1 | 0.09 |
| A1 | 53-55 | 6.4 | 11.4 | 6.1 | <0.1 | 0.11 | <5 | 11 | 2.9 | 0.02 | <0.1 | 0.30 |
| At the margin of the plume in a zone of aerobic treatment | | | | | | | | | | | | |
| B3 | *13-15 | 11.8 | 11.9 | 7.3 | 2.0 | 0.11 | 12 | 10 | 24 | <0.07 | 5.5 | 0.015 |
| B2 | 22-25 | 8.4 | 11.9 | 7.3 | 6.8 | <0.05 | <5 | 11 | 1.0 | <0.02 | 2.3 | 0.012 |
| B1 | 32-34 | 5.0 | 11.3 | 7.0 | 0.6 | <0.05 | <5 | 14 | 0.7 | <0.02 | 0.4 | 0.07 |
| At the margin of the plume in a renovated zone | | | | | | | | | | | | |
| C3 | *14-16 | <0.2 | 9.3 | 6.9 | 3.1 | 0.16 | <5 | 7 | 2.4 | <0.02 | 9.6 | <0.001 |
| C2 | 22-24 | <0.2 | 10.8 | 6.5 | 3.7 | <0.05 | <5 | 11 | 0.7 | 0.02 | 5.0 | <0.001 |
| C1 | 30-32 | <0.2 | 10.3 | 7.0 | 1.6 | 0.11 | <5 | 10 | 3.5 | <0.02 | 3.5 | <0.001 |

* Identifies the water table; nd - Not determined.

## Analyses of Samples

At each sampling interval, three replicate microcosms of each treatment were analyzed. Each bottle was sampled by purging the volatile compounds onto a stainless steel trap (11.5 cm by 0.4 cm I.D.) packed with Tenax/silica gel (2:1). The bottles were purged with nitrogen at a flow rate of 50 mL/min for 15 min. To keep the aquifer material well slurried, the bottles were placed on a rotary shaker during purging. The purging efficiencies were at least 99% for each of the alkylbenzenes and at least 90% for the chlorinated compounds.

The traps were analyzed by Method 624, modified for GC/FID, "Methods for Organic Chemical Analysis of Municipal and Industrial Wastewater," EPA-600/4-82-057, July 1982. The Tenax trap containing each purged sample was desorbed for 8 min at 180°C onto the head of a 1.8 m by 2 mL I.D. glass column packed with 60/80 Carbopack C/0.2% Carbowax 1500. The Hewlett Packard 5880A gas chromatograph was temperature programmed at 40°C for 8 min, then heated at 8°C/min to 220°C and held for 15 min. The internal standard method of calibration and quantitation was used with 4-bromofluorobenzene as the internal standard.

## RESULTS AND DISCUSSION

## Characterization of Waters at Traverse City

Geochemical analyses of the water samples collected at Traverse City revealed waters of four distinct characters (Tables 2-5): (1) the heart of the plume; (2) an anaerobic zone of biological activity; (3) an aerobic zone of biological activity; and (4) a pristine or renovated zone.

The heart of the plume contained high concentrations of methane and alkylbenzenes with no detectable oxygen. The heart of the plume was surrounded by an anaerobic zone of biological activity with greatly reduced concentrations of alkylbenzenes and no oxygen. This anaerobic zone was surrounded by an aerobic region with detectable oxygen and even greater reductions in alkylbenzene concentrations. A renovated or pristine zone surrounded the aerobic zone of treatment with high concentrations of oxygen and no detectable alkylbenzenes.

Table 2.  Geochemistry of Site A--In a Zone of Anerobic Biological Activity.

| Sample | Position | Electrode potential (mV) | Dissolved oxygen (mg/L) | Methane (mg/L) | Total alkyl-benzenes (μg/L) |
|---|---|---|---|---|---|
| A4 | Just below water table | +160 | 4.5 | 0.78 | 394 |
| A3 | Saturated zone | -150 | <0.1 | 10.60 | 2,250 |
| A2 | Saturated zone | -170 | <0.1 | 0.09 | 90 |
| A1 | Just above clay | -170 | <0.1 | 0.30 | 6.4 |

Table 3.  Geochemistry of Site B--A Zone of Active Aerobic Biological Activity.

| Sample | Position | Electrode potential (mV) | Dissolved oxygen (mg/L) | Methane (mg/L) | Total alkyl-benzenes (μg/L) |
|---|---|---|---|---|---|
| B3 | Just below water table | +230 | 5.5 | 0.015 | 11.8 |
| B2 | Saturated zone | +185 | 2.3 | 0.012 | 8.4 |
| B1 | Just above clay | +230 | 0.4 | 0.070 | 5.0 |

Table 4.  Geochemistry of Site C--A Renovated Zone.

| Sample | Position | Electrode potential (mV) | Dissolved oxygen (mg/L) | Methane (mg/L) | Total alkyl-benzenes (μg/L) |
|---|---|---|---|---|---|
| C3 | Just below water table | +235 | 9.6 | <0.001 | <0.02 |
| C2 | Saturated zone | +230 | 5.0 | <0.001 | <0.02 |
| C1 | Just above clay | +280 | 3.5 | <0.001 | <0.02 |

Table 5.    Geochemistry of Site D--In the Plume Above the
            Interdiction Field.

| Sample | Position | Electrode potential (mV) | Dissolved oxygen (mg/L) | Methane (mg/L) | Total alkyl-benzenes (μg/L) |
|--------|----------|--------------------------|--------------------------|----------------|------------------------------|
| D4 | Just below water table | -115 | <0.1 | 7.9 | 11,820 |
| D3 | Saturated zone | -90 | <0.1 | 16.1 | 3,150 |
| D2 | Saturated zone | -130 | <0.1 | 1.1 | 43 |
| D1 | Just above clay | +136 | 0.1 | 0.04 | 32 |

## Degradation of Organic Compounds in Microcosms

Material for aerobic and anaerobic fate studies was collected from three locations in the plume (Fig. 1). The anaerobic fate study was performed on aquifer material from Site A at the 32 to 34 ft depth (the anaerobic zone of biological activity). Material from Site B at the 32 to 34 ft depth (the aerobic zone of biological activity) and Site C at the 22 to 24 ft depth (the pristine or renovated zone) was collected for aerobic fate studies.

Concentrations of the five substituted aromatic compounds at time zero and at various incubation times are shown in Table 6. Each of the four alkylbenzenes studied exhibited a biological fate with removals observed in both aerobic and anaerobic aquifer material. The disappearances were quite rapid for all compounds in each of the three geochemical zones studied, whether the anaerobic zone of biological activity, the aerobic zone of biological activity, or the pristine material. By the end of 8 weeks of incubation of the compounds in the anaerobic aquifer material, the concentration of benzene, toluene, $m$-xylene, and $o$-xylene had been removed one order of magnitude. The biotransformation of the four compounds in the aerobic zone of biological activity occurred even more rapidly than from the anaerobic material. By the end of 2 weeks of incubation, the concentrations for all four compounds were decreased by two orders of magnitude. Similar losses of two orders of magnitude were seen in material from the pristine aquifer zone. By contrast, the decreases in the concentrations of the four alkylbenzenes in the autoclaved samples were not as great. At the end of 4 weeks of incubation, the concentrations for benzene, toluene, $m$-xylene, and $o$-xylene were 57%, 61%, 46%, and 49% of the original concentrations. Previous work in this laboratory has seen no significant loss of compounds such as benzene, toluene, or the xylenes in autoclaved controls. Concentrations of these compounds in previous studies remained approximately 100% of the original concentrations after 4 weeks of incubation. The reasons for the greater loss in this study have not yet been determined. However, sorption to the solids is a possibility, and sorption/desorption studies are currently underway.

Table 6. Behavior of Alkylbenzenes and Chlorobenzene in Aquifer
Material from an Aviation Gasoline Plume.

| Subsurface material | Elapsed time in weeks | Benzene | Toluene | m & p-Xylene | o-Xylene | Chloro-benzene |
|---|---|---|---|---|---|---|
| | | | (μg/L porewater) | | | |
| A3-anaerobic | 0 | 450 | 420 | 440 | 410 | 560 |
| | 4 | 12 | 56 | 78 | 41 | 38 |
| | 8 | 6 | 40 | 17 | 6 | 34 |
| B2-aerobic | 0 | 450 | 420 | 390 | 390 | 524 |
| | 2 | 2 | 2 | 1 | 2 | 54 |
| | 4 | 5 | 5 | 2 | 1 | 50 |
| C1-aerobic | 0 | 420 | 380 | 370 | 370 | 470 |
| | 2 | 4 | 3 | 3 | 3 | 100 |
| | 4 | 1 | 3 | 1 | 1 | 106 |
| C1-autoclaved | 0 | 420 | 380 | 370 | 370 | 470 |
| | 2 | 380 | 290 | 200 | 190 | 325 |
| | 4 | 240 | 230 | 170 | 180 | 208 |

Chlorobenzene was also biotransformed in the three types of
materials studied. Decreases of one order of magnitude were seen
for chlorobenzene in both the aerobic and anaerobic zones of
biological activity. Removals of 80% of the original concentrations were seen in the pristine material. Loss of chlorobenzene in
the autoclaved samples was not as great, with 44% of the compound
remaining after 4 weeks of incubation.

To check the maintenance of methanogenic conditions in the
microcosms, headspace concentrations of methane were measured
immediately before sampling for the alkylbenzenes. Methane was
found in all living samples with concentrations ranging from 50 to
100 ppb. No methane was found in the headspace of the autoclaved
samples.

Trichloroethylene and 1,1,1-trichloroethane were the chlorinated aliphatic hydrocarbons added with the alkylbenzenes to the
three geochemical types of aquifer material. Initial concentrations for each were approximately 550 μg/L in the pore water of the
microcosms. At the end of 8 weeks of incubation, no biotransformation was seen for either of the chlorinated solvents. Concentrations in the autoclaved samples remained the same as the initial
concentrations.

Previous work has shown that trichloroethylene and 1,1,1-
trichloroethane are not biologically transformed in aerobic subsurface environments (Wilson and McNabb, 1983). However, in
anaerobic subsurface environments, trichloroethylene undergoes a
reductive dehalogenation forming the dichloroethylenes and vinyl
chloride as daughter products (Bouwer and McCarty, 1983; Parsons
et al., 1984; Parsons and Lage, 1985). These compounds are more
mobile in subsurface environments than trichloroethylene, and, in
the case of vinyl chloride, more toxic.

In anaerobic subsurface materials 1,1,1-Trichloroethane is transformed to 1,1-dichloroethane (Barrio-Lage et al., 1986). This compound also has a nonbiological fate whereby 1,1-dichloroethylene is formed from the abiotic dehydrochlorination of 1,1,1-trichloroethane (Vogel and McCarty, 1986). This reaction can occur in conditions of dilute aqueous solutions at neutral pH and 20°C which are common to many ground waters. Gradual removals of 1,1,1-trichloroethane from both living and autoclaved samples should be seen from this abiotic process.

Previously, alkylbenzenes were not thought to be biolgically transformed without molecular oxygen (Young, 1984) or oxygen-containing substituents (Evans, 1977; Braun and Gibson, 1984). However, field evidence of biotransformation of alkylbenzenes was seen in methanogenic landfill leachate (Reinhard et al., 1984) with preferential removals of o-, m-, p-xylene over other petroleum hydrocarbons. Laboratory studies of four alkylbenzenes commonly found as indicators of gasoline contamination of ground water showed the biological removal of those compounds in methanogenic aquifer material (Wilson et al., 1986). Radiolabeled carbon dioxide ($^{14}CO_2$) was found as a product of transormation after $^{14}C$-toluene was added to the methanogenic microcosms. Biotransformations of the three xylene isomers were seen in a laboratory column study, with the xylenes being used as sole carbon and energy sources by the bacteria (Kuhn et al., 1985). Mixed-culture studies using methanogenic consortia actively metabolize both toluene and benzene to carbon dioxide. The incorporation of oxygen into the benzene ring during the anaerobic oxidation of those compounds to carbon dioxide was shown to come from water by the use of $^{18}O$-labeled water (Vogel and Grbić-Galić, 1986).

CONCLUSIONS

The fate of four alkylbenzenes and three chlorinated hydro-carbons was studied in aquifer material from three geochemical regions of an aquifer contaminated with aviation gasoline. All four of the alkylbenzenes were biologically transformed in each of the materials studied. The removals of benzene, toluene, m-xylene, and o-xylene were quite rapid in aquifer material from the anaero-bic zone of biological activity, the aerobic zone of biological activity and the pristine region with no lag times required before transformation began. Removals of one order of magnitude were seen at the end of 8 weeks of incubation in material from the anaerobic zone of biological activity. The removals of the alkylbenzenes at the end of 2 weeks of incubation in material from the aerobic zone of biological activity and the pristine zone were two orders of magnitude. No removals of 1,1,1-trichloroethane or trichloro-ethylene were seen. Chlorobenzene concentrations were reduced one order of magnitude in both aerobic and anaerobic zones of treatment and by 80% in the material from the pristine area adjacent to the plume.

Both natural aerobic and anaerobic in situ biorestoration of contaminated gasoline plumes can occur. Aerobic restoration of aquifers contaminated with petroleum products (Wilson and Ward, 1986) is frequently accomplished with oxygen being the greatest limiting constituent.

The anaerobic transformations seen at this site and confirmed by laboratory studies provide an attractive alternative to aerobic restoration. The removals of the alkylbenzenes in the anaerobic material were quite rapid and favorably compare to removals in the aerobic zone of treatment.

Oxygen concentrations in ground water are limited by solubility; also, oxygen is frequently depleted in heavily contaminated ground waters and soils because of the excessive oxygen demand. Anaerobic transformations of alkylbenzenes can enhance in situ biorestoration in oxygen depleted regions of a plume. Naturally-occurring aerobic and anaerobic biorestoration can remediate ground waters contaminated with petroleum products and significantly increase the reliability of existing reclamation methods.

## SUPPORT

This study was supported by the United States Air Force through Interagency Agreement RW57930615-01-1 with the U.S. Environmnetal Protection Agency.

## DISCLAIMER

Although the research described in this article has been supported in part by the U.S. Environmental Protection Agency under assistance agreement number CR-811146 to the University of Oklahoma and under in-house programs, it has not been subjected to the Agency's peer and administrative review and, therefore, does not necessarily reflect the views of the Agency, and no official endorsement should be inferred.

The use of trade or product names in this chapter is for iden-tification purposes only and does not constitute endorsement by the U.S. Geological Survey.

## REFERENCES

Barrio-Lage, G., F. Z. Parsons, R. S. Nassar, and P. A. Lorenzo. 1986. Sequential dehalogenation of chlorinated ethenes. Environ. Sci. Technol. 20:96-99.

Braun, K. and D. T. Gibson. 1984. Anaerobic degradation of 2-aminobenzoate (anthranilic acid) by denitrifying bacteria. Appl. Environ. Microbiol. 48:102-107.

Bouwer, E. J., P. L. McCarty. 1983. Transformations of 1- and 2-carbon halogenated aliphatic organic compounds under methanogenic conditions. Appl. Environ. Microbiol. 45(4):1286-1294.

Coleman, W. E., J. W. Munch, R. P. Streicher, H. P. Ringhand, and F. C. Kopfler, 1984. The identification and measurement of components in gasoline, kerosine, and no. 2 fuel oil that partition into the aqueous phase after mixing. Arch. Environ. Contam. Toxicol. 13:171-178.

Evans, W. C. 1977. Biochemistry of the bacterial catabolism of aromatic compounds in anaerobic environments. Nature. 270:17-22.

Federal Register. 1984. 49(114):24334.

Grbić-Galić, D., 1986. Anaerobic production and transformation of aromatic hydrocarbons and substituted phenols by ferulic acid-degrading BESA-inhibited methanogenic consortia. FEMS Microbiol. Ecol. 38:161-169.

Kuhn, E. P., P. J. Colberg, J. L. Schnoor, O. Wanner, A. J. B. Zehnder, and R. P. Schwarzenbach. 1985. Microbial transformations of substituted benzenes during infiltration of river water to groundwater: laboratory column studies. Environ. Sci. Technol. 19:961-968.

Parsons, F., and G. B. Lage. 1985. Chlorinated organics in simulated groundwater environments. J. Amer. Water Works Assoc. 77(5):52-59.

Reinhard, M., N. L. Goodman, and J. F. Barker. 1984. Occurrence and distribution of organic chemicls in two landfill leachate plumes. Environ. Sci. Technol. 18:953-961.

Twenter, F. R., T. R. Cummings, and N. G. Grannemann. 1985. Ground-water contamination in East Bay Township, Michigan. U.S. Geological Survey. Water-resources Investigations Report 85-4064.

U.S. Environmental Protection Agency. *In* Methods for Organic Chemical Analysis of Municipal and Industrial Wastewater. Longbottom, J.E. and J.J. Lichtenberg, eds. U.S. Governmental Printing Office: Washington, D.C. Purgeable Method 624, pp 624-1 to 624-12. EPA-600/4-82-057.

Vogel, T. M., and D. Grbić-Galić. 1986. Incorporation of oxygen from water into toluene and benzene during anaerobic fermentative transformation. Appl. Environ. Microbiol. 52:200-202.

Vogel, T. M., and P. L. McCarty. 1986. Rate of abiotic formation of 1,1-dichloroethylene from 1,1,1-trichloroethane in groundwater. Submitted to J. Contaminant Hydrol.

Wilson, B. H., G. B. Smith, and J. R. Rees. 1986. Biotransformations of selected alkylbenzenes and halogenated aliphatic hydrocarbons in methanogenic aquifer material: a microcosm study. Environ. Sci. Technol. 20:997-1002.

Wilson, J. T., and J. F. McNabb. 1983. Biological transformation of organic pollutants in groundwater. EOS (Trans. Am. Geophys. Union). 64:505-506.

Wilson, J. T., and C. H. Ward. 1986. Opportunities for bioreclamation of aquifers contaminated with petroleum hydrocarbons. Developments in Industrial Microbiol. In press.

Young, L. Y. *In* Microbial Degradation of Organic Compounds. Gibson, D.T., ed., Marcel Dekker, New York, 1984, pp. 487-523.

SECTION III

CHEMISTRY OF SURFACE WATER

# CHAPTER 9

# HYDROLOGIC AND CHEMICAL FLUX IN LOCH VALE WATERSHED, ROCKY MOUNTAIN NATIONAL PARK

Jill Baron, National Park Service, Fort Collins, Colorado

Owen P. Bricker, U.S. Geological Survey, Reston, Virginia

## ABSTRACT

Annual hydrologic and chemical budgets are described for 1984 and 1985 for Loch Vale Watershed, Rocky Mountain National Park, Colorado. Our objective in this paper is to define major physical and chemical processes that control hydrologic and chemical flux in an alpine-subalpine watershed. Between 60 and 70% of annual water input to Loch Vale enters as snow and this constitutes the major influence on hydrology. Snowmelt, a period of 2 months, is the major event of the hydrologic year. The melt process is also very important in governing the flow of solutes through Loch Vale. Annually 70% of the flux of $SO_4^{2-}$ out of the watershed occurs during the snowmelt period. Another important process influencing the movement of ions is the freezing of lake surfaces during winter, which causes a two-fold concentration of solutes in lake water. Biological uptake and immobilization of $NO_3^-$, $NH_4^+$, and $SO_4^{2-}$, and weathering and leaching of base cations from bedrock, talus, and soils also affects ionic movement.

## INTRODUCTION

In this paper we describe a watershed in the southern Rocky Mountains of the United States and explore in some detail the variation observed over a 2-year period in the hydrologic and biogeochemical processes believed to exert major influences on the fluxes of elements. We will demonstrate that the annual pattern of water movement, as it relates to temperature and atmospheric deposition, is the single most important factor in determining chemical flux. Other processes that affect water chemistry include

*Chemical Quality of Water and the Hydrologic Cycle*, Robert C. Averett and Diane M. McKnight (Eds.) © 1987 Lewis Publishers, Inc., Chelsea, Michigan. Printed in the United States of America.

weathering of bedrock and secondary materials (Johnson, 1984), soil processes such as cation exchange and anion adsorption (Reuss and Johnson, 1986), and biological processes such as nutrient uptake and release (Krug and Frink, 1983; Binkley and Richter, in press) and decomposition (Binkley and Richter, in press). These will be discussed relative to their influence on surface-water chemistry of Loch Vale Watershed.

Lentic systems in temperate mountainous areas are characterized by swiftly flowing streams along steep elevational gradients. Because the ultimate source of water for these systems is atmospheric, and the opportunity for subsurface storage on steep mountain slopes is low, the patterns of water input become the major factor in determining chemical flux. This has been noted in other systems, mountainous or otherwise, where streamflow is directly related to deposition (Bond, 1979; Kunkle and Comer, 1972; Verry, 1975).

There can be large increases in the concentration of solutes of surface water caused by springtime warming and melting of the snowpack in mountainous areas. Atmospherically deposited cations and anions within the snowpack migrate as the pack weathers and ages (Tranter et al., 1985) and they will emerge with the first meltwater (Wright, 1983). This process causes an initial pulse of solute to enter the surface waters, followed by very dilute snow water. During the snowmelt period there can be a flushing of surficial material such as organic detritus or minerals as well as a flushing of soil solution as meltwater courses through terrestrial systems. This can introduce soluble organic complexes, nutrients (Lewis and Grant, 1979; Likens et al., 1977), and cations from the soil water (Reuss and Johnson, 1986). A third factor influencing springtime surface waters is the melting of lake ice. The composition of ice is relatively pure and its melting in the spring can release a sizeable volume which will dilute existing waters (Gunn and Keller, 1984; Canfield et al., 1983).


DESCRIPTION OF THE STUDY AREA

Loch Vale Watershed is a northeastern-facing drainage located on the Front Range of Colorado in Rocky Mountain National Park (Fig. 1). Elevation in the 660 ha watershed ranges from 3,962 m at the Continental Divide to 3,109 m at the outlet of the lowest of three lakes. The bedrock is biotite gneiss and schist, with intrusions of Silver Plume granite (Cole, 1977). The Loch Vale Watershed is 81% exposed bedrock and talus containing the primary minerals biotite, quartz, orthoclase, and a plagioclase near the sodic end of the oligiclase range. The soils under spruce and fir forest range from undeveloped to well-developed alfisols containing an argyllic horizon. Alluvial and bog soils are found adjacent to stream channels and in places where lack of physical relief permits their development (Walthall, 1985). Roughly 11% of Loch Vale Watershed is an alpine ridge physically separated from the valley part of the watershed by almost 300 m of rock walls and talus.

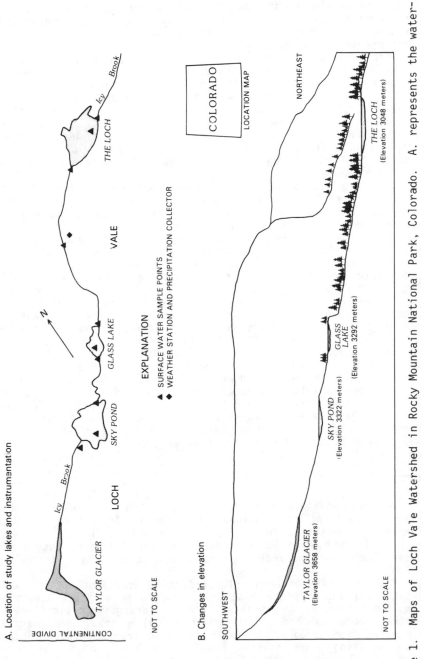

A. Location of study lakes and instrumentation

CONTINENTAL DIVIDE

TAYLOR GLACIER

Icy Brook

LOCH

SKY POND

GLASS LAKE

N

VALE

THE LOCH

Icy Brook

EXPLANATION

▲ SURFACE WATER SAMPLE POINTS
◆ WEATHER STATION AND PRECIPITATION COLLECTOR

NOT TO SCALE

B. Changes in elevation

COLORADO

LOCATION MAP

SOUTHWEST

TAYLOR GLACIER
(Elevation 3658 meters)

SKY POND
(Elevation 3322 meters)

GLASS LAKE
(Elevation 3292 meters)

THE LOCH
(Elevation 3048 meters)

NORTHEAST

NOT TO SCALE

Figure 1.  Maps of Loch Vale Watershed in Rocky Mountain National Park, Colorado.  A. represents the water-shed incross-section, and B. shows the watershed in plan view.

Because of this vertical distance, and because high wind velocities cause most moisture to evaporate or sublimate, it is thought that the processes occurring on the ridges do not contribute much to surface-water composition below.

Three vegetation types are found in the study area. The alpine ridges and the valley area above 3,300 m have sparse patches of alpine tundra interspersed with boulders or bedrock outcrops. From 3,300 m (treeline) to the lowest part of the drainage is an old-growth forest of Englemann spruce, Douglas fir and limber pine. Tree cores strongly suggest few major disturbances, such as fire, have occurred within the last 450 yr in the watershed; however, forest stands along the valley periphery are periodically knocked down by avalanches. Approximately 0.5% of the watershed supports wet sedge meadows.

The three lakes within the study area are connected by Icy Creek, a perennial stream. Sky Pond and Glass Lake are located above treeline and are surrounded by talus and sparse tundra. The lowest lake, The Loch, is partially surrounded by forest, and receives water and chemical input from forest and meadow soils, as well as from Icy Creek. Icy Creek is first order until just below treeline, where it is joined by Andrews Creek; it remains second order into The Loch. None of the three lakes stratify thermally due to the seasonally rapid flow of water and the strong winds which promote mixing. Sky Pond flushes three times per year; Glass Lake flushes 26 times per year; and The Loch flushes 19 times per year. Flushing is partially related to the rapid movement of water through Loch Vale Watershed as well as to the low average lake volumes, 61,000, 26,000, and 121,700 $m^3$ for The Loch, Glass Lake, and Sky Pond, respectively. During November through April no water flows through the watershed and the lake basins become closed.

Average precipitation ranges from 75 to 100 cm/yr, increasing with elevation (Marr, 1967). Our data from the higher elevations at Loch Vale Watershed suggest 60 to 70% of the annual precipitation falls as snow. Summer and autumn inputs often occur as thunderstorms, some of which are severe.

## METHODS

Wet precipitation has been collected weekly since September 1983 as part of the National Atmospheric Deposition Program. Samples are analyzed for $Ca^{2+}$, $Mg^{2+}$, $K^+$, $Na^+$, $H^+$, $SO_4^{2-}$, $NO_3^-$, $PO_4^{3-}$, $Cl^-$ and specific conductance at the Illinois State Water Survey laboratory according to methods described in Pedon et al. (1986).

Surface water has been collected at 10 locations in Loch Vale Watershed (Fig. 1) since 1982 on a seasonally variable schedule. Sampling has varied from weekly and daily springtime collections to monthly summertime and bimonthly winter collections. Grab samples are taken from streams. Lakes are sampled from a raft or through the ice using a peristaltic pump. Samples are placed in acid-washed 2-L linear polyethylene (lpe) containers, transported to the laboratory, and filtered (0.45-μm Nucleopore) into smaller, acid-washed lpe bottles for shipment to the Denver Central

Laboratory of the U.S. Geological Survey. Unfiltered samples are used for determination of pH (Corning electrode #476182) and specific conductance (Fisher model 09-327) in the field. Chemical determinations are made for the major anions and cations, $Fe^{3+}$, $Mn^{2+}$, $Al^{3+}$, $Br^-$, $F^-$, $SiO_2$, color, dissolved organic carbon (DOC), and alkalinity using standardized methods (Feltz and Anthony, 1984).

Precipitation volume is measured midway up the watershed with a Belfort weighing bucket rain gage (Fig. 1). Outflow from Loch Vale Watershed is gaged continuously with a Parshall flume located at the outlet of the lowest lake. Stilling well stage heights, measured with a float, are recorded electronically (Omnidata one-channel Datapod recorder) and calibrated to a backup Stevens Type F Recorder. Stage heights were converted to flow using the equation

$$Q = 4WHa^{1.522}W^{0.026}$$

where $Q$ = discharge, $W$ is throat width of flume, and $H_a$ is height of water in feet. Evaporation during 1984 and 1985 was calculated using a temperature-based model derived from Linacre (1977). Temperature was recorded every 15 min at a Remote Area Weather Station (Fig. 1). Seismic refraction studies were conducted by Zavodil (1984) to determine the presence of areas of fractured rock and depth to bedrock. This was done to assess the probability of water seeping out of Loch Vale Watershed as groundwater, or of water being stored in subsurface soil or rock reservoirs using techniques developed by Redpath (1973). The studies revealed very shallow soils and till overlaying continuous bedrock, suggesting minimal loss of water to either deep or shallow reservoirs.

The soil chemistry and the estimates of available soil pools of cations were derived from Walthall (1985). Estimates of nitrogen uptake by lake algal communities were derived from data collected by McKnight et al. (1986) and Redfield relationships (Redfield, 1958). Potential conversion rates of sulfate to organic and ester-bonded sulfur were obtained from S-35 labelled soil incubations at the University of Georgia using the methods of Swank et al. (1984).

RESULTS AND DISCUSSION

The total precipitation for 1984 and 1985 was 111 cm (a total input to the watershed of 7.3 million $m^3$) and 108 cm (a total input to the watershed of 7.2 million $m^3$), respectively, as measured at the weather station (Table 1). The 1984 snow year contributed 70.69 cm, or 64% of the yearly precipitation total. The remaining 40.21 cm or 36% of the precipitation fell as rain. The pattern was similar in 1985, with 66.58 cm (68%) falling as snow, and 30.91 cm or 32% falling as rain.

Outflow was 5.6 million $m^3$ in 1984 and 4.2 million $m^3$ in 1985, with outflow occurring only during April through December. Loch Vale Watershed is icebound from January to early April with little to no water movement between lake systems. Evaporation was calculated to be 5.165 million $m^3$ in 1984, and 4.078 million $m^3$ in 1985, or roughly equal to the measured outflow for both years.

Table 1.    Hydrologic Balance for Loch Vale Watershed for 1984 and
            1985.    Values are in Million Cubic Meters.

| Water parameter | 1984 | | 1985 | |
|---|---|---|---|---|
| | Value | Adjusted by Cl-balance | Value | Adjusted by Cl-balance |
| Deposition (D) (measured) | 7.337 | 9.905 | 7.162 | 7.162 |
| Evaporation (E) (calculated) | 5.165 | 5.165 | 4.078 | 4.078 |
| Outflow (O) (measured) | 5.996 | 5.996 | 4.152 | 4.401 |
| D - (E+O) | -3.824 | -1.256 | -1.068 | -1.317 |
| Percent deposition as snow | 64% | | 68% | |

There is a large discrepancy in the water budget for 1984 when it is developed using measured values of deposition and outflow and a calculated value for evaporation.  Chloride was used to balance the water budget, on the assumption that it is unreactive to biologic, geologic or soil processes (Likens et al., 1977).  The amount of chloride leaving the watershed should, therefore, equal the amount that enters.  The loss of chloride from Loch Vale Watershed in 1984 (921.2 kg/yr) was 35% greater than the measured inputs of that ion.  Increasing the volume of precipitation by 35% accounted for 67% of the discrepancy, still leaving 33% of the water inputs unaccounted for.

The remaining uncertainty could be caused by three factors. Undermeasurement of snow input is the source of some of the error in the water budget.  Alter-shielded rain gages measure 50 to 60% of total precipitation depths in winter at wind speeds up to 8 m/s (Goodison et al., 1981). Wintertime wind speeds between 8 to 13 m/s are common in Loch Vale Watershed, so catch efficiencies can possibly be as low as 30 to 40% of total precipitation. If 70% of annual precipitation is snow, and snow is underestimated by up to 60%, it could cause a sizeable error in annual precipitation values.

Accurate values for evaporation are difficult to obtain (Winter, 1981).  Furthermore, the estimated errors associated with different methods of measuring or computing evaporation are determined by comparing one method with another (Winter, 1981). The temperature-based model we used to calculate evaporation (Linacre, 1977) has not been compared with other commonly used techniques.  Our model suggests that annually half of the water that enters Loch Vale Watershed is evaporated.  This compares with the concentrations of a conservative ion, chloride, which in 1984 enterd the watershed at 2.62 µeq/L, but left the watershed in concentrations of 5.92 µeq/L (figures are annual averages; the concentration factor is 2.26). Values for 1985 give a concentration factor of 1.85. The error associated with this evaporation technique for Loch Vale Watershed may be tentatively assigned the value of ±26%.

The error associated with Parshall flumes is ±5% (Winter, 1981). We consider there to be an additional 5% error associated with water bypassing the flume through an adjacent meadow, so the total uncertainty assigned to flow values is ±10%.

Surface waters are extremely dilute, with solutes varying seasonally between 297 and 535 µeq/l (Table 2). On an annual basis silica is the most abundant of the dissolved species. Anions are found in the relationship $HCO_3^- > SO_4^{2-} > NO_3^- > Cl^-$. The most abundant cation is $Ca^{2+}$, followed in order by $Na^+$, $Mg^{2+}$, $K^+$, $NH_4^+$, and $H^+$. Annual average water composition does not reflect the pronounced seasonal chemical variability that occurs in conjunction with the hydrologic cycle. A hydrologic year can be divided into three parts, based upon flow and chemistry (Fig. 2). Phase I occurs during the summer and fall when there is decreasing flow (from 45 $ft^3/s$ to 5 $ft^3/s$) and dilute water. The lowest concentrations of all of the major ions are during Phase I (Table 2) and the variability, as characterized by the standard deviation around each mean, is also low. Phase I can be punctuated by storm events which cause flow to rise. The effect of storms on water chemistry is discussed in Chapter 3 of this book.

The winter snowpack begins to accumulate in October, and this marks the onset of the second phase of a hydrologic year. The lakes in Loch Vale Watershed freeze over, with the lake ice cover gradually thickening throughout the winter to about one meter. Flow at the outlet of Loch Vale Watershed decreases, and ceases by December. Chemical concentrations in the remaining water rise during winter, and most ions reach their maximum concentration at this time of the year (Table 2). Nitrate and ammonium ions are the exceptions, for reasons discussed below. The standard deviations around the means of these wintertime values are large because of the arbitrary initiation of Phase II at the beginning of ice formation. Solute concentrations are still very low at this time, in contrast to the majority of Phase II when concentrations are high. There are two reasons why the water becomes so concentrated during these months, lack of flushing and ice formation. Approximately half the volume of lake water freezes during the winter, creating an ice layer one meter or more in thickness which causes winter surface water ionic concentrations to be approximately twice summer ionic concentrations (Table 2).

Phase III encompasses the snowmelt period in Loch Vale Watershed, the period of greatest chemical and physical change (Fig. 2). The streams begin to flow near the end of April. For a short period this causes laminar flow under the ice because the stream water at 0°C is less dense than lake bottom water at 4°C. During the snowmelt period three distinct processes occur: flushing, pulses of solute inputs from the snowpack, and dilution.

Table 2.  Concentrations (µeq/L) of the Loch Surface Water for the Period 5/11/82 to 6/03/85.  Presented are Means (Std Dev) for the Total Sample Period as well as Means (Std Dev) for the Different Hydrological Groupings (Described in Text and Fig. 2).

| | Mean<br>n=44 | I<br>7/15 to 9/30 | II<br>10/1 to 4/30 | III<br>5/1 to 7/17 |
|---|---|---|---|---|
| $Ca^{2+}$ | 75.35<br>(28.94) | 50.30<br>(8.21) | 108.15<br>(18.14) | 72.90<br>(19.78) |
| $Mg^{2+}$ | 21.40<br>(7.41) | 14.04<br>(3.48) | 28.21<br>(6.13) | 21.63<br>(5.11) |
| $K^+$ | 5.37<br>(2.30) | 3.69<br>(0.70) | 6.03<br>(1.51) | 6.08<br>(2.26) |
| $Na^+$ | 26.10<br>(12.18) | 17.32<br>(3.84) | 40.19<br>(11.63) | 22.71<br>(11.03) |
| $NH_4^+$ | 2.77<br>(2.22) | 1.54<br>(1.74) | 2.21<br>(1.17) | 3.74<br>(2.68) |
| $H^+$ | 0.55<br>(0.37) | 0.28<br>(0.13) | 0.57<br>(0.28) | 0.66<br>(0.42) |
| $NO_3^-$ | 16.61<br>(7.26) | 12.30<br>(5.80) | 15.84<br>(10.47) | 19.16<br>(5.08) |
| $SO_4^{2-}$ | 38.73<br>(12.70) | 28.27<br>(4.20) | 51.33<br>(8.89) | 38.63<br>(9.29) |
| $Cl^-$ | 5.92<br>(1.97) | 4.84<br>(2.62) | 6.75<br>(1.50) | 6.18<br>(1.64) |
| $HCO_3^-$ | 46.51<br>(29.15) | 57.32<br>(37.75) | 87.23<br>(27.68) | 33.32<br>(8.08) |
| $H_4SiO_4$ | 146.45<br>(40.61) | 107.62<br>(14.84) | 189.11<br>(27.56) | 142.45<br>(54.70) |

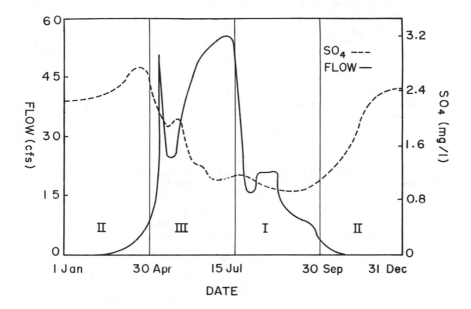

Figure 2.    A hypothetical water year in Loch Vale Watershed.   Lines
are derived from several years data.   Phase I represents
the least concentrated water chemistry ($SO_4$ is used as
an example) and low flow or decling flow from the snowmelt period.   Phase II represents wintertime conditions:
high chemical concentrations under lake ice, and little
to no flow in the stream system.   Phase III is the
period of snowmelt.   After an initial pulse of ions from
the snowpack, concentrations decline as flow rises,
peaks, and also begins to decline.

The concentrated bottom water in the lakes is flushed.   This is
augmented by the winter's accumulated atmospheric deposition, which
migrates through the snowpack and enters surface waters, resulting
in a temporary increase in solute concentration, particularly of
sulfate (Fig. 2), nitrate, ammonium, and hydrogen.   Johannessen et
al. (1980) found that the first 10% of the snowmelt can contain 50%
of the solute content of the winter pack, and this flushing can
cause a temporary drop in surface water pH if the snow solution is
acidic.   In Loch Vale Watershed the surface water pH minimum (5.7)
occurs when flow increases from 3.8 to 4.4 $ft^3/s$, representing
approximately 1 to 2% of the annual snowmelt.   The nitrogen species
and hydrogen ion reach their highest mean values during this phase
(Table 2), which can last from several days to two weeks.
Approximately 70% of the total annual flux of sulfate passes
through Loch Vale Watershed during this snowmelt period,
corresponding well with our calculations that 64 to 68% of the
annual precipitation enters Loch Vale as snow.

After the soluble ions leave the snowpack, there follows a long period of melting which is responsible for diluting solutes in the surface water to very low concentrations. This period lasts until mid-July in Loch vale Watershed, at which time the hydrologic cycle returns to Phase I, which is characterized by low flow and dilute waters.

Annual budgets (kg/ha) of deposition inputs and streamwater outputs for the major ionic species (Table 3) suggest that the ions can be grouped into three categories. The first category consists of the atmospherically deposited nutrients nitrate, ammonium, and possibly sulfate, which are utilized biologically so that some proportion is retained in Loch Vale on an annual basis. The second category is made up of base cations and other species which leave Loch Vale in greater amounts than their atmospheric input loadings. The watershed is a net exporter of calcium, magnesium, sodium, potassium, silica and alkalinity due to weathering of the bedrock and talus. The third category is made up of ions passing through the watershed with little or no change in total amount. Chloride follows this pattern. Another anion that might fall in this category is sulfate. In the 1985 chemical budget sulfate inputs balanced outputs, raising the possibility that there is no sink for additional sulfate in Loch Vale. The budget figures for 1984 did not show this pattern, rather there was a 30% retention of sulfate within Loch Vale. The errors associated with the water budget may be responsible for the uncertainty in sulfate flux values (LaBaugh and Winter, 1984).

Table 3.    Annual Budgets for Major Cations and Anions for 1984 and 1985 in Loch Vale Watershed. Values (kg/ha) have been Normalized to the Balanced Chloride Budget. In 1984 This Caused a 35% Increase in Input Values. In 1985 This Caused a 6% Increase in Output Values.

| Species | 1984 | | | 1985 | | |
| | In kg/ha | Out kg/ha | Out/In | In kg/ha | Out kg/ha | Out/In |
| --- | --- | --- | --- | --- | --- | --- |
| $Ca^{2+}$ | 2.61 | 7.55 | 2.90 | 2.46 | 8.21 | 3.34 |
| $Mg^{2+}$ | 0.57 | 1.36 | 2.40 | 0.42 | 1.24 | 2.95 |
| $Na^+$ | 1.07 | 2.76 | 2.59 | 0.79 | 2.69 | 3.41 |
| $K^+$ | 0.31 | 1.25 | 4.02 | 0.40 | 1.03 | 2.58 |
| $NH_4^+$ | 1.88 | 0.37 | 0.20 | 1.21 | 0.11 | 0.09 |
| $SO_4^{2-}$ | 10.68 | 8.25 | 0.77 | 8.18 | 9.30 | 1.14 |
| $NO_3^-$ | 10.23 | 7.14 | 0.70 | 8.25 | 6.49 | 0.79 |
| $Cl^-$ | 1.39 | 1.39 | 1.00 | 1.21 | 1.21 | 1.00 |
| $SiO_2$ | - | 11.84 | - | - | 9.68 | - |
| pH | 5.01 | 6.33 | - | 5.06 | 6.27 | - |

The physical movement of water is responsible for much of the transport of atmospherically-deposited ions through Loch Vale Watershed. This is easily seen if one examines nitrate and ammonium. Both of the nitrogen compounds are present in their highest concentrations during the snowmelt period because they are physically transported directly from the deposited snowpack to surface waters without any chance to interract with ecosystem components. In the summertime the situation is reversed. Nitrogen is in such biological demand that the summer season offsets the springtime "leakiness" such that the annual nitrogen budget shows net retention within Loch Vale (Table 4). Aquatic primary productivity has been approximated to be responsible for 1.25 kg/ha/yr N uptake (based on data from McKnight et al., 1986). This means the minimal terrestrial N accumulation is 0.70 kg/ha/yr. This does not take into account the substantial amount of internal N cycling which occurs.

There is a 70% sulfate flux in response to snowmelt. The remaining 30% enters Loch Vale Watershed during the ice-free times of year. Soil samples collected and analyzed for rates of microbial conversion of $SO_4^{2-}$ to ester-bonded S and organically-bound S show there is the potential for 4.225 kg/ha/yr to be converted to these quasi-stable forms (Fitzgerald, unpubl. data). If similar microbial conversion of sulfur occurs in situ in the watershed soils it could result in Loch Vale Watershed serving as a net sink for atmospheric sulfur, even though the clay minerals more commonly associated with sulfate adsorption are largely absent (Walthall, 1985).

Table 4.   Nitrogen Budget for Loch Vale Watershed.

|  | In | Out | Retained |
|---|---|---|---|
| $NH_4$-N | 1015.5<br>(1.55) | 199.7<br>(0.34) | 815.8<br>(1.21) |
| $NO_3$-N | 1488.3<br>(2.31) | 1038.6<br>(1.57) | 449.7<br>(0.74) |
| Total | 2503.8<br>(3.86) | 1238.3<br>(1.91) | 1265.5<br>(1.95) |
| Approx. aquatic n-uptake based on primary productivity |  |  | 840.0<br>(1.25) |
| Minimum terrestrial n-uptake |  |  | 425.5<br>(0.70) |

Values are in kg/yr.   Values in parentheses are kg/ha.

While biological activity may be controlling the flux of nitrogen compounds and perhaps sulfate, different processes control the flux of base cations through Loch Vale. All four cations ($Ca^{2+}$, $Mg^{2+}$, $K^+$, $Na^+$) leave the watershed in greater quantity than they enter via atmospheric deposition, indicating internal sources. The metamorphic and granitic parent material is rich in bases, and weathering can serve as a primary source of cations. The soils of the watershed also contribute base cations.

## SUMMARY

Clearly, the pattern of precipitation plays a dominant role in the annual hydrology of Loch Vale Watershed. The important annual event is the melting of 7 months of accumulated precipitation. Storm events in summer or autumn are minor compared to the peak flow of 60 $ft^2$/s or greater that is produced regularly in the spring by snowmelt.

The hydrology, in turn, exerts a large influence on water chemistry. It appears from the sulfate budget that solutes entering Loch Vale Watershed in snow do not have an opportunity to interract with the ecosystem before they are flushed through with the snowmelt. During the ice-free months, however, ecosystem processes become more important. Conversion of sulfate to ester-bonded sulfur and organic S complexes may be important processes affecting the movement of sulfate during the summer. Biological uptake and transformation of nitrate and ammonium ions plays a very important role in the flow of nitrogen through Loch Vale in the ice-free months. More work is needed to determine nitrogen dynamics during other times of the year.

Loch Vale Watershed is a net source for cations. The cations are weathering from bedrock and leaching from soils in volumes that increase their output 2.5 to 4.0 times above the input. Increases in alkalinity and dissolved silica reflect weathering processes. Alkalinity increases from negligible amounts in deposition (it is not measured as pH averages 5.0) to an average of 47.51 µeq/L in the surface waters, and the silica from negligible amounts in deposition to 146.45 µeq/L in the surface waters.

## ACKNOWLEDGEMENTS

We thank the field technicians who have worked so hard in collecting the data that went into this manuscript, particularly, Steve Zary, Sarah Spaulding and Brian Olver. We also are grateful for the use of unpublished data from Diane McKnight and John Fitzgerald. This project was funded by the National Park Service and the U.S. Geological Survey as part of the National Acid Precipitation Assessment Program.

DISCLAIMER

The use of trade or product names in this chapter is for iden-
tification purposes only and does not constitute endorsement by the
U.S. Geological Survey.

REFERENCES

Binkley, D. and D. Richter. Nutrient cycles and H+ budgets of
forest ecosystems. Advances in Ecological Research (in press).

Bond, H. W. 1979. Nutrient concentration patterns in a stream
draining a montane ecosystem in Utah. Ecology. 60:1184-1196.

Canfield, D. E., R. W. Bachmann, and M. V. Hoyer. 1983.
Freeze-out of salts in hard-water lakes. Limnol. Oceanog.
28:970-977.

Cole, J. C. 1977. Geology of East-Central Rocky Mountain National
Park and vicinity, with emphasis on the emplacement of the
Precambrian Silver Plume Granite in the Longs Peak-St. Vrain
batholith. Ph.D. Diss., Univ. Colorado, Boulder. 344 pp.

Feltz, H. R. and E. R. Anthony, eds. 1984. 1985 water quality
laboratory services catalog. U. S. Geological Survey Open File
Report 84-171. Reston, Virginia.

Goodison, B. E., H. L. Ferguson and G. A. McKay. 1981. Measurement and data analysis. Pages 191-274 In D. M. Gray and D. H.
Male, eds., Handbook of snow-principles, processes, management
and use. Pergamon Press, Willowdale, Ontario.

Gunn, J. M. and W. Keller. 1984. Effects of ice and snow cover on
the chemistry of nearshore lake water during spring melt.
International Glaciological Society. pp. 208-212.

Johannessen, M., A. Skartziet, R. F. Wright. 1980. Stream water
chemistry before, during and after snowmelt. In D. Dradlos and
A. Tollan, eds., Ecological inpact of acid precipitation.
Proceedings of International Conference. Sandefjord, Norway,
SNFF Project. Oslo, Norway.

Johnson, N. M. 1984. Acid rain neutralization by geologic
materials. Pages 37-54. In O. P. Bricker, ed., Geological
aspects of acid deposition. Acid precipitation series, Vol. 7,
Butterworth Publishers, Ann Arbor Science Book, Boston.

Krug, E. C. and C. R. Frink. 1983. Acid rain on acid soil: a new
perspective. Science. 221:520-525.

Kunkle, S. H. and G. H. Comer. 1972. Suspended, bed, and dissolved sediment loads in the Sleepers River, Vermont. Soil and Water Conservation Research Division, U. S. Department of Agriculture, ARS 41-188. 31 pp.

LaBaugh, J. W. and T. C. Winter. 1984. The impact of uncertainties in hydrologic measurement on phosphorus budgets and empirical models for two Colorado reservoirs. Limnol. Oceanogr. 29:322-339.

Lewis, W. M. and M. C. Grant. 1979. Relationships between stream discharge and yield of dissolved substances from a Colorado mountain watershed. Soil Science. 128:353-363.

Likens, G. E., F. H. Bormann, R. S. Pierce, J. S. Eaton, and N. M. Johnson. 1977. Biogeochemistry of a forested ecosystem. Springer-Verlag, New York. 146 pp.

Linacre, E. T. 1977. A simple formula for estimating evaporation rates in various climates, using temperature data alone. Agricul. Meteorol. 18:409-424.

McKnight, D., M. Brenner, R. Smith, and J. Baron. 1986. Seasonal changes in phytoplankton populations in lakes in Loch Vale, Rocky Mountain National Park. U.S. Geological Survey Water-Resources Investigations Report 86-4101, U.S. Geological Survey, Denver, Colo. 64 pp.

Marr, J. W. 1967. Ecosystems of the east slope of the front range of Colorado. Univ. of Colorado Studies Series in Biology #8. Univ. of Colorado Press, Boulder.

National Atmospheric Deposition Program. 1986. Quality Assurance Report: NADP/NTN Deposition Monitoring. Field operations July 1978 through December 1983. D. S. Bigelow ed., Natural Resource Ecology Laboratory, Colorado State University, Fort Collins. 113 pp.

Peden, M. E., S. R. Bachman, C. J. Brennan, B. Demir, K. O. James, B. W. Kaiser, J. M. Lockard, J. E. Rothert, J. Sauer, L. M. Skowron, and M. J. Slater. 1986. Development of standard methods for the collection and analysis of precipitation. Illinois State Water Survey Contract Report 381, Champaign, Illinois.

Redfield, A. C. 1958. The biological control of chemical factors in the environment. American Scientist. 46:205-221.

Redpath, B. B. 1973. Seismic refraction exploration for engineering site investigations. Technical Report E-73-4. U. S. Army Engineer Waterways Experiment Station. Livermore, California. 41 pp.

Reuss, J. O., and D. W. Johnson. 1986. Acid deposition and the acidification of soils and surface waters. Ecological Studies 59. Springer-Verlag, New York. 120 pp.

Swank, W. T., J. W. Fitzgerald, and J. T. Ash. 1984. Microbial transformation of sulfate in forest soils. Science. 223:182-194.

Tranter, M., P. Brimblecombe, T. D. Davies, C. E. Vincent, P. W. Abrahams, and I. Blackwood. 1985. The composition of snowfall, snowpack and meltwater in the Scottish Highlands-evidence for preferential elution. Atmospheric Environment. 20:1-9.

Verry, E. S. 1975. Streamflow chemistry and nutrient yields from upland-peatland watersheds in Minnesota. Ecology. 56:1149-1157.

Walthall, P. M. 1985. Acidic deposition and the soil environment of Loch Vale watershed in Rocky Mountain National Park. Ph.D. Diss., Colorado State Univ., Fort Collins. 148 pp.

Winter, T. C. 1981. Uncertainties in estimating the water balance of lakes. Wat. Res. Bull. 17:82-115.

Wright, R. F. 1983. Input-output budgets at Langtjern, a small acidified lake in southern Norway. Hydrobiologia. 101:1-12.

Zavodil, D. 1984. Report on geophysical investigation in Loch Vale, Rocky Mountain National Park. Unpublished Report. 6 pp.

# CHAPTER 10

ROLE OF ANAEROBIC ZONES AND PROCESSES
IN STREAM ECOSYSTEM PRODUCTIVITY

Clifford N. Dahm and Eleonora H. Trotter, University of
New Mexico, Albuquerque, New Mexico

James R. Sedell, USDA-Forest Service, Corvallis, Oregon

ABSTRACT

The character of flowing waters in North America has changed
dramatically since the days of early exploration and colonization.
Channel complexity, debris accumulations, beaver activity, riparian
vegetation, and the extent and structural diversity of the
floodplain have often been greatly reduced. These features all
affect the ability of a stream or river to trap, retain, and store
organic materials. Historically, accumulations of organic matter
that resulted in anaerobic interstitial waters were widespread and
commonplace in streams. Concentrations of ammonia, organic
nitrogen, organic carbon, organic phosphorus, phosphate, and
numerous metals are much greater within these zones. These regions
of the sediment represent important areas of nutrient regeneration
and potential sites of eventual nutrient introduction into stream
waters. The overall productivity of stream ecosystems is enhanced
by the greater nutrient availability represented by these regions
where anaerobic processes occur and nutrients are recycled into
soluble, mobile forms.

INTRODUCTION

Rivers and streams have served as magnets for colonization and
utilization throughout history. Available water, rich soils, wood,
grazing areas, and a source of transportation have drawn people to
their banks and riparian zones. The exploration and settlement of
the Americas by Europeans followed major waterways. Human activity
remains concentrated in these areas (Tiner, 1984).

*Chemical Quality of Water and the Hydrologic Cycle*, Robert C. Averett and Diane M. McKnight (Eds.) © 1987 Lewis Publishers, Inc.,
Chelsea, Michigan. Printed in the United States of America.

Industrialized society has developed numerous methods and machines that enable people to fundamentally alter the shape, form, and functioning of streams and rivers. In North America, these various impacts can be traced back as far as initial exploration and settlement. Few if any streams or rivers in the United States have remained totally untouched (Sedell and Luchessa, 1982).

Almost without exception, the first impact of Europeans on North American streams and rivers involved fur trappers. Beavers, in particular, were a major impetus for exploration and a major economic boon to the New World. Throughout the United States, the first European explorers were inevitably trappers. The exploitation of the resource was intense. Seton (1929) estimated that primal beaver populations in North America ranged from between 60 and 400 million or between 6 and 40 animals for each km of stream and river. Recent estimates place modern populations at 1 to 2 million animals, or approximately 0.1 to 0.2 animals per km (Denney, 1952).

The overall impact on floodplain landscapes has been substantial if one considers that populations of the prime animal agent of geomorphic change have been reduced one to two orders of magnitude. For example, Dobyns (1981) has suggested that arroyo formation, stream incision and increased flooding followed closely on the heels of beaver removal with subsequent dam failure and cattle introduction in the Gila River drainage of New Mexico. These features have been accepted for many years as ones natural to floodplain landscapes in the arid west. The diaries and reports of the early trappers, however, repeatedly record waters rich with beavers, lush with riparian vegetation, and running clear and cold where now they are channelized, silt-laden, and intermittent (Loyola, 1939; Cleland, 1950; Hastings and Turner, 1965; Weber, 1971; Dobyns, 1981). Beavers, a key geomorphic agent in streams and rivers, were greatly reduced in number throughout the United States at least a century ago.

A second widespread impact of human beings on streams and rivers began once permanent settlements were established. Waterways were a primary means to transport and market the agricultural output and natural resources of a region. Sedell and Luchessa (1982) documented the clearing of snags, boulders, and debris dams from almost every region of the United States. For more than a century, large wood and other obstructions have been systematically removed from rivers and streams (see Reports of the Secretary of War, Chief of Engineers, 1875 to 1914, for a detailed account of locations, maps, and numbers of logs removed). As one example, Triska (1984) has carefully documented the historical role of wood debris on a 400 to 500 km stretch of the Red River, Louisiana. Wood debris choked the channel. Exposed wood covered 80 to 120 km of channel and in one area 225 km of channel were affected. After 70 yr of intense channel and riparian zone clearing, a wide, unobstructed, meandering channel was produced by 1904. Today, this morphology may mistakenly be viewed as typical of pristine lowland rivers. Sedell and Froggatt (1984) have similarly reconstructed the historical changes to the Willamette River, Oregon. From 1870 to 1950, over 58,500 snags and streamside trees were pulled from a 114 km reach of the river. A pristine riparian forest originally extended 1.5 to 3.0 km on either side of the main channel. Now, after 80 yr of intense activity within the riparian zone, there exists one channel, few downed trees and a four-fold decrease in shoreline length. The shoals, multiple channels, oxbow lakes, and extensive marshes, wetland, sloughs, and backwaters are gone. Similar scenarios during the same time period can be documented in lotic ecosystems throughout North America.

A third impact of major historical importance was the introduction of domestic grazers to riparian ecosystems. Livestock commonly concentrate their foraging and calving near streams. In more xeric climates the end result of heavy grazing pressure is often the change from perennial to intermittent flow (Winegar, 1977; Stabler, 1985). Arroyo cutting, increased sediment loads, and channel incision often result (Dobyns, 1981). Most of the public lands of the western United States have been managed now for nearly a century as areas for livestock production. The riparian vegetation, channel geomorphology, hydrology, and stream processes have adjusted in numerous ways to the increased utilization by domestic animals.

Fourth, the use of rich alluvial soils for agriculture and the building of homes on the floodplain demanded a massive effort to control the course and quantity of flow in streams and rivers. Channelization and the building of levees, dikes, revetments and dams have attempted to restrict flow into well-defined main channels and irrigation ditches. Flood control has released vast areas of the floodplain for development of farmlands and cities. Containment of even the smallest of streams is now routinely practiced for irrigation, flood control, and livestock watering. Many of the lotic ecosystems of North America are now carefully regulated and are not representative of conditions under which native riparian vegetation and stream and river processes evolved (Fenner et al., 1985; Bradley and Smith, 1986).

Finally, the depression of the 1930's helped to further imprint human activities on many of the most remote and undisturbed streams remaining in the United States. Works Progress Administration and Civilian Conservation Corps crews participated in massive "clean up" efforts designed to remove large debris accumulations from streams throughout the United States (Sedell and Luchessa, 1982). The Federal Flood Control Act of 1936 made funds available for the clearing of almost any size stream anywhere in the United States. Debris dams, snags and brush were fastidiously removed. Thousands of miles of stream channel were made open and unobstructed.

Historical records help to reconstruct the primal conditions of riparian zones and streams. Two common themes reverberate throughout these records: 1) streams today are generally less retentive of organic matter and 2) riparian and stream ecosystem interactions have been widely uncoupled. Whether the focus is on beaver trapping, channel clearing, flood control, grazing, agricultural activity, or urbanization, human influence has acted to reduce long-term retention and storage of organic and inorganic material within the stream and floodplain and to diminish the extent of interaction between the riparian zone and the channel. A dynamic, heterogenous landscape has been increasingly molded into a more static, homogeneous one by human forces.

There are several reasons we stress the importance of historic reconstruction in studying stream-riparian interactions. First, Margalef (1960) proposed that the successional stage of a given terrestrial ecosystem determined the successional stage of a segment of stream flowing through it. This concept suggests a discontinuous series of stream habitats making up a waterway. The river continuum hypothesis (Vannote et al., 1980) and its refinement (Minshall et al., 1983) suggest that there are continuous gradients of structure and function from headwaters to mouth in a river system. Both ideas necessarily have, to some extent, been based on the structure and function of present-day stream and riparian systems. To understand underlying relationships in the floodplain between stream and riparian systems, it is instructive to know the historical backdrop upon which the present conditions came to exist. Sedell and Froggatt (1984) pointed out that even on major rivers there historically was extensive morphological heterogeneity produced by large quantities of woody debris. Within a large floodplain there would be much exchange between stream and riparian vegetation. Also, picture frequent positional changes of geomorphic surfaces, numerous types of channels (narrow or broad), and a variety of flow characteristics-- stagnant to free flowing. This is a profoundly different landscape from present-day floodplains and has important implications for the development of any theory concerning continua or discontinua of structure and function, and stream-riparian exchange.

Second, nutrient-spiralling theory (Elwood et al., 1983; Newbold et al., 1982) states that where there are mechanisms of retention, then the nutrient spirals become tighter; if there are few mechanisms enhancing retention, then the nutrient spirals grow longer. According to the river continuum hypothesis, large woody debris is an important structural component in headwater streams, but diminishes in importance in larger streams. Would spiralling length as a result then be shorter in headwater streams than in larger streams? The historical research of Sedell and Froggatt (1984) argues against making such predictions concerning the role of woody debris or spiralling length based on stream order.

Third, Hynes (1983) and Grimm and Fisher (1984) have pointed out the importance of interstitial waters as source areas for organic matter and key regions of stream metabolism and nutrient cycling. Hynes (1983) forcefully emphasized our lack of knowledge about conditions in the hyporheic zones of streams, drawing attention to the interstitial portion of the stream and the chemical and biological processes occurring there. Grimm and Fisher (1984) showed that a seemingly autotrophic desert stream was actually heterotrophic when metabolism within the sediment and interstitial waters was included in the estimates of stream respiration. Historical research suggests that conditions which would promote interaction and exchange between the stream and the interstitial waters of the sediment and riparian soils have been systematically removed or reduced throughout North America.

In summary, historic reconstructions repeatedly show that the floodplain landscapes of both small and large river systems were more complex and heterogeneous in space and time than they are presently. Redistribution of energy and nutrients was frequent and dependent on physical factors, such as debris dams, and animal agents, such as beaver. It is time to frame riparian-stream interactions in a landscape context. Risser et al. (1984) asked the question, "What formative processes, both historical and present, are responsible for the existing pattern in a landscape?" Efforts to produce a unified theory of stream ecology, research on riparian-stream interactions, and the concept of nutrient spiralling point the way to processes which could have shaped the rich, dynamic mosaic in the historic floodplain landscape.

The purpose of this paper is to present data on interstitial nutrient chemistry from anaerobic zones of stream sediment where geomorphic and hydrologic processes have created conditions conducive to organic matter accumulation and storage. Such zones, we hypothesize, were much more prevalent in historic floodplains because of the abundance of woody debris and beavers, and the less intense use by human populations. These data will be compared to aerobic interstitial waters and stream water. Also, the biomass of periphyton at the sediment-water interface above these various interstitial zones of the sediment will be compared. Finally, discussion will focus on how geomorphic patterns affect the spatial and temporal distribution of aerobic and anaerobic zones in streams and how these conditions might affect the riparian plant community.


METHODS AND SITES

In this study two sampling sites were used. Sulfur Springs is a second-order stream in the Coast Range of western Oregon. It is a small tributary of Knowles Creek which is a tributary of the Siuslaw River. A debris torrent blocked a portion of the channel and resulted in a section with lower gradient and debris dams. Beavers have occupied the area and have further enhanced the retentive characteristics of the reach. The second site is a reach of the headwaters of the Rio Cebolla in the Jemez Mountains of north-central New Mexico. This third-order stream has historically had extensive beaver activity. The study area includes one of the two remaining beaver colonies on the stream.

Samples for interstitial chemistry were taken with a modified 30-mL syringe. The syringe was marked at 5-cm length and the tip was enlarged to 2-cm diameter. The syringe was inserted into the sediment to a depth of 5 cm and the plunger pulled back to extract a slurry of sediment and water. The sediment and water mixture was immediately placed in sealed centrifuge tubes and centrifuged at 5,000 rpm for 3 min. The liquid supernatant was then decanted into a filtration system and filtered through GF/F filters that had been fired overnight at 450°C. The filtered sample was placed in an ice chest and transported to the lab for analyses. A subsample of 1 mL was also reacted with Ferrozine in the field for subsequent determination of ferrous iron in the lab (Stookey, 1970).

Samples of benthic algae were collected using a circular cork borer with a 2.5-cm diameter. Total chlorophyll was measured using the method of Parsons et al. (1984). Organic carbon content of the sediment below the algal mats was determined by ashing at 550°C for 4 hr and measuring weight loss.

Nitrate, ammonia, and phosphate were measured using automated nutrient analysis as described by Strickland and Parsons (1972). Dissolved organic nitrogen was calculated by subtracting the ammonia concentration from the Kjeldahl nitrogen value. Kjeldahl nitrogen was determined by automated ammonia analysis after sample digestion (Strickland and Parsons, 1972). Total phosphorus was measured as phosphate by automated nutrient analysis following persulfate and sulfuric acid digestion (Rand, 1976). Dissolved organic carbon (DOC) was measured as $CO_2$ after persulfate oxidation using the method of Menzel and Vaccaro (1964). Methane was measured using gas chromatography following the procedure of Lilley et al. (1983).

RESULTS

Comparisons of nutrient concentrations in anaerobic and aerobic zones of interstitial water were made at six sites in Sulfur Creek, Oregon. In addition, stream water samples were taken immediately above the interstitial samples at each site. Of the six interstitial waters samples, three were aerobic and three were anaerobic. Anaerobic samples occurred in the sediments of ponds behind debris dams or beaver dams. Aerobic samples were found at 5 cm depth in pools either upstream or downstream of the debris and beaver dams and in sediments immediately downstream of one of the debris dams in a free-flowing reach of stream.

Concentrations of ammonia, nitrate, and dissolved organic nitrogen (DON) are listed in Table 1. The reduced forms of nitrogen, ammonia and DON were present in higher concentrations within the anaerobic interstitial waters. Average ammonia values were 1.887 mg/L in the anaerobic samples, 0.334 mg/L in the aerobic samples and 0.023 mg/L in the stream water. Average DON values were were 4.014 mg/L in the anaerobic samples, 1.279 mg/L in the aerobic samples and 0.116 mg/L in the stream water. Interstitial waters collected in these various depositional environments were routinely at least one order of magnitude above stream values for ammonia and DON. Anaerobic zones had the highest levels of ammonia and DON. Going from the stream water to anaerobic interstitial waters at 5-cm depth in the sediment, a two orders of magnitude increase sometimes occurred. Nitrate concentrations showed the reverse trend. Average nitrate concentrations were 0.014 mg/L in the anaerobic samples, 0.058 mg/L in the aerobic samples and 0.100 mg/L in the stream water.

Dissolved organic carbon (DOC) and methane concentrations at the Sulfur Springs sites are given in Table 2. The average concentration of DOC was 37.7 mg/L as carbon for the anaerobic samples, 12.5 mg/L for the aerobic samples, and 1.8 mg/L for the stream water. A strong gradient existed between the stream water and the interstitial waters, particularly in or near the anaerobic zones. Methane samples were collected from the water immediately above the sediment-water interface. Substantial methanogenesis was occurring in the anaerobic regions of interstitial water as seen by the highly supersaturated levels of methane in waters in the very retentive reach of the stream. Average methane values were 0.278 mg/L in these waters compared to 0.005 mg/L in the water in pools upstream and downstream of the debris and beaver dams.

Table 1. Oxygen Conditions and Nutrient Concentrations in mg/L for Ammonia, Nitrate, and Dissolved Organic Nitrogen in the Stream and Interstitial Waters of Sulfur Creek, Oregon.

| Site | $O_2$[1] | $NH_4$ | $NO_3$ | DON |
|---|---|---|---|---|
| Upstream pool | Aerobic | 0.275 | 0.125 | 0.813 |
| Upper debris dam pond | Anaerobic | 1.100 | 0.030 | 2.363 |
| Lower debris dam pond | Anaerobic | 4.100 | 0.013 | 6.000 |
| Immediately below debris dam | Aerobic | 0.413 | 0.000 | 2.175 |
| Old pond | Anaerobic | 0.460 | 0.000 | 3.680 |
| Downstream pool | Aerobic | 0.313 | 0.050 | 0.850 |
| Stream water[2] | Aerobic | 0.023 | 0.100 | 0.116 |

[1]Based on the presence or absence of ferrous iron.
[2]Composite of six samples from above the six sites where interstitial waters were collected.

Table 2. Oxygen Conditions and Nutrient Concentrations in mg/L for DOC and Methane in the Stream and Interstitial Waters of Sulfur Creek, Oregon.

| Site | $O_2$[1] | DOC | $CH_4$[2] |
|---|---|---|---|
| Upstream pool | Aerobic | 8.2 | 0.002 |
| Upper debris dam pond | Anaerobic | 17.0 | 0.387 |
| Lower debris dam pond | Anaerobic | 82.6 | 0.320 |
| Immediately below debris dam | Aerobic | 20.6 | 0.128 |
| Old pond | Anaerobic | 13.4 | - |
| Downstream pool | Aerobic | 8.8 | 0.008 |
| Stream water[3] | Aerobic | 1.8 | - |

[1]Based on the presence or absence of ferrous iron.
[2]Methane samples were collected just above the sediment-water interface and not from the interstitial zone.
[3]Composite of six samples from above the six sites where interstitial waters were collected.

Phosphorus concentrations, both as phosphate and total phosphorus, were also much higher in interstitial water than stream water (Table 3). Phosphate levels averaged 0.638 mg/L in the anaerobic samples, 0.467 mg/L in the aerobic samples, and 0.007 mg/L in the stream water. Total average phosphorus concentrations were 0.792 mg/L in the anaerobic waters, 0.642 mg/L in the aerobic waters, and 0.026 mg/L in the stream water. Increases of one order of magnitude in concentration commonly occurred for phosphate and total phosphorus between stream water and interstitial waters in depositional areas of the stream. The differences in concentration between the aerobic and anaerobic zones were proportionally smaller for phosphate and total phosphorus than for DOC, ammonia, and DON. All interstitial waters, however, were greatly enriched over stream levels for both phosphate and total phosphorus. Abundant accumulations of benthic algae were often noted in stream reaches with extensive debris accumulations, beaver activity, and large amounts of organic matter storage. Stream reaches in these areas which were well-lit had especially large mats of benthic algae. Sampling was carried out in the Rio Cebolla, a meadow stream with beaver ponds, to quantify the amount of chlorophyll in various reaches (Table 4). Samples were collected from the reach immediately downstream of a beaver dam (less than 10 m below), the edges of the beaver pond, and upstream of the pond (10 m upstream). Sediments at 5 cm depth throughout the pond were anaerobic. The chlorophyll values showed a striking pattern with approximately a ten-fold increase from above the pond to within the pond. In the reach immediately below the pond, the chlorophyll levels were nearly 20 times those in the upstream reach.

Table 3. Oxygen Conditions and Nutrient Concentrations in mg/L for Phosphate, and Total Phosphorus in the Stream and Interstitial Waters of Sulfur Creek, Oregon.

| Site | $O_2$[1] | $PO_4$ | TP |
|---|---|---|---|
| Upstream pool | Aerobic | 0.350 | 0.563 |
| Upper debris dam pond | Anaerobic | 0.375 | 0.475 |
| Lower debris dam pond | Anaerobic | 0.820 | 1.000 |
| Immediately below debris dam | Aerobic | 0.600 | 0.863 |
| Old pond | Anaerobic | 0.720 | 0.900 |
| Downstream pool | Aerobic | 0.450 | 0.500 |
| Stream water[2] | Aerobic | 0.007 | 0.026 |

[1]Based on the presence or absence of ferrous iron.
[2]Composite of six samples from above the six sites where interstitial waters were collected.

Table 4. Sediment Organic Content (%) and Total Chlorophyll
($\mu g/cm^2$) from above, in, and below a Beaver Pond in
the Rio Cebolla, New Mexico.

| Site | n | Sediment organic content (%) | Total chlorphyll ($\mu g/cm^2$) |
|------|---|------------------------------|----------------------------------|
| Above pond | 3 | 2.6 +- 2.0 | 0.79 +- 0.24 |
| In pond | 4 | 7.6 +- 2.4 | 7.35 +- 3.27 |
| Below pond | 2 | 5.5 +- 0.1 | 14.64 +- 1.37 |

## DISCUSSION

Historical research presents a long-term perspective from which
to begin to consider the temporal and spatial heterogeneity of
streams and riparian zones. The changing mosaic along streams
provides a template on which various processes occur and different
biotic communities are distributed. How do the important roles of
large organic debris, beaver activity, complex channel morphology,
and enhanced retention and storage of organic detritus affect
stream processes and the plant community of the riparian zone?

An important effect is the ability of the stream to retain,
store, and process organic inputs. Efficient retention of organic
inputs not only influences the quantity of organic matter available
but can also dictate the pathway by which much of this material is
decomposed. Anaerobic decomposition occurs when oxygen
availability does not meet the metabolic demands of the decomposer
population. Reaches of stream with debris dams, beaver ponds,
sloughs, backwaters, and side channels create areas where
widespread anaerobic conditions potentially can occur within the
interstitial sediments. Anaerobic sediments are more likely to
exist where accumulations of organic material and fine-grained
sediments occur in regions of relatively restricted exchange
between surface and interstitial waters. The extent to which this
type of environment is found in streams has not been investigated,
but the data presented in this paper show that nutrient concentrations, the chemical form of various nutrients, and the rate and
pathways of nutrient cycling can be shifted dramatically in these
environments (Tables 1-3). The overall extent of anaerobic
interstitial waters will be closely linked to physical, hydrologic,
and geomorphic process within each drainage.

Light, nitrogen, and phosphorus are common factors limiting the
rate of primary production in streams (Peterson et al., 1983;
Gregory, 1980; Elwood et al., 1983; Triska et al., 1983; Grimm and
Fisher, 1986). The structure of the riparian zone strongly influences the amount of light reaching the stream. Primary production in highly-shaded streams is commonly light-limited (Gregory,
1980). Where shading becomes less of a problem, a nitrogen or
phosphorus limitation often is present (Peterson et al., 1983;
Grimm and Fisher, 1986). Anaerobic interstitial waters are greatly

enriched in both nitrogen and phosphorus when compared to the overlying water of the stream (Tables 1 and 3). The gradient between the aerobic sediment-water interface and the anaerobic zone supports a flux of nitrogen and phosphorus to the interface. The predicted response would be increased rates of primary production in these zones where light is not a limiting factor. In many ways, this response would be similar to stream enrichment experiments where direct addition of nutrients to the stream stimulates increased photosynthetic activity and standing crops from the periphyton community (Gregory, 1980; Triska et al., 1983; Elwood et al., 1983; Petersen et al., 1983; Grimm and Fisher, 1986). The difference is that the increased supply of nutrients is supported by anaerobic processing within certain zones of the stream sediments and floodplain, which are then mobilized to the benthic algal community by turbulent diffusion and the concentration gradient which arises. The standing crop of algae indicated by the total chlorophyll data (Table 4) supports this scenario, but direct measurements of the rates of primary production would be a better test.

Changing conditions of oxygen and nutrient concentrations may also be key variables influencing the structure and composition of the riparian plant community. Increased nutrient concentrations and changes in chemical forms and ratios in anaerobic zones and around aerobic-anaerobic interfaces present both potentially advantageous and deleterious conditions for various plants. What effects do increased concentrations of ammonia, DON, DOC, and phosphorus, but low concentrations of nitrate and oxygen, have on the riparian plant community?

Until recently, most research on riparian-stream interactions has focused on contributions from plants to the stream. Large organic debris (Bilby, 1981; Cummins, 1980; Keller and Swanson, 1979; Keller and Tally, 1979; Sedell and Froggatt, 1984; Swanson et al., 1982; Triska, 1984; and Triska et al., 1982), plant nutrient contribution to stream nutrient economy (Bormann and Likens, 1967; Elwood et al., 1983; Minshall, 1978; Petersen and Cummins, 1974), stream retention of organic inputs (Speaker et al., 1984), shading and buffering roles of riparian vegetation (Aho, 1976; Brown and Kryger, 1970; Gray and Edington, 1969; Lowrance et al., 1984a; Nabhan, 1985), reclamation of ephemeral streams (Heede, 1981), and the relationship of terrestrial and stream succession (Fisher, 1983; Margalef, 1960; Molles, 1982; Vannote et al., 1980) have been considered.

Recently, however, the effects of stream hydrology and geomorphology on riparian plant community composition and distribution have become an active arena for research. Floods and stream meandering create geomorphic surfaces such as sand and gravel bars, swales and natural levees which, in part, determine plant species distribution on the floodplain (McKee and Swanson, personal communication; Hupp and Ostercamp, 1985). These geo-morphic surfaces also establish the template upon which nutrient dynamics and oxygen utilization and resupply are superimposed. A number of possible responses can be hypothesized to occur within the riparian plant community to this mosaic of geomorphic surfaces, oxygen availability, and nutrient dynamics. For example, oxygen levels in the soils and sediments may be an important variable in structuring riparian plant communities with their ability to handle various periods of anoxia as a determining role in plant distributions (e.g. Crawford, 1966). Nutrient chemical form and cycling rates may also be key elements in the structure and composition of the riparian plant community, although limited research in these veins has yet been reported. Possible responses of riparian plants to changing nutrient conditions associated with anaerobic zones or near aerobic-anaerobic interfaces that deserve consideration include: 1) total nitrogen and phosphorus content within the plant; 2) the nitrogen to phosphorus ratio within the vegetation; 3) plant allocation of photosynthate to either vegetative growth or sexual reproduction; 4) changes in phenology with different nutrient levels or ratios; and 5) compensatory ability of the plant to respond to herbivory. These possible responses to shifts in the form and concentration of plant nutrients may also influence the structure and composition of plants in the riparian zone.

Many riparian plants are anatomically and physiologically adapted to withstand low $O_2$ conditions found in saturated soils. Anatomical adaptations in many woody riparian species include well-developed stem lenticels which allow them to grow in low $O_2$ conditions (Koslowski, 1984). Air moves through the lenticels by diffusion to root aerenchyma, and toxic products of respiration are volatilized and released via the lenticels (Chirkova and Gutman, 1971; Hook and Scholtens, 1978). Many flood-adapted species are extremely sensitive to anoxia (Koslowski, 1984) but the well-developed aerenchyma-lenticel system insures an oxygen supply to roots growing in anaerobic conditions. Riparian plants are also adapted physiologically to varying periods of flooding or anoxia. Some species resistant to $O_2$ deficiency such as *Salix alba* can decrease respiration rates and detoxify metabolic products by volatilization, root exudation, or secondary metabolism (Chirkova, 1978). McMannon and Crawford (1971) showed an inverse relationship between flood tolerance and production of alcohol dehydrogenase in roots. Armstrong (1975, 1978) reported that although sensitive to anoxia, roots of wetland plants growing in permanently anaerobic soils maintained a fairly steep $O_2$ gradient from lenticel to root meristem, where respiration requires much $O_2$. Armstrong (1971) and Hook et al. (1971) showed that roots actually leak $O_2$ to the rhizosphere. Therefore, two processes, respiration and root exudation of $O_2$, maintain the $O_2$ gradient necessary for survival in anaerobic soil. Oxygen leakage into the rhizosphere could confer at least three advantages on plants. First, oxidation of the rhizosphere creates a nutrient-rich aerobic-anaerobic interface at the site of mineral nutrient uptake. Second, oxygen stimulates the activity of nitrogen-fixing bacteria and fungal symbionts (Armstrong, 1978), and third, stimulation of mycorrhizae may lead to increased $NH_4^+$ (Rygiewicz et al., 1984), the form of inorganic nitrogen most abundant in reducing environments (see Table 1). Another physiological adaptation to flooding is the rapid production of adventitious roots above the anaerobic zone (Koslowski, 1984). The placement of these roots may enable plants to take advantage of the nutrient enrichment occurring near anoxic interstitial water. Riparian plants clearly possess anatomical and physiological adaptations to overcome low dissolved $O_2$ concentrations.

Riparian plants can use nutrient supplies in the interstitial waters of the riparian zone. Asmussen et al. (1979) showed that $NO_3N$ streamflow outputs from fertilized row-crops were much less after passing through the riparian zone. They concluded that $NO_3^-$ uptake within the riparian zone between the field and stream was responsible. Lowrance et al. (1984b) showed that riparian plants were highly retentive of N and moderately retentive of Ca, P and Mg.

Nutrient distributions in the riparian zone can also affect growth rate, species representation, distribution, and reproduction. Woody species such as willow (*Salix* sp.) use both nitrate nitrogen and ammonia nitrogen (Smirnoff et al. 1984, W.C. Martin and R.H. Waring, personal communication). Ingestad (1979a, 1979b) showed that concentrations of mineral nutrition in willow tissue varied depending on the rate at which nutrients are supplied. Therefore, willow can track changes in mineral nutrient levels in the soil, and can use ammonia nitrogen as well as nitrate nitrogen. An increase in mineral nutrients favors rapid growth rates. Waring et al. (1985) showed that under different light and nutrient regimes, the leaf chemistry of a willow clone (*Salix aquatica*) differed. Total N of the high light, high nutrient willows exceeded 5% compared with 1.91% for the high light, moderate nutrient willows. Growth rate was also affected (16.1% per day for high nutrient willows compared with 5.5% for moderate nutrient willows, both under high light). An increase in the N:P ratio available also favors rapid vegetative growth instead of flowering (Salisbury and Ross, 1985; D. Marshall, personal communication). Plants growing in areas of preferential nitrogen enrichment could, therefore, spread very rapidly by vegetative growth. The ability of many woody riparian plants to sprout vigorously when physically damaged by abrasion during a flood may be linked to an abundant nutrient supply. Barnes (1985) attributed the survival of woody riparian species on a frequently flooded island to its ability to spread clonally. Seedling establishment on this island was poor, and clonal growth ensured the spread of established individuals, insuring high immediate fitness (Handel, 1985).

Rapid vegetative growth may also help riparian plants to weather periods of moderate herbivory. Larsson et al. (1986) showed that willow grown under high light and high nutrient conditions had high phenolic glycoside concentrations and were not chosen by defoliating beetles. Belovsky (1981) showed that of two species of birch (*Betula* sp.), the faster growing one was able to produce more ramets and recovered more rapidly under moose herbivory. Klein (1977) showed that browsing by hare on birch stimulated production of basal shoots, which contain high resin levels and are less preferred than twigs at the top of the tree. Red willow (*Salix lasiandra*) maintained high growth rates and increased in basal diameter when exposed to beaver herbivory (Kindschy, 1985).

In summary, riparian plants possess anatomical and physiological adaptions to withstand and exploit anaerobic, waterlogged soils. Many of these plants can take up nitrate and ammonia nitrogen and their growth rate is positively correlated with nutrient levels available to them. Therefore, riparian plants may be favored if they tap pockets of nutrient enrichment such as those occurring in the retentive reaches of the two study streams discussed here. Such nutrient uptake potentially affects plant growth rates, clonal spread or sexual reproduction, and response to herbivory. Herbivory might also stimulate changes in plant reproductive strategies by favoring vegetative growth. Changing nutrient conditions and herbivory are largely unstudied forces in determining the riparian plant distribution in floodplain landscapes.

A conceptual model of how changes in the floodplain may affect oxygen distributions, nutrient cycling pathways, and the structure, functioning, and productivity of stream and riparian ecosystems is given in Table 5. Historical changes in the floodplain landscape are predicted to have decreased the input, retention, storage, and processing of energy present as organic matter. This in turn affects dissolved $O_2$ concentrations within the soil and sediment interstitial waters. Changing oxygen conditions direct nutrient cycling pathways which determine the quantity and type of nutrients present. Nutrient concentrations, chemical speciation, and elemental ratios can potentially help to regulate such processes as benthic primary production, within plant nutrient allocation, and growth rates of riparian plants. These elements of production within the stream and riparian ecosystems can then influence plant response to herbivory, plant reproductive strategies, the composition, distribution, and architecture of the plant community, litter quantity, quality, and type, and overall stream productivity. In general, we hypothesize that highly retentive zones in the floodplains of streams and rivers have been diminished due to protracted and ever-increasing anthropogenic intervention in these areas. Nutrient chemistry, geomorphology, hydrology, and the structure and functioning of streams, rivers, and riparian zones throughout North America are therefore often substantially altered from the primal conditions under which biotic adaptation and evolution once proceeded.


## CONCLUSIONS

Anaerobic zones are widespread in retentive reaches of streams where organic matter is efficiently captured, stored, and processed. These regions of the stream and riparian zone are enriched in dissolved ammonia, organic nitrogen, organic carbon, and phosphorus. The historical legacy of ever-increasing anthropogenic utilization of the riparian zone has decreased the amount of large organic debris, the extent of beaver activity, the complexity of riparian habitat, and the diversity of off-channel habitats. Systematic removal of these features has decreased the extent and overall importance of anaerobic pathways of nutrient cycling for numerous riparian areas throughout North America.

Table 5.   A Generalized Conceptual Representation of how Changing Geomorphic Conditions within the
Floodplain Landscape can Affect Oxygen Concentrations, Nutrient Cycling Pathways, Stream
Productivity, and the Riparian Plant Community.

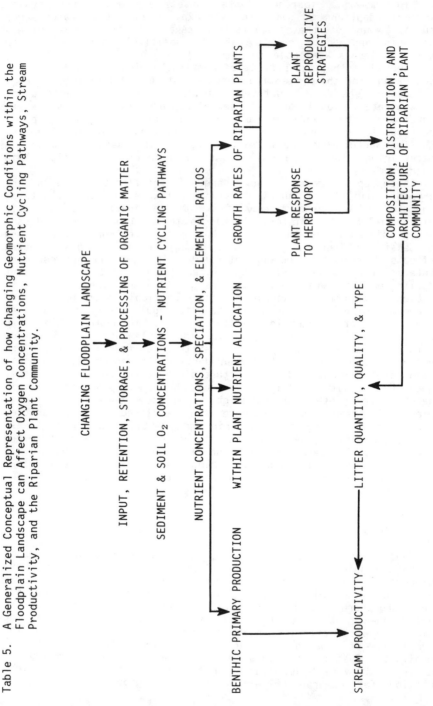

These regions can be important source areas for soluble carbon, nitrogen, and phosphorus to the stream ecosystem and the riparian plant community. Rates of benthic primary production, within plant nutrient allocation, and the structure and growth rates of riparian plant communities are often directly linked to the concentration, speciation, and cycling rates of limiting nutrients in these environments. The interplay of geomorphology, hydrology, and biotic activity is a key variable forming and maintaining anaerobic conditions where enriched sources of nutrients are found.

## ACKNOWLEDGEMENTS

We would like to thank the following organizations and individuals for their help in this study. Funding was provided by the U.S. Forest Service under contract PNW-80-281 and the National Science Foundation under grant DEB 81-12455. Most of the chemical analyses were run by the Central Chemistry Analytical Laboratory in the Forestry Sciences Laboratory of the U.S. Forest Service in Corvallis, Oregon. Methane analyses were run by Marv Lilley of the School of Oceanography, University of Washington, Seattle, Washington. Matt Stebleton assisted in the chlorophyll analyses and Diane Marshall provided numerous ideas and references concerning riparian plant communities. Joanne McEntire assisted in manuscript preparation.

## DISCLAIMER

The use of trade or product names in this chapter is for iden- tification purposes only and does not constitute endorsement by the U.S. Geological Survey.

## REFERENCES

Aho, R.S. 1976. A population study of the cutthroat trout in an unshaded and shaded section of stream. Masters Thesis. Oregon State University, Corvallis.

Armstrong, W. 1971. Radial oxygen losses from intact rice roots as affected by distance from the apex, respiration, and waterlogging. Physiologia Plantarum. 25:192-197.

Armstrong, W. 1975. Waterlogged soils. p. 184-216. *In* J.R. Etherington, ed. Environment and Plant Ecology. John Wiley, London.

Armstrong, W. 1978. Root aeration in wetland conditions. p. 269-298. *In* D.D. Hook and R.M.M. Crawford, eds. Plant life in anaerobic environments. Ann Arbor Science Publishers, Ann Arbor.

2222222

222222222222222222222222

222222222222222222222222

22222222222222222222222

2222222222222222222222222

2222222222222222222222222

22222222222222222222222222

22222222222222222222222222

222222222222222222222222222222222222

Asmussen, L.E., J.M. Sheridan, and C.V. Booram, Jr. 1979. Nutrient movement in streamflow from agricultural watersheds in the Georgia Coastal Plain. Transactions of the Society of Agricultural Engineering. 22:809-815, 821.

Barnes, W.J. 1985. Population dynamics of woody plants on a river island. Canadian Journal of Botany. 63:647-655.

Belovsky, G.E. 1981. Food plant selection by a generalist herbivore: the moose. Ecology. 62:1020-1030.

Bilby, R.E. 1981. Role of organic debris dams in regulating the export of dissolved and particulate matter from a forested watershed. Ecology. 62:1234-1243.

Bradley, C.E., and D.G. Smith. 1986. Plains cottonwood recruitment and survival on a prairie meandering river floodplain, Milk River, southern Alberta and northern Montana. Canadian Journal of Botany. 64:1433-1442.

Bormann, F.A., and G.E. Likens. 1967. Nutrient cycling. Science. 155:424-429.

Brown, G.W., and J.T. Kryger. 1970. Effects of clear-cutting on stream temperature. Water Resources Research. 6:1133-1139.

Chirkova, T.V. 1978. Some regulatory mechanisms of plant adaptation to temporal anaerobiosis. p. 137-154. In D.D. Hook and R.M.M. Crawford, eds. Plant life in anaerobic environments. Ann Arbor Science, Ann Arbor.

Chirkova, T.V., and T.S Gutman. 1971. Physiological role of branch lenticels of willow and poplar under conditions of root anaerobiosis. Fiziol. Rast. (English translation). Plant Physiology. 19:352-359.

Cleland, R.G. 1950. This reckless breed of men: The trappers and fur traders of the southwest. Alfred A. Knopf, New York, 361 pp.

Crawford, R.M.M. 1966. The control of anaerobic respiration as a determining factor in the distribution of the genus *Sececio*. Journal of Ecology. 54:403-413.

Cummins, K.W. 1980. The multiple linkages of forests to streams. p. 191-198. In Richard H. Waring, ed. Forests: fresh perspectives for ecosystem analysis. Proceedings of the 40th Annual Biology Colloquim. Oregon State University Press, Corvallis, Oregon.

Denney, R.N. 1952. A summary of North American beaver management, 1946-1948. State of Colorado Game and Fish Department, Current Report 28, 58 pp.

Dobyns, R.N. 1981. From Fire to Flood: Historic Human Destruction of Sonoran Desert Riverine Oases. Ballena Press Anthropological Papers No. 20, 222 pp.

Elwood, J.W., J.D. Newbold, R.V. O'Neill and W. Van Winkle. 1983. Resource spiralling: an operational paradigm for analyzing lotic ecosystems. *In* T.D. Fontaine and S.M. Bartell, eds. The dynamics of lotic ecosystems. Ann Arbor Sciences. Ann Arbor.

Fenner, P., W.W. Brady, and D.R. Patton. 1985. Effects of regulated water flows on regeneration of Fremont cottonwood. Journal of Range Management. 38:135-138.

Fisher, S.G. 1983. Succession in streams. p. 7-27. *In* J.R. Barnes and G.W. Minshall, eds. Stream Ecology. Plenum Press. New York.

Gray, G.R.A., and J.M. Edington. 1969. Effect of woodland clearance on stream temperature. Journal of Fisheries Board of Canada. 26:399-403.

Gregory, S.V. 1980. Effects of light, nutrients, and grazing on periphyton communities in streams. Ph.D. thesis, Oregon State University, 150 pp.

Grimm, N.B., and S.G. Fisher. 1984. Exchange between interstitial and surface water: Implications for stream metabolism and nutrient cycling. Hydrobiologia. 111:219-228.

Grimm, N.B., and S.G. Fisher. 1986. Nitrogen limitation in a Sonoran Desert stream. Journal of North American Benthological Society. 5:2-15.

Handel, S.N. 1985. The intrusion of clonal growth patterns on plant breeding systems. American Naturalist. 125:367-384.

Hastings, J.R., and R.M. Turner. 1965. The changing mile: An ecological study of vegetation change with time in the lower mile of an arid and semiarid region. University of Arizona Press, Tucson. 317 pp.

Heede, B.H. 1981. Rehabilitation of disturbed watersheds through vegetation treatment and physical structures. *In* David M. Baumgartner, ed. Interior west watershed management, proceedings of a symposium, Spokane, 1980. Washington State University, Pullman.

Hook, D.D., C.L. Brown, and P.P. Kormanik. 1971. Inductive food tolerance in swamp tupelo (*nyssa sylvatica*) var biflora (Walt. Sarg.). Journal of Experimental Botany. 22:78-89.

Hook, D.D., and J.R. Scholtens. 1978. Adaptations and flood tolerance of tree species. p. 299-332. *In* D.D. Hook and R.M.M. Crawford, eds. Plant life and anaerobic environments. Ann Arbor Science, Ann Arbor.

Hupp, C.R. and W.R. Ostercamp. 1985. Bottomland vegetation distribution along Passage Creek, Virginia, in relation to fluvial land forms. Ecology. 66:670-681.

Hynes, H.B.N. 1983. Groundwater and stream ecology. Hydrobiologia. 100:93-99.

Ingestad, T. 1979a. Nitrogen stress in birch seedlings. II. N, K, P, Ca, and Mg nutrition. Physiologia Plantarum. 45:148-157.

_____. 1979b. Mineral nutrient requirements of *Pinus silvestris* and *Picea abies* seedlings. Physiologia Plantarum. 45:137-147.

Kindschy, R.R. 1985. Response of red willow to beaver use in Southeastern Oregon. Journal of Wildlife Management. 49:26-28.

Keller, E.A., and F.J. Swanson. 1979. Effects of large organic material on channel form and fluvial processes. Earth Surface Process. 4:361-380.

Keller, E.A. and T. Tally. 1979. Effects of large organic debris on channel form and fluvial processes in the coastal redwood environment. p. 169-197. *In* D.D. Rhodes and G.P. Williams, eds. Adjustments of the fluvial system. Proceedings of the 10th Annual Geomorphology Symposium. State University of New York, Binghamton.

Koslowski, T.T. 1984. Plant responses to flooding of soil. Bioscience. 34:266-275.

Klein, D.R. 1977. Winter food preferences of snowshoe hares (*Lepus americanus*) in interior Alaska. Proceedings of the International Congress of Game Biology. 13:266-275.

Larsson, S., A. Wiren, L. Lundgren, and T. Ericsson. 1986. Effects of light and nutrient stress on leaf phenolic chemistry in *Salix dasyclados* and susceptibility to *Galerucella lineola* (Coleoptera). Oikos. 47:205-210.

Lilley, M.D., J.A. Baross, and L.I. Gordon. 1983. Reduced gases and bacteria in hydrothermal fluids: the Galapagos spreading center and 21°N east Pacific Rise. p. 411-449. *In* P.A. Rona, K. Bostrom, L. Laubier and K.L. Smith, Jr., eds. Hydrothermal processes at seafloor spreading centers. Plenum Press, New York.

Lowrance, R.R., R.L. Todd, and L.E. Asmussen. 1984a. Nutrient cycling in an agricultural watershed: I. Phreatic movement. Journal of Environmental Quality. 13:22-27.

____. 1984b. Riparian forests as nutrient filters in agricultural watersheds. BioScience. 34:374-377.

Loyola, M. 1939. The American occupation of New Mexico, 1821-1852. New Mexico Historical Review. 14:34-75.

Margalef, R. 1960. Ideas for a synthetic approach to the ecology of running waters. Internationale Revue de Gesamten. Hydrobiologie. 45:133-153.

Marshall, D. Personal communication.

Martin, W.C., and R.H. Waring. Personal communication.

McKee, and Swanson. Personal Communication.

McMannon, M., and R.M.M. Crawford. 1971. A metabolic theory of flooding tolerance: the significance of enzyme distribution and behavior. New Phytologist. 70:299-204.

Menzel, D.W., and R.F. Vaccaro. 1964. The measurement of dissolved organic and particulate carbon in seawater. Limnology and Oceanography. 9:138-142.

Minshall, G.W. 1978. Autotrophy in stream ecosystems. BioScience. 28:767-771.

Minshall, G.W., R.C. Petersen, K.W. Cummins, T.L. Bott, J.R. Sedell, C.E. Cushing, AND R.L. Vannote. 1983. Interbiome comparison of stream ecosystem dynamics. Ecological Monographs. 53:1-25.

Molles, M.C., Jr. 1982. Trichopteran communities of streams associated with aspen and conifer forests: long-term structural change. Ecology. 63:1-6.

Nabhan, G.P. 1985. Riparian vegetation and indigenous southwestern agriculture control of erosion, pests, and microclimate. p. 232-236. *In* Riparian ecosystems and their management: Reconciling Conflicting Uses. U.S. Department of Agriculture, Forest Service General Technical Report RM-120.

Newbold, J.D., R.V. O'Neill, J.W. Elwood, and W. Van Winkle. 1982. Nutrient spiralling in streams: implications for nutrient limitation and invertebrate activity. The American Naturalist. 120:628-652.

Parsons, T.R., Y. Maita, and C.M. Lalli. 1984. A manual of chemical and biological methods for seawater analysis. Pergamon Press, Elmsford, 173 pp.

Petersen, R.C., and K.W. Cummins. 1974. Leaf processing in a woodland stream. Freshwater Biology. 4:343-368.

Petersen, B.J., J.E. Hobbie, T.L. Corliss, and K. Kriet. 1983. A continuous-flow periphyton bioassay: tests of nutrient limitation in a tundra stream. Limnology and Oceanography. 28:583-591.

Rand, M.C. 1976. Standard methods for the examination of water and waste water. 14th ed. American Public Health Association, Washington, D.C.

Reports of the Secretary of War. Reports of the Chief of Engineers, 1875-1914: *In* House Executive Documents, Sessions of Congress, U.S. Government Printing Office, Washington, D.C. (annual reports).

Risser, P.G., J.R. Karr, and R.T.T. Forman. 1984. Landscape ecology: directions and approaches. Illinois Natural History Survey Special Publication Number 2. Illinois Natural History Survey, Champaign.

Rygiewicz, P.T., C.S. Bledsoe, and R.J. Zasoki. 1984. Effects of ectomycorrhizae and solution pH on [15N] uptake by coniferous seedlings. Canadian Journal of Forestry Research. 14:893-899.

Salisbury, J.R., and C.W. Ross. 1985. Plant physiology, 3rd ed. Wadsworthy Publishing Company, Inc., Belmont.

Sedell, J.R., and J.L. Froggatt. 1984. Importance of streamside forests to large rivers; the isolation of the Willamette River, Oregon, U.S.A., from its floodplain by snagging and streamside forest removal. Verhandlyngen Internationale Vereinigung Limnologie. 22:1828-1834.

Sedell, J.R., and K.J. Luchessa. 1982. Using the historical record as an aid to salmonid habitat enhancement. *In* N.B. Armantrout, ed. Proceedings of a symposium on acquisition and utilization of aquatic habitat inventory information, American Fisheries Society, Bethesda. pp. 210-223.

Seton, E.T. 1929. The beaver. *In* Lives of game animals, vol. 4, part II, Doubleday, Doron, and Company, Inc., Garden City. pp. 441-501.

Smirnoff, N., P. Todd, and G.R. Stewart. 1984. The occurrence of nitrate reduction in the leaves of woody plants. Annals of Botany. 54:363-374.

Speaker, R., K. Moore, and S. Gregory. 1984. Analysis of the process of retention of organic matter in stream ecosystems. Verhandlungen Internationale Vereinigung Limnologie. 22:1835-1841.

Stabler, D.F. 1985. Increasing summer flow in small streams through management of riparian areas and adjacent vegetation: a synthesis. p. 206-210. *In* Riparian ecosystems and their management: reconciling conflicting uses, First North American Riparian Conference, U.S. Department of Agriculture, Forest Service General Technical Report RM-120.

Stookey, L. 1970. Ferrozine - a new spectrophotometric reagent for iron. Analytical Chemistry. 42:779-781.

Strickland, J.D.H., and T.R. Parsons. 1972. A practical handbook of seawater analysis, Bulletin 167, Fisheries Research Board of Canada, Ottawa, 310 pp.

Swanson, F.J., R.L. Fredriksen, and F.M. McCorison. 1982. Material transfer in a western Oregon forested watershed. p. 233-265. *In* R.L. Edmonds, ed. Analysis of coniferous forest ecosystems in the western United States. Hutchinson Ross Publishing Company, Stroudsburg.

Tiner, R.W. 1984. Wetlands of the United States: current status and recent trends. U.S. Department of the Interior, Fish and Wildlife Service, 59 pp.

Triska, F.J., J.R. Sedell, and S.V. Gregory. 1982. Coniferous forest streams. p.292-332. *In* R.L. Edmonds, ed. Analysis of coniferous forest ecosystems in the western United States. Hutchinson Ross Publishing Company, Stroudsburg.

Triska, F.J., V.C. Kennedy, R.J. Avonzino, and B.N. Reilly. 1983. Effect of simulated canopy cover on nitrate uptake and primary production by natural periphyton communities. *In* Proc. Savannah River Ecol. Lab. Ecol. Symp. 7, pp. 129-159.

Triska, F.J. 1984. Role of wood debris in modifying channel geomorphology and riparian areas of a large lowland river under pristine conditions: a historical case study. Verhandlungen Internationale Vereinigung Limnologie. 22:1876-1892.

Vannote, R.L., G.W. Minshall, K.W. Cummins, J.R. Sedell, and C.E. Cushing. 1980. The river continuum concept. Canadian Journal of Fisheries and Aquatic Sciences. 37:370-377.

Waring, R.H., A.J.S. McDonald, S. Larsson, T. Ericsson, A. Wiren, E. Arwidsson, A. Ericsson, and T. Lohammar. 1985. Differences in chemical composition of plants grown at constant relative growth rates with stable mineral nutrition. Oecologia. 66:157-160.

Weber, D.J.    1971.    The Taos trappers:    The fur trade in the far
    southwest, 1540-1846.    University of Oklahoma Press, Norman.
    263 pp.

Winegar, H.H.    1977.    Camp Creek channel fencing - plant, wildlife,
    soil, and water response.    Rangeman's Journal.    4:10-12.

# CHAPTER 11

LONGITUDINAL DISPERSION OF TRACE METALS
IN THE CLARK FORK RIVER, MONTANA

E. D. Andrews, U.S. Geological Survey, Denver, Colorado

ABSTRACT

The Clark Fork River drains the west-central part of Montana.
A large deposit of copper ore was discovered near the headwaters in
1880 and was mined until 1982. Efforts to retain the mill tailings
in ponds often were limited or unsuccessful. As a result, an
estimated 100 million tons of tailings material was supplied to the
Clark Fork River. Samples of fine-grained bed sediment collected
from a reach of 400 river kilometers show that concentrations of
arsenic, cadmium, copper, lead, and zinc decrease downstream,
although at different rates. There is no evidence that an
appreciable quantity of any trace metal is being dissolved from the
sediment particles and is entering the fluid phase. The longitudinal distribution of trace metals associated with fine-grained bed
sediment in the Clark Fork River appears to be solely the result of
physical mixing of mill tailings with floodplain material deposited
prior to mining.

INTRODUCTION

An investigation of river meanders and floodplain evolution in
a river that has received a large quantity of mine tailings during
the past 100 years was begun in the fall of 1984. This is a
continuation of work by other investigators who have used the
chemical characteristics of sediment particles to study the long-
term evolution of river channels (e.g., Ritchie et al., 1975; and
Lewin et al., 1983). Rates of geomorphic processes typically are
extremely variable and slow compared to available lengths of
record. Furthermore, the spatial extent frequently is large.
Attempts to simplify systems by studying just one landform, a
single process, or a small plot for a few years may remove
essential complexity and may lead to erroneous conclusions.

*Chemical Quality of Water and the Hydrologic Cycle*, Robert C. Averett and Diane M. McKnight (Eds.) © 1987 Lewis Publishers, Inc.,
Chelsea, Michigan. Printed in the United States of America.

Field experiments involving tracer particles have been used with some success to study the downstream movement of sediment particles--entrainment, transport, and redeposition--from one channel deposit to another.  For some geomorphic questions such as the evolution of a floodplain during a century or more, the injection of marked sediment particles into a river has notable limitations.  An alternative approach is to study rivers that have received wastes that are rich in trace metals from mines during a relatively long period of time.

The  Clark Fork River drains the west-central part of Montana (Fig. 1) and is a tributary to the Columbia River.  A very large deposit of copper ore was discovered at Butte in 1880 (Hutchinson, 1979).  Between 1880 and 1982, more than $400 \times 10^6$ tons of ore was mined.  The ore was ground to a slime (particle size less than 0.04 mm) and roasted with carbon.  Efforts to retain the mill tailings in ponds often were limited or unsuccessful.  As a result, an estimated $100 \times 10^6$ tons of material were supplied to the Clark Fork River during the period of mining that ended in 1982.  In spite of processing, the mill tailings still contained significant concentrations of five trace metals:  arsenic, cadmium, copper, lead, and zinc.  Concentrations of these trace metals in the mill tailings are 10 to 100 times the expected background values. Because of the large trace-metal concentrations, the mill tailings are distinguished easily from other alluvial sediments in the basin.  Moore and Johns (1984) investigated that distribution of trace metals in floodplain deposits along the Clark Fork River and found significant enrichment compared to background values, especially within the first 60 km downstream from the mouth of Warm Springs Creek.

The Clark Fork River flows through a series of alluvial valleys which are separated by narrow, bedrock canyons.  Where free adjustment of the channel is not constrained, the river has an irregular, meandering course.  Bed sediment is predominantly gravel and cobbles throughout the entire study reach from the mouth of Warm Springs Creek to the confluence with the Flathead River.  The coarse bed sediment is entrained and transported and only discharges near the bankfull stage and greater.  Estimated mean annual transport of gravel- and cobble-sized material is small, less than 3%, compared to the mean annual load of suspended sediment.  Since the end of mining, 1982, suspended-sediment concentrations typically are less than 150 mg/L.  An estimated mean annual suspended-sediment load of $3.4 \times 10^5$ tons per year was transported past the Clark Fork River below Missoula gage, number 12353000, during the period October 1978 to September 1986. Approximately 55% of this suspended sediment or $1.9 \times 10^5$ tons per year was smaller than 0.062 mm in diameter.  At this gage, the mean annual runoff is 160 $m^3$/s.  Peak river discharges are the result of snowmelt runoff.  Variation of annual peak flows is relatively small; the estimated flood discharge with an exceedance probability of 1% is only 1.8 times the mean annual peak discharge. Flood peaks, however, are usually sustained, and the bankfull discharge is typically exceeded for 2 to 3 weeks per year.

Figure 1.   The Clark Fork River basin, Montana.

METHODS

Forstner and Wittmann (1983) summarized the results of many studies concerning heavy metal transport in aquatic environments. Commonly, the concentration of trace metals associated with the solid phase increases rapidly with decreasing particle size. Therefore, it is essential that temporal and areal differences in particle size be minimized to the extent possible. Jenne et al. (1980) discuss the collection and separation of fluvial sediment for heavy metal analysis.

During the summer of 1984, fine-grained bed sediment was sampled at 21 locations along the Clark Fork River from the downstream edge of the tailings ponds (mouth of Warm Springs Creek) to just below the mouth of the Flathead River, 396 km downstream. In addition, a sample of fine-grained bed sediment was collected from the five largest tributaries to the Clark Fork River study reach. Fine-grained bed sediment was collected at low flow from very small deposits located around and behind gravel and cobble particles in the slack water along the river's edge. Only the top ~1 cm of bed-sediment was taken from each deposit to ensure that the underlying material, which was frequently organic rich and reduced, was excluded from the sample. Approximately 1 to 2 kg of fine-grained bed sediment were collected at each location by sampling several tens of deposits along a few hundred meters of the river channel.

The composite sample was washed through an 0.062-mm nylon sieve, and the retained fraction was discarded. The filtrate was agitated vigorously and then allowed to stand until particles larger than 0.016 mm had settled to the bottom of the container. Particle fall velocities and, thus, settling time, were determined by Stoke's law assuming sediment-free water. In fact, the filtrates typically had suspended-sediment concentrations in excess of 10 g/L. As a result, the fall velocity of particles larger than 0.016 mm was overestimated. From the remaining water and sediment mixture two 250-mL samples were collected. These samples were packed in ice until the end of the day when they were centrifuged, the excess water decanted, and the remaining sediment frozen. Subsequently, the bed sediment was freeze dried.

A split of the bed sediment was dissolved completely in hydrofluoric and sulfuric acids. A second split was treated with a warm mixture of 0.6 m hydrochloric and 0.3 m nitric acids for 24 h. The concentration of major and trace elements, except arsenic, in both extracts were determined with an induction-coupled plasma instrument. Arsenic concentrations were determined by graphite-furnace atomic absorption. The total organic-carbon content of each bed-sediment sample also was determined.

Bed-sediment samples were collected from the Clark Fork River (4) and from the tributary Blackfoot River (1), separated into size fractions, dissolved completely, and the concentrations of copper, zinc, and manganese determined (Wilhelm, 1986). Different particle sizes were separated by resuspending the samples and centrifuging for the appropriate interval. Because this separation procedure was performed in a laboratory with much lower sediment concentration than had been possible during the field separation, the accuracy of this second size fractionation is

much improved.   The laboratory separation showed that the field
samples contained an appreciable quantity of material, 40 to 50%
larger than 0.016 mm in diameter.   The size distributions of the
five bed-sediment samples, however, are similar.
    The U.S. Geological Survey operates several gaging stations in
the Clark Fork River basin.   The principal gaging stations in the
study area are shown in Figure 1.   Mean daily water discharge is
determined at each gage.   In addition, various characteristics of
water chemistry, pH, suspended-sediment concentration, and concentrations of dissolved major and trace elements have been determined
at the Clark Fork River below Missoula gage, number 12353000,
during the period from October 1978 to September 1985.   These
measurements are published annually by the U.S. Geological Survey
in a series titled "Water Resources Data for Montana."


RESULTS AND DISCUSSION

    Concentrations of the five trace metals studied (arsenic,
cadmium, copper, lead, and zinc, plus aluminum, iron, and manganese
in the bed sediment of the Clark Fork River and major tributaries
determined by total and partial digestion) are listed in Table 1.
Several observations indicate that the five trace metals studied
primarily are associated with ferromanganese material on the
particle surfaces (Jenne, 1968, 1977).   Concentrations of arsenic,
cadmium, copper, and zinc in fine-grained bed sediment determined
by the total and partial digestions essentially were identical.   In
the instance of lead, the partial digestion recovered about 70% of
the material found in the total digestion.   Therefore, it may be
concluded that, except for lead, an insignificant amount of the
trace metals are bound in silicate minerals.   Concentrations of the
five trace metals also are well correlated with concentrations of
manganese and, to a lesser extent, iron.   The concentration of
total organic carbon in the bed-sediment samples was nearly
constant throughout the study reach and, therefore, was not
correlated with the decreasing downstream concentration of trace
metals.   Ferromanganese coatings frequently form on the surface of
river sediment where normal to slightly alkaline pH conditions
prevail (Davison and Seed, 1983; Robinson, 1983).   The pH has been
measured at streamflow gaging stations located 35 and 217 km
downstream from the Warm Springs tailings ponds, and values ranged
from 7.5 to 8.6 throughout the year regardless of discharge.   The
most highly enriched bed-sediment sample was collected 21.2 km
downstream from the tailings ponds.   Analysis of this sample by
scanning electron microscopy and x-ray elemental mapping have shown
that copper and zinc are primarily associated with manganese-rich
material on the surface of quartz and feldspar particles (J. J.
Fitzpatrick, U.S. Geological Survey, oral commun., 1986).   Concentrations of arsenic, cadmium, and lead were not large enough for an
accurate determination by energy dispersive x-ray analysis.
    The distribution of copper with bed-sediment particle size is
shown in Figure 2 for five samples.   Samples were collected from
the Clark Fork River at 21.2, 89.2, 168.3, and 299.6 km downstream
from the tailings ponds, and a single sample was collected from the

Table 1.    Concentrations of Trace Metal Associated with Fine-Grained Bed Material in the Clark Fork River and Major Tributaries.

| Location river kilometer | Arsenic mg/kg | | Cadmium mg/kg | | Copper mg/kg | | Lead mg/kg | |
|---|---|---|---|---|---|---|---|---|
| | Total | Partial | Total | Partial | Total | Partial | Total | Partial |
| 14.3 | 165 | 164 | 9.3 | 7.3 | 1,290 | 1,300 | 173 | 117 |
| 21.2 | 199 | 194 | 9.7 | 10 | 2,490 | 1,410 | 179 | 136 |
| 34.8 | 151 | 195 | 8.7 | 11 | 1,660 | 1,540 | 213 | 151 |
| 48.1 | 100 | 80 | 7.3 | 5.9 | 1,620 | 1,080 | 170 | 116 |
| 78.4 | 60 | 62 | 7.3 | 6.9 | 1,700 | 990 | 139 | 89.8 |
| 89.2 | 39 | 26 | 4.8 | 3.3 | 1,000 | 641 | 100 | 62.2 |
| 94.1 | 46 | 53 | 4.8 | 17 | 1,050 | 747 | 111 | 67.9 |
| 104.4 | 44 | 11 | 1.7 | 5.9 | 650 | 680 | 100 | 63.4 |
| 115.7 | 54 | 52 | 3.3 | 3.7 | 400 | 418 | 112 | 77.2 |
| 130.7 | 69 | 50 | 3.5 | 3.5 | 420 | 428 | 116 | 84.9 |
| 140.8 | 49 | 51 | 4.1 | 2.2 | 335 | 345 | 95 | 36.8 |
| 153.4 | 40 | 38 | 2.4 | 2.7 | 305 | 305 | 87 | 52.1 |
| 168.3 | 33 | 35 | 3.4 | 1.1 | 325 | 321 | 79 | 43.5 |
| 181.5 | 35 | 38 | 3.1 | 1.9 | 333 | 345 | 80 | 51.9 |
| 207.1 | 18 | 20 | 2.8 | 4.3 | 225 | 230 | 54 | 30.2 |
| 222.4 | 15 | 19 | 2.2 | 1.7 | 245 | 231 | 62 | 30.1 |
| 228.4 | 19 | 20 | 2.0 | 3.3 | 325 | 353 | 62 | 37.0 |
| 264.9 | 17 | 21 | 1.3 | 2.6 | 212 | 221 | 45 | 20.9 |
| 299.6 | 8.5 | 17 | 1.2 | <.1 | 121 | 107 | 34 | <.5 |
| 387.7 | 17 | 23 | 1.2 | 2.1 | 235 | 245 | 57 | 27.4 |
| 399.7 | 9.4 | 4 | <.5 | .79 | 93 | 101 | 24 | 1.4 |

Major Tributaries

| Location | Arsenic mg/kg | | Cadmium mg/kg | | Copper mg/kg | | Lead mg/kg | |
|---|---|---|---|---|---|---|---|---|
| | Total | Partial | Total | Partial | Total | Partial | Total | Partial |
| Little Blackfoot River | 3.2 | 17 | .7 | .8 | 25 | 27.5 | 31 | 4.2 |
| Flint Creek | 126 | 128 | 1.5 | .7 | 48 | 51 | 165 | 124 |
| Rock Creek | 5.4 | 14 | <.5 | <.1 | 10 | 12 | 6 | <.5 |
| Blackfoot River | 4.8 | 6.4 | <.5 | .3 | 19 | 17 | 9 | <.5 |
| Bitterroot River | 3.0 | 5 | <.5 | <.1 | 30 | 29 | 24 | <.5 |

Table 1. Concentrations of Trace Metal Associated with Fine-Grained Bed Material in the Clark Fork River and Major Tributaries--Continued.

| Location river kilometer | Zinc Total (mg/kg) | Zinc Partial (mg/kg) | Aluminum Total (Weight percent) | Aluminum Partial (Weight percent) | Iron Total (Weight percent) | Iron Partial (Weight percent) | Manganese Total (Weight percent) | Manganese Partial (Weight percent) |
|---|---|---|---|---|---|---|---|---|
| 14.3 | 1,660 | 1,580 | 5.9 | 1.2 | 3.9 | 2.2 | 0.80 | 0.60 |
| 21.2 | 1,770 | 1,770 | 5.7 | 1.4 | 3.7 | 2.7 | .90 | .70 |
| 34.8 | 1,850 | 1,880 | 6.0 | 2.0 | 3.9 | 2.7 | .40 | .36 |
| 48.1 | 1,460 | 1,380 | 5.9 | 1.3 | 3.5 | 1.9 | .30 | .30 |
| 78.4 | 1,380 | 1,390 | 5.8 | 1.7 | 3.2 | 2.1 | .30 | .33 |
| 89.2 | 1,030 | 1,030 | 5.5 | 1.7 | 3.0 | 1.9 | .20 | .17 |
| 94.1 | 1,130 | 1,090 | 5.4 | 1.4 | 3.1 | 1.8 | .20 | .22 |
| 104.4 | 560 | 1,130 | 5.2 | 1.6 | 2.8 | 1.8 | .20 | .36 |
| 115.7 | 900 | 916 | 5.3 | 1.4 | 2.8 | 1.6 | .20 | .21 |
| 130.7 | 940 | 916 | 5.4 | 1.1 | 2.9 | 1.3 | .20 | .23 |
| 140.8 | 830 | 836 | 5.3 | 1.4 | 2.7 | 1.5 | .20 | .16 |
| 153.4 | 800 | 761 | 5.8 | 1.1 | 2.6 | 1.3 | .15 | .15 |
| 168.3 | 325 | 780 | 5.2 | 1.2 | 2.6 | 1.4 | .10 | .11 |
| 181.5 | 900 | 873 | 5.3 | 1.2 | 2.6 | 1.4 | .20 | .17 |
| 207.1 | 690 | 685 | 5.1 | .83 | 2.3 | 1.1 | .11 | .07 |
| 222.4 | 540 | 489 | 5.2 | .90 | 2.6 | 1.1 | .07 | .05 |
| 228.4 | 760 | 740 | 5.2 | 1.3 | 2.6 | 1.5 | .08 | .07 |
| 264.9 | 610 | 613 | 4.2 | 1.1 | 2.3 | 1.5 | .10 | .11 |
| 299.6 | 330 | 300 | 4.7 | .95 | 2.1 | 1.1 | .07 | .05 |
| 387.7 | 540 | 527 | 4.4 | 1.3 | 2.6 | 1.7 | .12 | .09 |
| 399.7 | 250 | 267 | 4.5 | 1.04 | 2.2 | 1.3 | .10 | .10 |
| **Major Tributaries** | | | | | | | | |
| Little Blackfoot River | 153 | 128 | 5.5 | 1.3 | 3.2 | 1.5 | .11 | .07 |
| Flint Creek | 560 | 542 | 4.6 | 1.5 | 2.6 | 1.4 | .30 | .31 |
| Rock Creek | 38 | 35 | 2.8 | 1.1 | 1.7 | 1.1 | .02 | .02 |
| Blackfoot River | 54 | 41 | 4.8 | .75 | 2.6 | 1 | .04 | .03 |
| Bitterroot River | 80 | 79 | 6.3 | 2 | 2.7 | 1.9 | .04 | .04 |

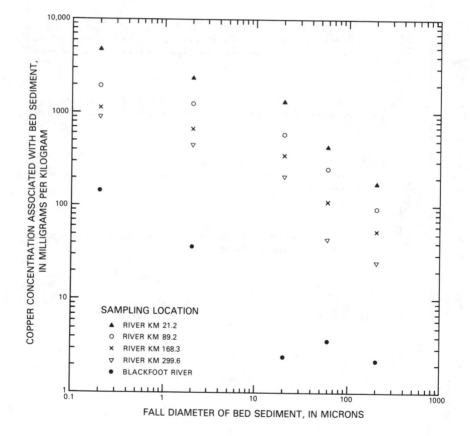

Figure 2.   Distribution of copper concentration with particle size
in selected bed-sediment samples from the Clark Fork and
Blackfoot Rivers, Montana.

Blackfoot River as a control.    The concentration of copper is largest for all particle size fractions in the sample collected closest to the tailings ponds and decreases downstream.    At a distance of 299.6 km downstream, the copper concentration in fine-grained bed sediment of the Clark Fork River is still an order of magnitude greater than that which occurs in the Blackfoot River. For all bed-sediment samples, copper concentration increases significantly with decreased particle size.    The rate at which copper concentration decreases with increasing particle size is remarkably similar, including the Blackfoot River sample.    Both zinc and manganese show similar distributions as those described for copper.    Increasing particle surface area per unit weight of material as particle size decreases is apparently a factor given the predominant association of the trace metals with the particle surface as described above.    Trace-metal concentrations, however, do not increase proportionately with increasing particle surface area.    For example, assuming flat, plate-like particles, surface area per unit weight of material increases by $(200/.2)^2 = 1 \times 10^6$ from the largest to smallest particle sizes shown in Figure 2. Therefore, it is concluded that copper, zinc, and manganese have the greatest abundance per unit surface area on the largest particle.    A detailed analysis of particle surfaces is currently being conducted in order to determine the degree to which the thickness and areal coverage of trace-metal rich coatings vary with particle size.

The longitudinal distribution of copper and lead in fine-grained bed sediment of the Clark Fork River are compared in Figure 3.    Within the first 20 km downstream from the Warm Springs tailings pond, the concentrations of trace metals in the bed sediment are large and are within the range of values determined for the mill tailings.    The location and concentration of trace metals in tributaries also are shown in Figure 3.    Except for Flint Creek, which drains a small mining district, trace-metal concentrations in the bed sediment of the Clark Fork River tributaries are indicative of background values, that is, trace-metal concentrations that occurred in bed sediment of the Clark Fork River prior to the mining.    The concentration of trace metals in the tributaries approximate values for their occurrence in shales worldwide as reported by Turekian and Wedepohl (1961).    The concentration of copper and lead in fine-grained bed sediment of the Clark Fork River decreases to approximately the background values 400 km downstream from Warm Springs Creek.

The concentration of copper and lead vary with distance downstream at different rates (Fig. 3).    The concentration of copper in bed sediment decreases downstream much more rapidly than the concentration of lead.    The longitudinal distributions of arsenic, cadmium, and zinc in the Clark Fork River are intermediate between the extremes represented by copper and lead.    Several possible chemical processes have been examined to explain these observations and all were judged to be unsatisfactory.    As noted previously, measured pH values in the reach investigated support the formation and stability of iron and manganese oxides. Furthermore, a comparison of dissolved arsenic, cadmium, copper,

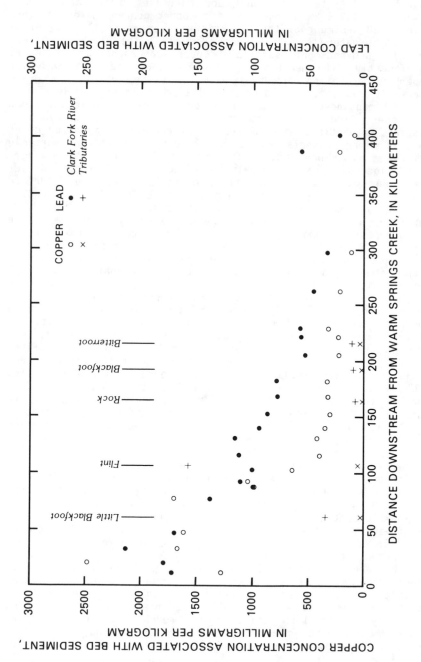

Figure 3.    Longitudinal distribution of copper and lead associated with bed sediment in the Clark Fork River and major tributaries, Montana.

lead, and zinc concentrations sampled at gaging stations, located at river kilometer 35 and 217, shows that appreciable quantities of these trace metals are not entering the dissolved phase.

The contribution of water and sediment by the tributaries has very little effect upon the longitudinal distribution of trace metals. The Bitterroot River contributes approximately 40% of the flow and 50% of the suspended-sediment load to its confluence with the Clark Fork River (Cartier, 1984). The longitudinal distribution of trace metals associated with fine-grained bed sediment of the Clark Fork River, however, does not show an appreciable dilution as a result of the advection of sediment with background trace-metal concentrations contributed by the Bitterroot River. Consequently, it was concluded that the mixing of bed sediment rich in mill tailings with sediment from the tributaries is not a significant process in determining the longitudinal distribution of trace metals in the Clark Fork River.

Longitudinal distributions of the five trace metals studied can be explained by mixing sediments rich in mill tailings with flood-plain material initially having background trace-metal concentrations. The volume of floodplain material stored along the Clark Fork River is a few hundred times the annual quantity of transported sediment. Furthermore, the exchange of sediment between the river and floodplain is large relative to the quantity of sediment supplied by tributaries. Hence, the tributary contributions have no appreciable effect. The concentration of copper in the mill tailings is nearly 80 times the background concentration, whereas, the concentration of lead in the tailings is only 12 times its background concentration. Thus, the admixture of an equal proportion of mill tailings and floodplain material causes a more rapid decrease in the copper concentration than the lead concentration. Fine-grained bed sediment at river kilometer 21 is composed of about 70% mill tailings. In contrast, fine-grained bed sediment at river kilometer 400 is composed of less than 1% mill tailings.

The floodplain of the Clark Fork River differs from the regional norm in one significant aspect. Throughout the Rocky Mountain region of the United States, floodplains predominantly are formed by lateral accretion to the point bar at the inside of a meander bend. Overbank (vertical accretion) deposits are relatively thin. For example, along channels tributary to the Clark Fork River, the thickness of vertically accreted material is only 5 to 20 cm compared to a bank height of 1 to 2 m. In contrast, overbank deposits along the Clark Fork River immediately downstream of the tailing ponds are relatively thick (1-1.5 m) and become thinner downstream. The relatively thick overbank deposits result, in part, from comparatively slow rates of meander migration. Comparisons of aerial photographs taken during 1955 and 1982 indicate channel migration rates of 10 to 20 cm/yr, or less than 1% of the channel width per year. The relatively slow rates of channel migration and sustained flood peaks in themselves probably are not sufficient to account for the quantity of overbank material deposited downstream from the tailings pond. Suspended-sediment concentrations probably were substantially greater in the past than under the present conditions.

The quantity of mill tailings deposited within and along the Clark Fork River can be estimated approximately from the measured trace-metal concentrations and the longitudinal distribution of floodplain material. As much as one-half of the mill tailings supplied to the Clark Fork River ($\sim50 \times 10^6$ tons) appears to have been incorporated into the river bed and floodplain. Although the trace metals are associated with fine-grained, and therefore, easily suspended particles, the net downstream flux of trace metals is not large. As noted previously, the bed sediment of the Clark Fork River is predominantly coarse gravel and cobbles. This material is entrained and transported by the flow only at discharges near the bankfull stage and greater. Consequently, the coarse bed material along the base of the river banks is eroded and undermines the banks at relatively larger discharges. As a result, the banks of the Clark Fork River are relatively stable and resistant to erosion except during floods. Thus, the fine-grained floodplain sediment (i.e., the vertical accretion deposit which contains the highest fraction of mill tailings) is eroded and transported downstream primarily during periods when the floodplain is inundated. Relatively large concentrations of suspended sediment and extensive flow across the floodplain, however, favor the redeposition of fine-grain sediment in the overbank areas. The relatively slow rate of meander migration plus hydraulic conditions tending to redeposit fine-grained sediment on the floodplain explain the relatively slow net downstream flux of trace metals associated with fine-grained sediment, as well as the rapid exchange of fine-grained sediment between the channel and floodplain.

## ACKNOWLEDGEMENTS

Sam Luoma, Howard Taylor and Robert Stallard contributed to this investigation by their thoughtful discussion of solid phase sampling and analysis. I have appreciated the assistance of J. N. Moore on several occasions and benefited from his extensive knowledge of the Clark Fork River geochemistry. Arthur Horowitz made the chemical analysis of total elemental composition. Judith McHugh, Ann Choquette, and Dennis Helsel collected the bed-sediment samples and prepared them for analysis.

## REFERENCES

Cartier, K. D. 1984. Sediment, channel morphology and streamflow characteristics of the Bitterroot River drainage basin, southwestern Montana. Unpublished master thesis, University of Montana, Missoula, Montana, 191 p.

Davison, W., and G. Seed. 1983. The kinetic of the oxidation of ferrous iron in synthetic and natural waters. Geochem. et Cosmochem. Acta, 47: 67-79.

Fitzpatrick, J. J. U.S. Geological Survey. 1986. Oral communication.

Forstner, U., and G. T. W. Wittmann. 1983. Metal pollution in the aquatic environment, 2nd ed. Springer-Verlag, Berlin, 486 p.

Hutchinson, T. C. 1979. Copper contamination of ecosystems caused by smelter activities, p. 451-502. *In* J. O. Nriaga, ed., Copper in the environment. John Wiley and Sons Inc., New York.

Jenne, E. A. 1968. Controls on Mn, Fe, Co, Ni, Ca, and Zn concentrations in soil and water. The significant role of hydrous Mn and Fe oxides p. 337-387. *In* R. F. Gould, ed., Trace inorganics in water advances in chemistry series, V. 73. Am. Chem. Soc., Washington, D.C.

Jenne, E. A. 1977. Trace element sorption by sediments and soil-sites and processes, p. 425-553. *In* W. Chappel and K. Petersen, eds., Symposium on molybdenum, V. 2. Marcel Dekker, New York.

Jenne, E. A., V. C. Kennedy, J. M. Burchard, and J. W. Ball. 1980. Sediment collection and processing for selective extraction and for total trace element analysis, p. 169-189. *In* R. A. Baker, ed., Contaminants and sediments V. 2. Ann Arbor Science Publishers, Ann Arbor.

Lewin, J., S. B. Bradley, and M. G. Macklin. 1983. Historical valley alluviation in mid-Wales. Geol. J. 18:331-350.

Moore, J. N., and C. Johns. 1984. Occurrence, distribution, and fractionation of metals in contaminated sediments originating from mining and smelting operations along the Clark Fork River, Montana. EOS Transactions. Am. Geophys. Union. 65(45):890.

Ritchie, J. C., P. H. Hawks, and J. R. McHenry. 1975. Deposition rates in valleys determined using fallout cesium -137. Geol. Soc. America Bull. 86:1128-1130.

Robinson, G. D. 1983. Heavy-metal adsorption by ferromanganese coatings on stream alluvium. Natural controls and implications for exploration. Chem. Geol. 38:157-174.

Turekian, K. K., and K. H. Wedepohl. 1961. Distribution of the elements in some major units of the earth's crust. Geol. Soc. America Bull. 72:702-706.

Wilhelm, S. R. 1986. Smelter-contaminated river sediments--Effects of particle size on heavy metal distribution, Clark Fork River, Anaconda to Missoula, MT. Unpublished senior thesis, Department of Geological and Geophysical Sciences, Princeton University, Princeton, N.J., 57 p.

# CHAPTER 12

# FLOODPLAIN STORAGE OF METAL-CONTAMINATED SEDIMENTS DOWNSTREAM OF A GOLD MINE AT LEAD, SOUTH DAKOTA

Donna C. Marron, U.S. Geological Survey, Denver, Colorado

## ABSTRACT

Strong associations between sediment and some contaminants in river systems cause the transport and storage of sediment to exert considerable control on the fate of those contaminants. Sediment can be stored in a variety of environments on the floodplains of meandering rivers. Stored sediment eventually is remobilized by meander migration. The effect of gold-mining activity in Lead, South Dakota, on metal concentrations of floodplain sediments downstream of Lead illustrates the role of sediment transport and storage processes in determining the distribution of contaminants associated with sediments. The introduction of a large volume of finely milled mine tailings into a river system has resulted in an extensive floodplain deposit of metal-contaminated sediment. Rough calculations of the amount of metal-contaminated sediment stored in floodplains indicate that at least one-third of the mine tailings discharged into the river system presently are stored along approximately 121 kilometers of the Belle Fourche River floodplain downstream of Lead. Measurements of meander-migration rates of channels in contaminated floodplain areas indicate that the metal-contaminated floodplain deposits will be a source of metals to adjacent streams for centuries.

## INTRODUCTION

### Background

Relations between sediments and some pollutants in river systems have received growing recognition in recent years. Several authors, among them Lewin et al. (1977), Alloway and Davies (1971), Macklin (1985), and Klimek and Zawilinska (1985) have noted

*Chemical Quality of Water and the Hydrologic Cycle*, Robert C. Averett and Diane M. McKnight (Eds.) © 1987 Lewis Publishers, Inc., Chelsea, Michigan. Printed in the United States of America.

associations between metals and sediments in rivers that received wastes from metal mining. Metals introduced into river systems by human activities other than metal mining are also commonly transported in association with sediments (Trefry et al., 1985; and Rutherford, 1977). Bopp et al. (1982) noted strong associations between sediments in the Hudson River estuary and radionuclides, polychlorinated biphenyls, and chlorinated-hydrocarbon pesticides. Clearly, an understanding of the fate of these pollutants in the river systems into which they have been introduced must be based on an understanding of the transport, storage, and remobilization of the sediments with which the pollutants are associated.

This report discusses aspects of sediment transport and storage that affect the movement of pollutants through meandering river systems. Meandering rivers are common in valleys with gentle gradients; they flow, for the most part, through a single channel with a sinuous course. Most meandering channels are located within valley flats called floodplains, which are created by the back-and-forth migration of the stream channel over long periods of time. During channel migration, sediment is eroded along outsides of meander bends, and is deposited along insides of meander bends. During stream flows that are large enough to overflow the channel banks, sediment is deposited on top of floodplain surfaces. The sinuosity of meandering rivers causes the distance measured along a river channel between two points to exceed the distance measured along the floodplain between the same two points. Distances measured along river channels are expressed in river kilometers; whereas, distances measured along floodplains are expressed in floodplain kilometers.

The continual exchange of sediment between river channels and their floodplains has an important effect on the movement of pollutants that are associated with sediments through river systems. The time scale over which sediment can be stored in floodplains increases with the size of the river and can be on the order of 1,000 years (Leopold et al., 1964). Aside from their potential to be remobilized by meander migration, contaminants associated with sediment that is stored in floodplains can affect the health of plants and animals that naturally inhabit riparian areas or that are placed in those areas by human activity. Contaminated sediment that is stored in floodplains also may affect water quality in alluvial aquifers.

A case study concerning the movement and storage of metal-contaminated sediment resulting from gold-mining activity in Lead, South Dakota (Fig. 1), in downstream river systems will be discussed as an illustration of more general points presented in this report. This case study is of particular interest because site conditions favor the chemical immobility of the metals associated with sediments, and because an estimate of the volume of contaminated sediment introduced into the river system can be used to approximate the proportion of that material that has been stored in floodplains downstream. Aspects of the sedimentology, chemical characteristics, and geometry of the floodplain deposits of metal-contaminated sediments downstream of Lead, South Dakota, will be discussed.

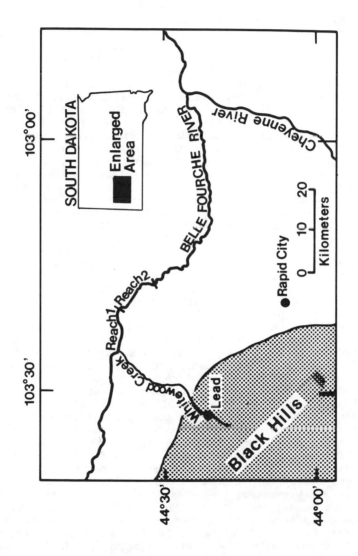

Figure 1.   Location of study area.

Transport, Storage and Remobilization of Sediment in
Meandering Rivers

The particulate load of rivers is transported as bedload or
suspended load. The bedload consists of particles that travel by
saltation, sliding, and rolling along the bed of a river. The
suspended load consists of particles that are entrained in the
turbulent eddies of a flowing river. Where the various sizes of
sediment are available for transport, the suspended load is likely
to consist of clay, silt, and fine to medium sand, and the bedload
is likely to consist of medium sand and coarser particles (Dunne
and Leopold, 1978). Data presently available suggests that the
suspended load comprises more than 90% of the particulate load in a
majority of rivers (Hadley et al., 1985). The finer particles that
are carried in suspension have more surface area; they are more
likely to be associated with many pollutants than the coarser
particles that are transported as bedload.

Suspended-sediment transport can be related to stream discharge
in a particular river, although the relation may differ at
different stages during a storm hydrograph or at different times of
the year. Data presently available indicate that annual sediment
transport is highly skewed, with a large proportion of the annual
sediment yields of rivers being moved during a small number of
floods (Meade, 1982; Hadley et al., 1985). Moderate-frequency
floods, which are likely to occur within the time scale of one year
to several years, are responsible for the majority of long-term
sediment transport (Wolman and Miller, 1960; Webb and Walling,
1982) because (1) extremely large floods that are particularly
effective in transporting sediment are rare and (2) frequently
occurring small floods have small sediment yields.

Point bars, overbank deposits, and the filled channels of
abandoned meanders are the main storage sites for sediment in the
floodplains of meandering rivers. Arcuate bars, called point bars,
are deposited on the insides of channel bends as the outsides of
the bends are eroded during meander migration. The channel
migration that occurred between 1910 and 1981 along the 30 km of
the Belle Fourche River floodplain downstream of Whitewood Creek is
shown in Figure 2. Point-bar deposits commonly are composed
predominantly of sediments that were transported as bedload (Allen,
1965). Overbank sediments are deposited by flood waters that have
risen over the banks of a river. Recently deposited point bars are
preferred sites for overbank deposition. Overbank deposits
commonly are composed predominantly of sediments that were
transported as suspended load (Allen, 1965). Channels that have
been abandoned because of meander cutoffs are filled by moderate
and large flows that allow water and sediment to enter the
abandoned channel after the cutoff has occurred. Both bedload and
suspended-load sediments are deposited in abandoned channels.

Controls on rates of meander migration are not well
understood. Meander-migration rates are known to be discontinuous
in time (Nanson and Hickin, 1986) and space (Brice, 1973). Hooke
(1980) and Nanson and Hickin (1986) found that meander-migration
rates generally increase with increasing river size. Daniel (1971)
and Hooke (1980) found that meander-migration rates decrease with

increasing bank cohesion or percentage of silt and clay in streambanks. Nanson and Hickin (1986) found that meander-migration rates decrease with increasing grain size of the sediment at the base of outer banks of meanders in gravel- and sand-bed rivers in western Canada. At the present time, historical information, dendrochronology, and time-sequential air photographs and maps provide the most reliable information on meander-migration rates in any particular stream system.

Reported rates of overbank deposition on floodplain surfaces vary widely. Data compiled by Bridge and Leeder (1979) show that rates of overbank deposition on floodplain surfaces with a period of record of 1,000 years or less are usually between 0.80 and 2.80 cm/yr, with reported rates ranging as high as 8.3 cm/yr and as low as 0.1 cm/yr. Some of these rates reflect increases in the sediment yield of rivers as a consequence of land use (Costa, 1975; Happ, 1945). Because the introduction of pollutants into a river system often either is directly related to, or is synchronous with, the disturbance of a watershed by land use, these high rates of overbank deposition are relevant to the discussion of the deposition of contaminated sediments on floodplain surfaces.

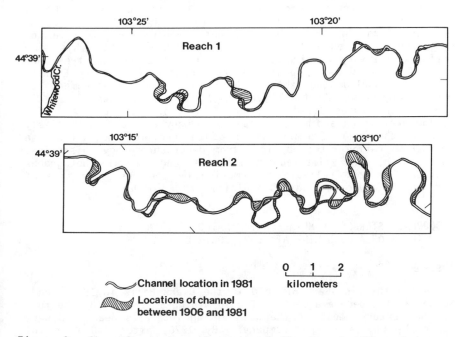

Figure 2. Channel changes along the Belle Fourche River between 1910 and 1981. Reach locations are on Figure 1.

The lateral extent of high rates of overbank deposition on floodplain surfaces is unclear. Overbank deposition during periodic flooding elevates floodplain surfaces over time. A higher elevation makes a floodplain surface less vulnerable to inundation by flood waters and to additional overbank deposition. In his study of floodplain formation along Beatton River in northeastern British Columbia, Nanson (1980) reports a mean sedimentation rate of 6.10 cm/yr on floodplain surfaces that are less than 50 years old, a mean sedimentation rate of 0.80 cm/yr on floodplain surfaces that are 50 to 250 years old, and negligible sedimentation on surfaces that are more than 250 years old. Deposits that imply an approximate overbank deposition rate of 1 cm/yr for metal-contaminated sediment on the floodplain of the Belle Fourche River in South Dakota (discussed hereafter) extend away from the channel over distances that in places are less than one-fourth the width of the meander belt. The thickness of metal-contaminated overbank sediments decreases with increasing distance from the stream channel; this pattern implies a smaller rate of overbank deposition in floodplain positions that are farther away from the stream channel. In estimating the volume of sediment stored on floodplain surfaces, care must be taken to recognize spatial variations in the rate of overbank deposition.

One final point in this general discussion of sediment transport and storage concerns the time period over which sediment is likely to remain in floodplains before being remobilized by meander migration. The dependence of meander-migration rates on the nature of sediment in streambanks suggests that this time period could vary considerably. A study of cottonwood-tree ages along 2.5 km of a floodplain in western North Dakota led Everitt (1968) to conclude that about half of that floodplain segment had been reworked in about 70 years. In contrast, Leopold et al. (1964) and studies cited by Schumm (1977) indicate that the time period over which sediment is stored in many floodplains is likely to be on the order of a millennium. Site-specific studies that use meander-migration rates obtained from dendrochronology, historical information, and time-sequential maps and aerial photographs provide the most reliable information on the residence time of sediment in any particular floodplain.

## FLOODPLAIN STORAGE OF METAL-CONTAMINATED SEDIMENTS DOWNSTREAM OF A GOLD MINE AT LEAD, SOUTH DAKOTA

### The Study Area

After the discovery of gold in the Black Hills in 1874, Lead, South Dakota, became a center of gold-mining activity. By the turn of the century, most of the gold-mining interests near Lead were controlled by a single company. By 1976, after 100 years of production, that company had milled 127,000,000 Mg of ore to obtain 893,000 kg of gold (Mg equals 1,000 kg or 1 metric ton) (Homestake Gold Mine, 1976).

The ore that was mined is a fine- to medium-grained schist consisting of silicates, iron and magnesium carbonates, and various metallic sulfides, particularly pyrrhotite (FeS), pyrite (FeS$_2$), and arsenopyrite (FeAsS) (Goddard, in press). Gold is associated with quartz, chlorite, and arsenopyrite in the ore body. Oxidized portions of the ore body were mined in the late 1800's and early 1900's, whereas more recent mining has taken place in unoxidized parts of the ore body.

In the early days of mining, the ore was crushed in stamp mills that yielded sand-size and finer particles. In subsequent years, rod-and-ball mills that yielded even finer particles were installed. In recent years, some of the sand-size tailings have been used to backfill mine shafts. A mercury-amalgamation procedure was used to extract gold from the milled rock until the early 1970's. Starting around the turn of the century, a cyanide-leaching procedure was employed for the further extraction of gold after the amalgamation procedure.

Prior to 1977, milled tailings were discharged into Whitewood Creek (Fig. 1) via a small tributary. Whitewood Creek is a meandering stream that has a drainage area of 105 km$^2$ at Lead. Approximately 43 floodplain kilometers below Lead, Whitewood Creek flows into the Belle Fourche River, which is a meandering river with a drainage area of 15,200 km$^2$ at a streamflow-gaging station that is 37 floodplain kilometers downstream of the mouth of Whitewood Creek. The Belle Fourche River flows into the Cheyenne River 121 floodplain kilometers downstream of the mouth of Whitewood Creek. The Cheyenne River carries a large sediment load from areas of badlands erosion. Discussions with local residents indicate that, prior to the cessation of mine-tailings discharge into Whitewood Creek, the water in Whitewood Creek and the Belle Fourche River downstream of Whitewood Creek flowed black due to the suspension of finely milled mine tailings. Approximately 100,000,000 Mg of the mine tailings were discharged into Whitewood Creek between the 1880's and 1977.

A reconnaissance study by Goddard (in press) indicates that floodplain sediments along Whitewood Creek, the Belle Fourche River, and the Cheyenne River contain above-background concentrations of a variety of metals, including arsenic, cadmium, silver, copper, and mercury. The difference between background concentrations and concentrations measured in the contaminated floodplain deposits was largest for arsenic, which ranged between 500 and 10,000 µg/gm in the floodplain deposits in contrast to concentrations of 5 to 50 µg/gm in uncontaminated alluvium, (Goddard, in press). Based on their distinctive orange-red color, the contaminated floodplain deposits along Whitewood Creek and the Belle Fourche River are visually distinguishable from premining alluvium. Dilution of the mine tailings on the Cheyenne River floodplain makes them chemically, but not visually, identifiable.

### Characteristics of Contaminated Floodplain Deposits
### Along Whitewood Creek and the Belle Fourche River

Sedimentology and Geometry

Three major types of metal-contaminated floodplain deposits occur on the floodplains of Whitewood Creek and the Belle Fourche River. Channel migration during the period of tailings discharge (Fig. 2) left metal-contaminated point-bar deposits on the insides of channel bends. Meander cutoffs that occurred during the period of tailings discharge resulted in the filling of abandoned channels with metal-contaminated sediment. The most widely distributed type of metal-contaminated floodplain deposit along Whitewood Creek and the Belle Fourche River consists of overbank sediments. The three deposit types differ from each other in their geometry and their textural characteristics.

Point-bar deposition was considerably more extensive along the Belle Fourche River than along Whitewood Creek during the period of tailings discharge. Where exposed, metal-contaminated point-bar sediments along the Belle Fourche River are about 1.5 m thick, consisting mostly of horizontal beds of sand and gravel, with some finer sediments included. The sediments generally become finer in texture toward the top of the deposit. The metal-contaminated point-bar deposits along the Belle Fourche River are located for the most part at elevations similar to those of the present river channel. A comparison of a map that was surveyed in 1904 and 1910 (Darton, 1919) and aerial photographs that were taken in 1981 indicate that approximately 2,740,000 m² of point bar were deposited along the 30 km of the Belle Fourche River floodplain immediately downstream of the mouth of Whitewood Creek between 1910 and 1981.

Abandoned meanders that were filled with metal-contaminated sediment during the period of tailings discharge occur in both the Whitewood Creek and the Belle Fourche River floodplains. The deposits along the lower part of Whitewood Creek have bases that are above the present level of Whitewood Creek, as a result of channel incision that occurred during the period of tailings discharge. Cross sections of these deposits show considerable textural variation, with grain sizes ranging mostly from gravel to sand. A permeability contrast between the contaminated alluvium and the material beneath it causes ground water to seep out of the bases of numerous filled channels along the lower part of Whitewood Creek. The amount of metal-contaminated sediment stored in abandoned channels downstream of Lead is small in comparison to the amount of metal-contaminated sediment stored as point bars and overbank deposits.

The tops of streambank cuts along Whitewood Creek and the Belle Fourche River typically expose 1 m or more of overbank deposits. Most of this sediment is horizontally bedded, consisting predominantly of fine sand, silt, and clay. Gravel layers occur in some places. Grain-size analyses of samples taken from soil pits dug in overbank deposits show a decrease in particle size with increasing distance downstream from Lead.

Soil pits that were dug along transects perpendicular to Whitewood Creek and the Belle Fourche River at five locations exposed more than 1 m of metal-contaminated overbank sediments, extending to distances that are mostly about 50 m, but range as far as 175 m away from the present channel location. The transects were located in areas where a comparison of the map that was surveyed in 1904 and 1910 (Darton, 1919) and aerial photographs that were taken in 1981 showed essentially no change in channel position. Auger holes along two of these transects showed the thickness of the metal-contaminated overbank sediments to range between 0.5 m and 1.5 m, and to average about 1.2 m. The thick overbank deposits are most common on floodplain surfaces on the insides of meander bends. More widespread sampling and chemical analysis indicate that surface contamination of floodplain deposits is common at a distance of about 150 m away from the present channel of the Belle Fourche River along the insides of meander bends (K. E. Goddard, U.S. Geological Survey, written commun., 1986).

Arsenic Concentration

The arsenic concentration of contaminated floodplain deposits along Whitewood Creek and the Belle Fourche River is used here to provide a measure of the degree to which the mine tailings discharged into Whitewood Creek are diluted by uncontaminated sediment as they travel downstream. Carbonates in the ore body and in other bedrock units in the watershed, plus the lime that is added to the mine tailings during milling, serve to prevent acid formation; they limit potential desorption of the metals associated with sediments in the study area into the ground water. Dissolved-metal concentrations in ground water generally are orders of magnitude lower than the particulate-metal concentrations of surrounding metal-contaminated soils in the floodplains of Whitewood Creek and the Belle Fourche River; this attests to the stable association of metals and sediments in many floodplain locations (Goddard, in press). Arsenic is of primary interest here, because it is the metal that is the most enriched in the contaminated floodplain sediments.

Total digestions, using hydrofluoric perchloric acid, were performed on at least four samples from soil pits in overbank sediments at one location along Whitewood Creek, and at three locations along the Belle Fourche River. Total digestions also were done on grain-size fractions, obtained using an air elutriator, of two samples from the Whitewood Creek site and two samples from the downstream-most site on the Belle Fourche River. In addition, 13 samples, for which total digestion data were obtained, were analyzed using a partial-digestion procedure that is not effective for sulfide minerals. All the metal analyses were obtained using inductively coupled plasma emission spectroscopy.

Arsenic concentrations obtained by the partial-digestion procedure were consistently more than 70% of the arsenic concentrations obtained using the total-digestion procedure for the 13 samples for which both types of data were obtained. This result suggests that much of the arsenic that is presently in the contaminated floodplain sediments no longer occurs in association with the primary sulfides that were the original source of the arsenic.

When comparing metal concentrations of sediments with different grain-size distributions, a standardization procedure to compensate for the greater association of metals with finer sediments commonly is employed (Horowitz, 1984). The usefulness of such a standardization procedure with regard to metal-contaminated overbank deposits along Whitewood Creek and the Belle Fourche River is unclear. Evidence of in situ oxidation suggests that much of the arsenic was transported to floodplains as primary sulfides in the mechanically crushed mine tailings. The extremely large absorptive capacity of the contaminated floodplain sediments with respect to arsenic (Wuolo, U.S. Geological Survey, written commun., 1986) suggests that, following its release by the oxidation of primary sulfides, the arsenic in the deposits may have been quickly reattached to the floodplain sediments in association with oxide and hydroxide minerals. Although analyses of grain-size fractions of floodplain sediments in the study area show a relation between grain size and arsenic concentration (Table 1), differences in metals concentrations of different grain-size fractions of the floodplain soils may reflect the efficiency with which the different grain sizes adsorb arsenic, rather than the arsenic concentration of the original material.

Table 1.    Arsenic Concentrations of Different Grain-Size Fractions of Overbank Deposits Downstream of Lead.

| River distance from Lead (km) | Onsite field description | Arsenic concentration, in µg for grain-size fractions | | | | |
|---|---|---|---|---|---|---|
| | | <4 | 4-16µ | 4-64µ | 16-64µ | >64µ |
| 15 | gray silt | 2,800 | -- | 750 | -- | 480 |
| 25 | orange-brown fine sand | 13,000 | 8,100 | -- | 4,000 | 3,800 |
| 160 | gray silt | 2,300 | 1,700 | -- | 810 | 490 |
| 160 | orange-brown fine sand | 2,700 | 2,400 | -- | 1,400 | 1,100 |

Despite considerable variability at a site, overbank-deposit arsenic concentrations that are not standardized for the effects of grain size do not appear to decrease with increasing distance downstream from Lead along the floodplains of Whitewood Creek and the Belle Fourche River (Fig. 3). This result is not surprising, because regional estimates of the amount of sediment likely to be carried in rivers the size of Whitewood Creek and the Belle Fourche River (Hadley and Schumm, 1961) are small in comparison to the tremendous volume of mine tailings that was discharged yearly. The lack of an adequate volume of uncontaminated sediment to dilute the metal-contaminated sediment appears to be reflected in the lack of significant downstream decreases in metals concentrations in contaminated sediments on the floodplains of Whitewood Creek and the Belle Fourche River.

## Floodplain Storage of Metal-Contaminated Sediments

The amount of metal-contaminated floodplain deposits resulting from the discharge of mine tailings over a 100-year period at Lead, South Dakota, is important from two perspectives. First, it provides a measure of the efficiency of floodplains of meandering streams in some environments in storing contaminated sediments. Second, it provides an illustration of the ability of contaminated sediment stored in floodplains to become a long-lived, nonpoint source of contaminants. The usefulness of these observations is perhaps questionable in environments such as the study site discussed here, where the biological effects of the metals associated with stored floodplain sediments are not clear. However, the transfer value of these observations of river systems to study areas where contaminants introduced into river systems are considerably more toxic, makes these observations more useful.

The mappable nature of the metal-contaminated floodplain deposits along the Belle Fourche River provides a basis to estimate the amount of mine tailings presently stored along the floodplain of the Belle Fourche River downstream of Whitewood Creek. Arsenic concentrations discussed previously suggest that the mine tailings were not significantly diluted during transport and deposition. Sediment storage in point bars and overbank deposits is considered here. The amount of metal-contaminated sediment stored in abandoned channels is not considered because it is minor in comparison to the amount of metal-contaminated sediment stored in point bars and overbank deposits in the floodplain of the Belle Fourche River.

A comparison of Darton's (1919) map, which was surveyed in 1904 and 1910, and aerial photographs that were taken in 1981, indicate that between 1910 and 1981, 2,740,000 m² of floodplain area were reworked by meander migration along a 30-km length of the Belle Fourche River floodplain (Fig. 2). This area of reworked floodplain is multiplied by four to extrapolate it over the 120-m length of the Belle Fourche River floodplain downstream of Whitewood Creek. The average thickness of the point-bar deposits is 1.5 m; an estimated 30% of the deposits by weight is too coarse

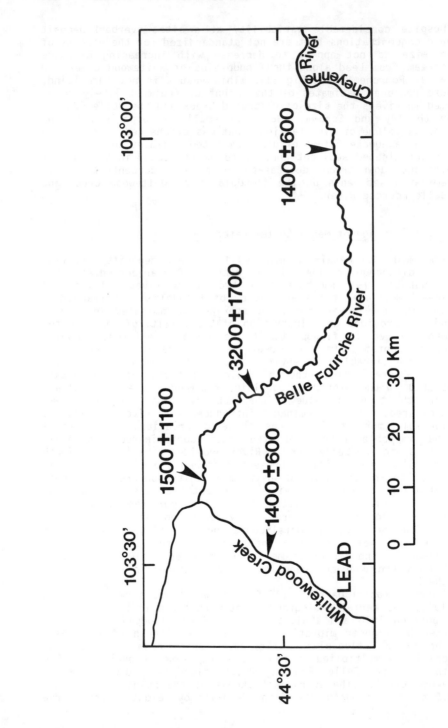

Figure 3.    Average arsenic concentrations (in µg/gm) of overbank sediments.

to have come from the mine tailings; and the bulk density of the deposits is assumed to be 1.6 gm/cm$^3$. Based on these observations and assumptions, an estimated weight of 18,000,000 Mg is obtained for mine tailings that presently are stored as point-bar deposits along the Belle Fourche River floodplain downstream of Whitewood Creek.

Observations made in soil pits and auger holes along the transects discussed previously indicate that downstream of Whitewood Creek, metal-contaminated overbank deposits averaging 1.2 m in thickness commonly extend 50 m away from the Belle Fourche River on the insides of meander bends. On a floodplain where the meander radius is usually about 500 m (Fig. 2), the area covered by a deposit that extends 50 m away from the channel on the inside of a meander bend is approximately equal to the product of the deposit width (50 m) and the channel length. The river distance from the mouth of Whitewood Creek to the mouth of the Belle Fourche River is 190 km. The bulk density of the deposits is assumed to be 1.6 gm/cm$^3$. Based on these observations and assumptions, an estimated weight of 18,000,000 Mg is obtained for mine tailings that are presently stored as overbank deposits along the Belle Fourche River downstream of Whitewood Creek.

The calculations presented are rough and likely to be revised. Thin overbank deposits that are farther than 50 m from the Belle Fourche River and filled abandoned meanders are not considered. The estimated amount of metal-contaminated overbank deposits along the Belle Fourche River downstream of Whitewood Creek is probably conservative because it is based on data collected along relatively stable river reaches; overbank deposition during an undetermined period of time prior to the study period along these reaches may have reduced the magnitude and frequency of further overbank deposition by elevating floodplain surfaces. Sediment storage along 43 km of the Whitewood Creek floodplain below Lead also is not considered. Despite these limitations, the approximate figures obtained clearly indicate that at least one-third of the mine tailings discharged over about 100 years at Lead are presently stored along 164 floodplain kilometers in a downstream direction.

Data discussed previously also can be used to estimate the length of time required for meander migration to remobilize the contaminated floodplain deposits along the Belle Fourche River downstream of Whitewood Creek. Data (Goddard, U.S. Geological Survey, written commun., 1986) indicate that arsenic contamination is common in floodplain sediments at a distance of 150 m from the present channel location along the insides of meander bends of the Belle Fourche River downstream of Whitewood Creek. On a floodplain where the meander radius is commonly on the order of 500 m (Fig. 2), the area covered by a deposit that extends 150 m away from the channel on the inside of a meander belt is approximately equal to one-half the product of the deposit width (150 m) and the channel length. The channel length of the Belle Fourche River downstream of Whitewood Creek is about 190 km. Therefore, an area of 14,250,000 m$^2$ of contaminated floodplain is estimated to occur along the Belle Fourche River downstream of Whitewood Creek. The extrapolated rate at which the Belle Fourche River floodplain downstream of Whitewood Creek has been reworked by meander migration between 1910 and 1981, based on a comparison of Darton's (1919) map, which was surveyed in 1904 and 1910, and aerial photographs that were taken in 1981, is about 150,000 m²/yr. At this rate, it would take about 100 years to rework the contaminated area of floodplain along the Belle Fourche River downstream of Whitewood Creek by meander migration. It is unlikely that the floodplain area that will be reworked by meander migration in the near future will coincide completely with the area affected by metal contamination. In addition, some of the remobilized contaminated sediment will be restored in floodplains downstream. Consequently, it is likely that the metal-contaminated floodplain deposits that resulted from the discharge of mine tailings at Lead will be a direct and indirect source of metals to the Belle Fourche River for centuries.

CONCLUSIONS

The transport and storage of sediment in river systems control the fate of contaminants in environments where contaminants are strongly associated with sediments. Contaminants that can travel in association with sediments include metals, radionuclides, polyclorinated biphenyls, and chlorinated-hydrocarbon pesticides. Most sediment is transported as suspended load; moderate streamflows with a recurrence interval of one to several years accomplish the majority of long-term sediment transport. In meandering river systems, sediment can be stored by point-bar deposition during meander migration, by overbank deposition, and by the filling of abandoned channels. Deposited sediment eventually is remobilized by meander migration. Sediment can be stored in floodplains for centuries or even millenia.

About 100,000,000 Mg of finely milled mine tailings were discharged into a small stream near Lead, South Dakota. The discharge of mine tailings resulted in an extensive floodplain deposit of metal-contaminated sediment that is visually distinguishable from premining alluvium along 160 floodplain kilometers downstream of Lead. Arsenic concentrations in the contaminated overbank sediments along Whitewood Creek and the Belle Fourche River do not decrease with increasing distance downstream of Lead. This pattern suggests that the amount of mine tailings discharged yearly was so large in relation to the background sediment load of these streams that the mine tailings were not diluted by uncontaminated sediment during transport and overbank deposition. Rough calculations of the amount of mine tailings that have been deposited along the Belle Fourche River floodplain indicate that at least one-third of the mine tailings discharged at Lead are presently stored within 160 floodplain kilometers of Lead. A comparison of the aerial extent of metal-contaminated sediments in the floodplain of the Belle Fourche River downstream of Whitewood Creek and the rate at which a segment of the Belle Fourche River floodplain was reworked by meander migration between 1910 and 1981, suggests that metal-contaminated floodplain sediments will be a source of arsenic to the Belle Fourche River for centuries.

ACKNOWLEDGMENTS

I acknowledge excellent field help from Sara L. Rathburn (U.S. Geological Survey) and useful discussions with Kimball F. Goddard (U.S. Geological Survey) and Sara L. Rathburn. Metals analyses that required total digestion were done by Paul H. Briggs (U.S. Geological Survey). Robert H. Meade (U.S. Geological Survey) and Diane M. McKnight (U.S. Geological Survey) provided helpful manuscript reviews.

REFERENCES

Allen, J. R. L. 1965. A review of the origin and characteristics of recent alluvial sediments. Sedimentology. 5:89-191.

Alloway, B. J., and B. E. Davies. 1971. Trace element content of soils affected by base metal mining in Wales. Geoderma. 5:197-208.

Bopp, R. F., H. J. Simpson, C. R. Olson, R. M. Trier, and N. Kostyk. 1982. Chlorinated hydrocarbons and radionuclide chronologies in sediments of the Hudson River and Estuary, New York. Environmental Science and Technology. 16:666-676.

Brice, J. C. 1973. Meandering pattern of the White River in Indiana--An analysis, p. 176-200. *In* M. Morisawa, ed., Fluvial geomorphology. Publications in Geomorphology. Binghamton, New York.

Bridge, J. S., and M. R. Leeder. 1979. A simulation model of alluvial stratigraphy. Sedimentology. 26:617-644.

Costa, J. E. 1975. Effects of agriculture on erosion and sedimentation in Piedmont Province, Maryland. Geological Society of America Bulletin. 86:1281-1286.

Daniel, J. F. 1971. Channel movement of meandering Indiana streams. U.S. Geological Survey Professional Paper 732-A, p. A1-A18.

Darton, N. H. 1919. Geologic Atlas of the United States, Newell Folio. U.S. Geological Survey, 1 volume.

Dunne, T., and L. B. Leopold. 1978. Water in environmental planning. W. H. Freeman, San Francisco. 818 p.

Everitt, B. L. 1968. Use of the cottonwood in an investigation of the recent history of a floodplain. American Journal of Science. 266:417-439.

Goddard, K. E. Composition, distribution, and hydrological effects of mine and mill wastes discharged to Whitewood Creek at Lead and Deadwood, South Dakota. U.S. Geological Survey Open-File Report, (in press).

Goddard, K. E. 1986. U.S. Geological Survey. Written communication.

Hadley, R. F., R. Lal, C. A. Onstad, D. E. Walling, A. Yair. 1985. Recent developments in erosion and sediment yield studies. UNESCO Technical Documents in Hydrology. 125 p.

Hadley, R. F., and S. A. Schumm. 1961. Sediment sources and drainage basin characteristics in Upper Cheyenne River Basin. U.S. Geological Survey Water Supply Paper 1531B, p. 137-198.

Happ, S. C. 1945. Sedimentation in South Carolina Piedmont Valleys. American Journal of Science. 243:113-126.

Homestake Gold Mine. 1976. 1876 Homestake Centennial 1976. Homestake Mining Company, Lead, South Dakota, 1 volume.

Hooke, J. M. 1980. Magnitude and distribution of rates of river bank erosion. Earth Surface Processes. 5:143-157.

Horowitz, A. J. 1984. A primer on trace metal-sediment chemistry. U.S. Geological Survey Open-file Report 84-709, 82 p.

Klimek, K., and L. Zawilinska. 1985. Trace elements in alluvia of the upper vistula as indicators of Paleohydrology. Earth Surface Processes and Landforms. 10:273-280.

Leopold, L. B., M. G. Wolman, and J. P. Miller. 1964. Fluvial processes in geomorphology. W. H. Freeman, San Francisco, 522 p.

Lewin, J., B. E. Davies, and P. J. Wolfenden. 1977. Interactions between channel changes and historic mining sediments, p. 353-367. *In* K. J. Gregory, ed., River Channel Change. John Wiley and Sons, Chichester.

Macklin, M. G. 1985. Flood-plain sedimentation in the upper Axe Valley, Mendip, England. Institute of British Geographers Transactions. 10:235-244.

Meade, R. H. 1982. Sources, sinks, and storage of river sediment in the Atlantic drainage of the United States. Journal of Geology. 90:235-252.

Nanson, G. C. 1980. Point bar and floodplain formation of the meandering Beatton River, northeastern British Columbia, Canada. Sedimentology. 27:3-29.

Nanson, G. C., and E. J. Hickin. 1986. A statistical analysis of bank erosion and channel migration in western Canada. Geological Society of America Bulletin. 97:497-504.

Rutherford, G. K. 1977. Anthropogenic influences of sediment quality at a source, p. 95-104. *In* H. Shear and A. E. P. Watons, eds., Workshop on the fluvial transport of sediment-associated nutrients and contaminants. Proceedings, 1976. Kitchener, Ontario.

Schumm, S. A. 1977. The fluvial system. John Wiley and Sons, New York, 333 p.

Trefry, J. H., S. Metz, R. P. Trocine, and T. A. Nelson. 1985. A decline in lead transport by the Mississippi River. Science. 230:439-441.

Webb, B. W., and D. E. Walling. 1982. The magnitude and frequency of fluvial transport in a Devon drainage basin and some geomorphological implications. Catena. 9:9-23.

Wolman, M. G., and J. P. Miller. 1960. Magnitude and frequency of forces in geomorphic processes. Journal of Geology. 68:54-74.

Wuolo, R. 1986. U.S. Geological Survey. Written communication.

# CHAPTER 13

# THE PARTICLE-SOLUTION CHEMISTRY OF LEAD IN ACIDIC LAKE SYSTEMS

Jeffrey R. White, Indiana University, Bloomington, Indiana

## ABSTRACT

The chemistry and transport of Pb is important because of its role as a toxicant to aquatic biota, as a sediment marker of anthropogenic activity, and in the Pb-210 form as a sediment-dating tool. Temporal and spatial variations in the partitioning of Pb between particles and solution (based upon sediment-trap derived values of distribution coefficient, $K_D$) were studied in acidic Darts Lake (Adirondack State Park, New York). The particle-solution chemistry of Pb was found to be strongly influenced by seasonal changes in pH (seasonally variable from 4.8 to 5.6) and retention of Al within the lake. The affinity of Pb for particles was most pronounced during stratification periods (winter and summer; log $K_D$ = 5.5). During spring snowmelt, Pb was more conservative with regard to surface interaction and retention in the lake (epilimnetic log $K_D$ = 4.1). Water column pools of lead were only 5% of surficial sediment pools (pool ratio = 0.05), supporting significant in-lake retention of Pb. Temporal changes in pH found in Darts Lake were less significant to the particle-solution chemistry of Zn and Mn, which are more conservative elements as suggested by lower average log $K_D$'s (3.7), and higher pool ratios (1.9 and 13, respectively). Sediment titration work supported the differences in affinities of trace metals for particles observed in Darts Lake. Results from this study may have implications for interpretation of sediment metal stratigraphy.

## INTRODUCTION

Recent studies of dilute acidic lake systems have reported elevated concentrations of trace metals and aluminum (Dickson, 1978; Johnson et al., 1981; Schofield, 1976; Henriksen and Wright, 1978; White and Driscoll, 1985). The acidification process has

*Chemical Quality of Water and the Hydrologic Cycle*, Robert C. Averett and Diane M. McKnight (Eds.) © 1987 Lewis Publishers, Inc., Chelsea, Michigan. Printed in the United States of America.

been shown to increase the mobilization of Al from the edaphic environment to the aquatic system (Johnson et al., 1981; Hooper and Shoemaker, 1985; Nordstrom and Ball, 1986). Trace-metal concentrations have also been shown to increase with decreasing surface water pH, and may be derived predominantly from enhanced mobilization from soils and sediments upon acidification (White and Driscoll, 1985; Schindler and Turner, 1982; Schindler et al., 1980; Henriksen and Wright, 1978).

The biological impact of metal ions in aquatic environments is linked to metal concentrations, pH, and the concentration of complexing organic and inorganic ligands. Toxicity of many trace metals to aquatic biota increases with decreasing pH, and decreasing ligand concentration (National Academy Press, 1985; Stumm and Morgan, 1981). In waters characteristic of clear, dilute, acidic lakes, pH often ranges from 4.5 to 5.5, and important metallic-ligand concentrations ($HCO_3^-$, $OH^-$, organic matter) are below levels yielding significant (>20%) complexation (Driscoll et al., 1984). A recent study involving American Flagfish (Hutchinson and Sprague, 1986) demonstrated that concentrations of Al, Zn, and Cu at approximately 57% of those typical for dilute acidic waters (pH = 5.8) resulted in complete reproductive failure in less than 7 days. Therefore, although trace metal concentrations in dilute acidic clear-water lakes are low in comparison to many lakes (e.g. systems close to population centers), low trace metal buffering capacity and low pH apparently greatly enhance the toxicity of these substances to the biological population.

The biogeochemical cycling of lead is of particular importance in dilute acidic systems not only because lead is a potential toxicant, but also since both stable and radioisotopic (Pb-210) forms of lead are used in sediment geochronology. There have been extensive paleolimnological investigations in recent years addressing the impacts of acidic deposition and atmospheric trace-metal inputs on the sediment record of dilute lakes (Galloway and Likens, 1979; Norton and Hess, 1980; Heit et al., 1981; Norton et al., 1980; Hanson et al., 1982; Nriagu et al., 1982; Kahl et al., 1984; Baron et al., 1986). These and other studies have relied on Pb-210 dating methods and have interpreted trace-metal stratigraphies as unbiased indicators of atmospheric trace-metal inputs to lakes and thus a measure of anthropogenic activity. The underlying assumptions in such chronological reconstructions are that: (1) diffusive transport of trace metals within the sediment is negligible, and (2) the rate of trace metal transport to the sediments is directly proportional to the rate of supply of atmospheric trace metals to the lake. Varifying such assumptions requires an understanding of the in-lake processing of trace metals.

Until recently, little information was available on the chemistry of trace metals within dilute acidic lakes. A number of studies have been conducted which use input-output budgets to determine changes in the chemistry of trace metals in lake systems (Beamish and VanLoon, 1977; Troutman and Peters, 1982; Davis and Galloway, 1982; Schut et al., 1986). Further information on the chemistry of trace metals in acidic systems has been derived from artificial acidification experiments using lake enclosures (Schindler et al., 1980; Hesslein et al., 1980; Jackson et al., 1980; Santschi et at., 1986). Paleolimnological studies have also been used to reconstruct historical changes in trace-metal chemistry of acidic lakes. The sediment record is an integrated signal of the inputs of trace metals to sediments, and provides no direct information on in-lake processes. Charles and Norton (1986) review the results from such studies. Thus, detailed studies of the cycling of trace metals within acidic lakes have been lacking.

A detailed study of the chemistry and cycling of Pb, Mn, Zn, Fe, and Al has recently been completed on a lake system in the North Branch Moose River area of the Adirondack Park in New York. This study has allowed examination of in-lake processes which affect the fate and transport of Al, Fe, dissolved organic carbon (DOC), and trace metals in dilute acidic systems. Results from this work have been published, in part, elsewhere (White and Driscol, 1985; White and Driscoll, 1987a; Driscoll and Schafran, 1984; Schafran and Driscoll, 1987; White and Driscoll, 1987b). The purpose of this chapter is to examine the particle-solution chemistry of Pb in acidic lakes, and to demonstrate how Pb differs significantly from other trace metals (Zn, Mn). Improving our understanding of how trace metals differ in their interaction with particles will help in the interpretation of sediment stratigraphy and advance modeling of trace-metal chemistry in aquatic systems.

METHODS

Site and Field Sampling Description

A detailed study of the Darts Lake system was conducted from October 1981 to November 1982. Darts Lake (43°47'N, 74°51'W) is located in the North Branch of the Moose River, in the Adirondack Region of New York State. The watershed (107 km$^2$) is forested except for small areas of exposed bedrock (predominantly granitic gneiss (Isachsen and Fisher, 1970)). Secondary growth hardwood vegetation predominates in soils composed of mainly thin till (<3m) (Newton et al., 1987). Detailed descriptions of regional forest vegetation (Cronan et al., 1987), Darts Lake hydrology (White and Driscoll, 1985), surficial geology (Newton et al., 1987), general water chemistry (Driscoll et al., 1987), and aluminum chemistry (Schafran and Driscoll, 1987) have recently been published.

The field sampling program used in the study of Darts Lake has been described in detail previously (White and Driscoll, 1985). The study was based on an inlet-outlet, water column, and sediment trap sampling program involving collections every 2 to 3 weeks for a period of 1 yr. Samples of the water column were collected from the pelagic sampling station (major lake depression 15 m maximum depth) at 2-m intervals. Sedimenting materials were collected in sediment traps (triplicate collectors set at 6 m and 14 m) located at the pelagic sampling station. Sediment trap design followed recommendations of Bloesch and Burns (1980) (aspect ratio = 9).

Sediment cores were also collected in triplicate from profundal sediments at the pelagic sampling station, using a 5-cm gravity coring device. Sediment from cores was sectioned into 1-cm intervals and analyzed for metal content (Al, Fe, Mn, Zn, Pb) to estimate surficial sediment pools.

## Analytical Methods

Field measurements included stream flow, pH, dissolved oxygen (Standard Methods, 1976), water column temperature (thermistor), monomeric Al forms (Barnes, 1975; Driscoll, 1984), and daily on-site climatological data. Other stream and water column chemical parameters monitored included dissolved organic carbon (DOC) (Menzel and Vaccaro, 1964) and dissolved metal species (Fe, Mn, Zn, Pb). To determine "dissolved" metal concentrations, high-speed centrifugation (5,720 G for 30 min) was used to separate particulate-bound metals from dissolved forms (Salim and Cooksey, 1981) and the supernatant acidified (0.5% v/v Ultrex $HNO_3$). Particulate material collected by sediment traps was analyzed for metal content (Al, Fe, Mn, Zn, Pb), particulate carbon (Menzel and Vaccaro, 1964), and suspended solids (Standard Methods, 1976). The acid-labile metal content of sediment trap collections was determined by short term (1 hr) acid digestion at pH 1.0. Dissolved concentrations of each metal were subtracted from total acid-leachable concentrations to determine acid-labile particulate fractions (White and Driscoll, 1985). The operationally defined acid-labile fraction probably reflects recently sorbed metals rather than metals present within a mineral matrix (Hsu, 1977). This study focused on acid-labile metals, since the acid-labile fraction is likely to be most involved in chemical cycling within the lake. Similar methods of acid digestion (including hydrogen peroxide treatment) were used to quantify the acid-labile pools of trace metals in surficial sediments. Total metal content of the sediment was determined by summing each individual fraction of the sequential extraction (White, 1984).

In addition to field sampling, laboratory studies were conducted to investigate the importance of pH in trace-metal sorption to particulates from the lake system. Batch adsorption experiments were performed using fresh wet surficial sediment collected from the pelagic sampling station (dry weight = 0.1 g $l^{-1}$). Suspensions of sediment in deionized water (ionic strength adjusted with $10^{-3}$ M $NaNO_3$) received 1 µmol $l^{-1}$ of trace metal and were adjusted with NaOH and $HNO_3$ to yield a range in pH from 3.0 to 6.0. After 24 hr, particulates were separated by centrifugation (5,720 G for 30 min) and supernatants analyzed for pH, DOC, Al, and trace-metal concentrations.

Trace-metal analyses were conducted by atomic absorption spectrophotometry using graphite furnace atomization. The total error associated with analyses of soluble concentrations of metal species and the fractional uncertainties for sediment trap collection and analysis are presented elsewhere (White and Driscoll, 1985; White and Driscoll, 1987b).

Computative Methods

The gross deposition rate of substances into sediment traps ($D_g$) was calculated from the mass of particulate-associated substances accumulated during the collection period, expressed on an areal basis:

$$D_g = C_m V/(A_t \ t) \tag{1}$$

where $D_g$ is equal to the gross deposition rate (mmol $m^{-2}d^{-1}$), $C_m$ is the sediment trap concentration (mmol $cm^{-3}$) of particulate-associated acid-labile metal, V is the sediment trap volume ($cm^3$), $A_t$ is the area of the sediment trap opening ($m^2$), and t is the collection period (d).

Another way of expressing the rate of deposition of trace metals from the water column is through a specific rate of deposition. The $D_g$ value is normalized to the water column pool of metal through which particulates are depositing. This value is then expressed on the basis of time, and referred to as the specific turnover rate (T, days):

$$T = M/D_g \tag{2}$$

where M (mmol $m^{-2}$) is the quantity of soluble metal per $m^2$ above the sediment trap, and $D_g$ is the gross deposition rate (Equation 1). The specific turnover rate, T, represents the time required to deplete the water column pool of soluble metal at an average rate of deposition ($D_g$). Nonconservative elements have relatively short specific turnover rates, while trace metals which are more conservative with respect to particle-solution interactions and adsorption are likely to exhibit higher values of T. The above computational methods are particularly useful since inputs of trace metals to Darts Lake were largely soluble species entering through the inlet. Atmospheric deposition directly to the surface of the lake represented a minor componenet of trace-metal input (Pb<10%; Zn<2%; Mn<0.5%). Particulate concentrations of trace metals in the inflow were characteristically less than 5% of the total trace-metal content, with suspended solids concentrations less than 1.0 mg $l^{-1}$ (White, 1984).

The distribution coefficient ($K_D$) is a parameter which is indicative of the intrinsic affinity of a metal for particulate matter in an aqueous system. It is a measure of the equilibrium mass distribution of an adsorbate (trace metal) between solution and solid phase adsorbent (particulates). Using sediment-trap data of particulate concentrations and concentrations of acid-labile trace metals, water column values of $K_D$ can be determined as follows:

$$K_D = C_m/C_p M_i \ . \tag{3}$$

$K_D$ ($cm^3 \ g^{-1}$) is a function of the concentration of particulate-bound trace metal ($C_m$), particulate concentration ($C_p$, total solids in g $cm^{-3}$ (Standard Methods, 1976)), and the volume-weighted mean concentration of soluble trace metal in the water column above the sediment trap ($M_i$; mmols $cm^{-3}$). The distribution coefficient was also calculated for the laboratory batch adsorption studies using the mass of particulate-bound metal associated with a known mass of adsorbent at equilibrium. Particulate-bound metal is determined by the difference between the soluble metal concentration added to suspension and the soluble metal concentration at equilibrium. This approach assumes that the trace metal adsorption process is not influenced significantly by background concentrations of particulate-bound trace metals present in the sediment being used as an adsorbent. The surficial sediment used in the laboratory experiments represented less than 0.2 μmols of acid-labile trace metals (Pb<0.1 μmol; Zn, Mn<0.2 μmol).

## RESULTS AND DISCUSSION

### Changes in Water Column Chemistry

Darts Lake is a dimictic system, exhibiting complete turnover in late April and again in October. Drainage input of solutes is likely to be controlled by output from Big Moose Lake upstream, since over 90% of the Darts Lake watershed is common with the Big Moose Lake watershed. Although it is not located in the upper portion of the North Branch of the Moose River watershed, Darts Lake exhibited significant changes in water chemistry over short time periods. Changes in pH within the lake occurred rapidly during spring snowmelt as concentrated mineral acidity (mainly nitric acid) was released from melting snowpack. The pH then increased significantly due to biological denitrification in the hypolimnion during stratified periods (winter and summer hypolimnion) and photosynthesis (summer epilimnion) (Fig. 1). The seasonal changes in acid-base chemistry of this lake are extensive and much of the variation in hydrogen ion and aluminum acidity can be attributed to changes in nitrate concentrations, rather than variations in sulfate, chloride, or organic anion concentrations (Driscoll and Schafran, 1984).

The variable most important to metal chemistry in dilute aquatic systems is pH. Temporal and spatial changes in pH observed in Darts Lake also have important implications for the chemistry of Al and trace metals. On an annual basis Darts Lake was a net sink for Al. However, significant seasonal variation was associated with the fluxes of Al within the lake system. Within the lake, aqueous Al was converted to particulate Al and deposited to the sediments, particularly during summer low-flow periods when pH was elevated (5.4-5.6) (Driscoll and Schafran, 1984). The nonconservative behavior of Al (and other elements, such as Fe, Mn, and Si) might significantly impact the fate of trace metals in the system, since metal oxides are excellent adsorbents for trace metals (James and Healy, 1972; Hohl and Stumm, 1976; Davis and Leckie, 1978; Millward and Moore, 1982). The range in pH observed in Darts Lake includes significant portions of the typical adsorption edges of a number of trace metals (Stumm and Morgan, 1981). The adsorption and affinity of trace metals on colloidal $Al(OH)_3$ exhibited the

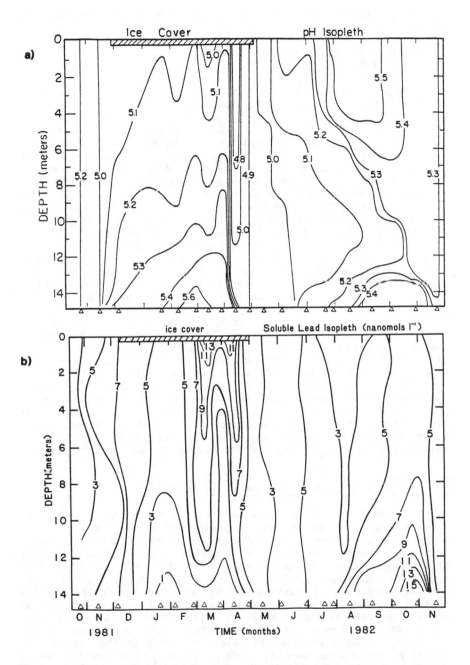

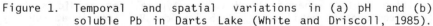

Figure 1. Temporal and spatial variations in (a) pH and (b) soluble Pb in Darts Lake (White and Driscoll, 1985).

following pattern with decreasing pH (Kinniburgh and Jackson, 1981):

metal:    Cu < Pb < Zn < Ni < Cd ~ Co
$pH_{50}$:   4.8    5.2   5.6   6.3    6.6

where $pH_{50}$ is the pH at which 50% of the soluble metal has been absorbed (based on $10^{-1.3}$ M absorbent and $10^{-3.9}$ M metal). Although the adsorbent and metal concentrations in Darts Lake are considerably lower than those used in laboratory experiments, trace-metal adsorption to hydrous oxides may be an important in-lake process. The temporal and spatial variations in pH observed in Darts Lake (Fig. 1) may, therefore, be significant to adsorption of Cu, Pb, and Zn on oxide particles in the system.

Chemical transformations within the water column of Darts Lake appeared to play a significant role in Pb cycling. Note that during winter stratification Pb concentrations declined in the hypolimnion, while in the vernal hypolimnion Pb concentrations were enriched, from 3 nmols $l^{-1}$ during turnover to 15 nmols $l^{-1}$ at late stratification, as a result of release of DOC from the sediment during periods of low dissolved oxygen (<0.1 mM $O_2$) (White and Driscoll, 1985) (Fig. 1). Concentrations of Fe were also highly enriched in the vernal hypolimnion (from 1 μmol $l^{-1}$ in the epilimnion to >28 μmols $l^{-1}$ in the hypolimnion), while Mn concentrations increased only slightly (from 1.2 to 2.2 μmols $l^{-1}$) (White and Driscoll, 1987a), and Zn concentrations exhibited no significant change (White and Driscoll, 1987b). Spring melt waters did exhibit increases in the concentration of Pb, Zn, and Mn, when pH was lowest (4.8-5.0). Pb was the most variable trace metal on a seasonal basis within the water column. This variability may be due partly to increased mobilization to the lake during the spring pH depression, and due to release from sediments during summer accumulation of DOC in the hypolimnion. In addition, metal scavenging by particulates may have contributed to a decrease in epilimnetic concentrations of Pb during summer stratification.

## The Role of In-Lake Retention of Al in Darts Lake

A dominant cation in the waters of Darts Lake is Al. Loss of Al from the water column during periods of elevated pH coincided with supersaturation with respect to hydrous aluminum oxide (gibbsite), suggesting that the formation of particulate-inorganic Al was responsible for the increase in depositional loss of Al from the Darts Lake water column (Driscoll and Schafran, 1984). The in-lake formation of particulate Al appeared to facilitate the vertical deposition of Pb, particularly in the upper waters. Al was a major constituent of material collected in the sediment traps (13 ± 14% on dry mass basis). The nature of the adsorbent surface is likely to be a function of water chemistry in the system. It should be noted that pure Al oxide particles would normally have a positive surface charge at pH values typical for acidic systems (electrophoretic mobility = +5.5 μm cm $V^{-1}$ $sec^{-1}$ at pH 5) (Letterman et al., 1982). Electrophoretic mobility results from particles collected in Darts Lake sediment traps were markedly different (of opposite charge and lower charge density) from synthetic Al oxides, showing no significant variation with pH, time, or depth (mean = -1.75 ± 0.16 µm cm $V^{-1}$ $sec^{-1}$) (White and Driscoll, 1985). The surface charge results from Darts Lake sediment traps suggest that the particulate matter is coated with organic surface films derived from DOC in the water (Hunter and Liss, 1979; Tipping et al., 1981; Davis and Gloor, 1981). It is likely that inorganic hydrous oxides (coatings) and adsorbed-coprecipitated organic matter constitute the solid matrix of particulate material in Darts Lake (Table 1).

Results from multiple linear regression of Pb deposition to sediment traps as a function of the deposition of particulate constituents demonstrated that Pb was strongly correlated with the deposition of Al (partial r = 0.85, p<0.004; Fig. 2) (White and Driscoll, 1985). Although there is an abundance of Fe and organic carbon in sedimenting material, particularly in the hypolimnetic trap, regression results suggest that these materials are not important to the deposition of Pb in Darts Lake. Cycling of redox-sensitive oxides within the lake would also be expected to play a role in trace-metal chemistry. The mass of Mn was very low in Darts Lake sediments (<2 µmols $g^{-1}$), and did not exhibit significant internal cycling (White and Driscoll, 1987a). In contrast, Fe was highly enriched within Darts Lake sediments (>1 mmol $g^{-1}$) (White, 1984) and in pore waters (>0.5 mM in nearby Big Moose Lake) (Gubala and White, Chapter 14). Extensive redox-controlled cycling of Fe was observed in the vernal hypolimnion of Darts Lake, however, no significant impact on trace-metal cycling was observed. Therefore, the seasonal retention of Pb within the lake was predominantly dependent on the formation of particulate Al within the water column. Multiple linear regression results for Zn and Mn deposition as a function of Al, Fe, and organic matter deposition indicated no significant relationships (except between $D_g$-Mn and $D_g$-Fe in the 14-m traps; r = 0.80, p<0.001). Sediment trap results suggest, therefore, that Pb is scavenged from the water column by Al, while Zn and Mn do not exhibit significant scavenging.

Table 1. Ratios of Elemental Components to Total Suspended Solids in Sediment Material from Darts Lake[1].

| Component | 6 m depth | | 14 m depth | |
|---|---|---|---|---|
| | $\bar{x}$ | C.V. | $\bar{x}$ | C.V. |
| Al | 4.8 | 106 | 3.0 | 53 |
| Fe | 2.2 | 177 | 4.3 | 91 |
| Mn | 0.05 | 113 | 0.03 | 85 |
| Organic C | 59.3 | 110 | 78.3 | 54 |

[1]Mean values ($\bar{x}$) are expressed as micromoles per milligram of suspended solids with coefficient of variation (C.V.). C.V. = (S.D./$\bar{x}$) x 100 (White and Driscoll, 1985).

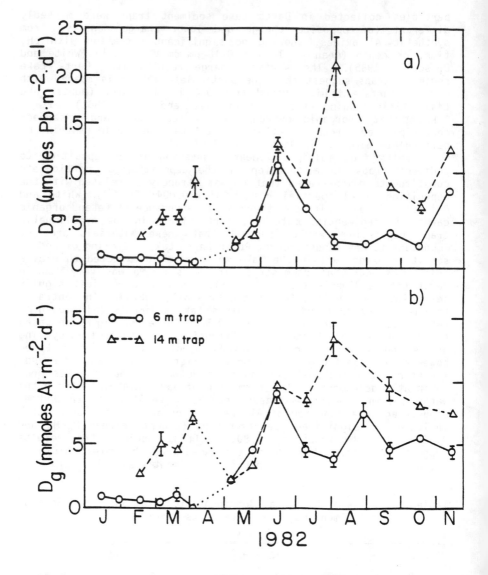

Figure 2.   Temporal changes in gross deposition of (a) Pb and (b) Al in Darts Lake (White and Driscoll, 1985).

## Temporal Trends in $K_D$-Pb in Darts Lake

An important measure of inherent affinity of trace metals for natural particulate matter in lakes is the distribution coefficient, $K_D$ (Equation 3). Changes in the value of $K_D$ suggest significant changes in the particle-solution behavior of a trace metal. In a natural system, major changes in water chemistry or the chemical nature of the adsorbent may contribute to a shift in the distribution coefficient. Values of $K_D$-Pb changed little over most of the study period in Darts Lake for either the upper waters (<6 m), or the entire water column (14 m) (Fig. 3) (note that the values are presented as logarithms, log $K_D$). One exception was a rapid decrease in log $K_D$, from 5.5 to 4.1, observed during spring snowmelt, when pH decreased to 4.8. Solution pH is an important controlling variable in the surface reactions of hydrous oxides (Stumm and Morgan, 1981). A decrease in solution pH has been shown to reduce adsorption of Pb on Al oxides (Hohl and Stumm, 1976), therefore the observed decrease in log $K_D$-Pb during snowmelt is consistent with the surface chemistry of metal oxides. The temporal trends in log $K_D$-Pb may also be explained by changes in the nature of the particulate material depositing through the water column. A decrease in the amount of oxide present per unit mass of particulate (adsorbent) during spring snowmelt may account for the decline in log $K_D$-Pb, and would be consistent with Al solubility (Fig. 3). However, only the sediment trap located in the upper waters exhibited a decrease in log $K_D$. This suggests that a change in water chemistry (the collection of acidic melt waters along the ice-water interface) is the more likely explanation. A decline in log $K_D$ during the spring period was also observed for Mn (White and Driscoll, 1987a) and Zn (White and Driscoll, 1987b). However, temporal trends in $K_D$ were generally more variable for these elements, due partly to significantly lower rates of deposition ($D_g$) to sediment traps (larger errors associated with flux determinations) (White, 1984).

## The Role of pH in Affecting $K_D$-Pb in Darts Lake

Batch acid-base titration experiments using surficial sediment from Darts Lake were conducted to investigate more carefully the role of pH in affecting particle-solution chemistry. Changes in the water chemistry of sediment suspensions as a function of pH (3-6) were extensive (Fig. 4). Note that soluble Pb, Zn, and Mn concentrations increased 2- to 4-fold with decreasing pH below ambient (pH~5.2) (Fig. 4). Low pH also dissolved Al from the sediment matrix (Fig. 4). However, at elevated pH (>5), Pb and Al were released from the sediment, while Zn and Mn concentrations decreased over the entire pH range studied. The actual metal concentrations observed depended on the sediment concentration and the amount of metal added, but the trends as a function of pH were independent (White, 1984). The increases in soluble Pb and Al at higher pH coincided with release of DOC from the suspended sediment, suggesting that DOC complexation of Pb and Al may enhance desorption (Fig. 4). Mn and Zn did not respond to the release in

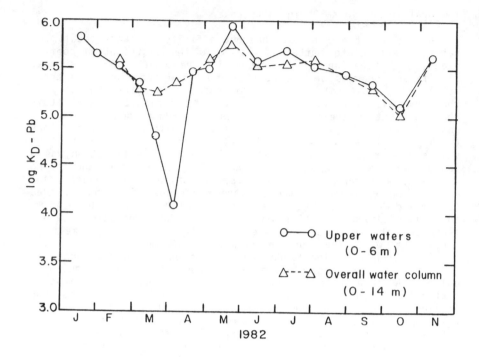

Figure 3.    Temporal changes in the distribution coefficient for Pb
            in Darts Lake as determined by sediment trap collec-
            tions.

DOC.  The mechanism responsible for release of DOC from sediment
particles with increasing pH is not well understood.  However,
partitioning of DOC on hydrous oxides has been shown to be a pH-
dependent process, exhibiting maximum absorption at pH~5.5 (Davis,
1982).
      Results from the batch titration experiments can also be
expressed in terms of $K_D$ values to demonstrate how trace metal
affinity for natural particles changes with pH.  In-lake values of
log $K_D$ based upon sediment trap collections are superimposed on
titration data (Fig. 5).  Note that Pb appeared to reach a maximum
in affinity for Darts Lake sediment particles at pH~5.5 (Figs. 4
and 5).  The maximum in $K_D$-Pb corresponds with the maximum in DOC
adsorption to hydrous Al oxides reported by Davis (1982).  In-lake
values of $K_D$-Pb were consistent with laboratory results (Fig. 5).
Similar analyses were conducted for Zn (White and Driscoll, 1987a)
and Mn (White and Driscoll, 1987a), and demonstrated that $K_D$'s
increased continuously as a function of pH, and the maximum
affinities of these elements for Darts Lake sediment particulates
occurred at pH values >6 to 6.5.  These field and laboratory
results using natural particulate material are corroborated by
trace metal adsorption to pure metal oxides (Hohl and Stumm, 1976;

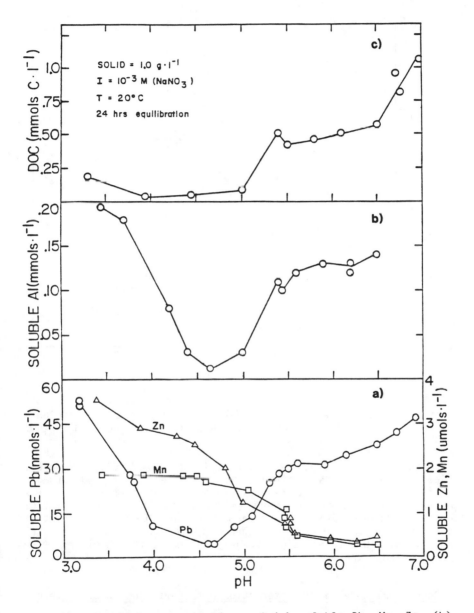

Figure 4.    Chemical concentrations of (a) soluble Pb, Mn, Zn, (b) soluble Al, and (c) DOC following acid-base titration of Darts Lake sediment.

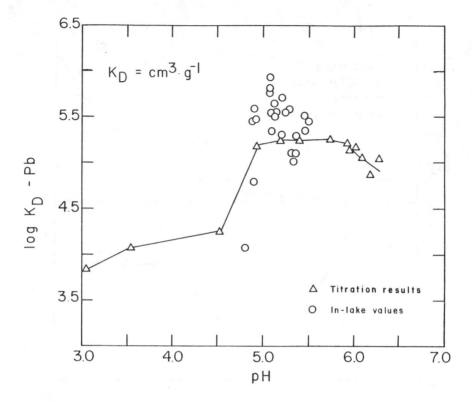

Figure 5.    The in-lake and experimental distribution coefficients of Pb as a function of the pH of the suspending medium.

Kinniburgh and Jackson, 1981; Millward and Moore, 1982). However, clean oxide systems (without the presence of natural organic matter) do not exhibit desorption of Pb at higher pH as was observed with natural sediments (Figs. 1 and 4). It appears from both laboratory adsorption studies and Darts Lake chemistry that the interactions of DOC with particulates are important to Pb partitioning but do not significantly influence the chemistry of Zn and Mn within the system.

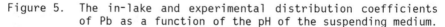

A Conceptual Model of Pb Interaction with Particles in Darts Lake

Based upon results from water column, sediment trap, and sediment titration analyses, Pb interactions with naturally occurring particles in Darts Lake are summarized by a simple conceptual model (Fig. 6). The nature of the substrate is likely to be variable in composition and mass, and might include algal cells, diatom frustules, clay particles, and other forms of detritus and suspended minerals. These surfaces may act as nucleation sites for adsorption-precipitation of Al, Fe, Mn, and organic matter upon/within which Pb adsorption-coprecipitation may occur (James and Healy, 1972; Hohl and Stumm, 1976; Davis and Leckie, 1978; Millward and Moore, 1982). Colloidal organic matter may also combine with Al, Fe, Mn, and trace metals through complexation, adsorption and coagulation reactions (Schwertmann, 1966; Kodama and Schnitzer, 1977; Davis and Gloor, 1981; Davis, 1982). Based upon the composition of sediment trap collections and electrophoresis results (White and Driscoll, 1985), the solid matrix dominant in the water column of Darts Lake was probably composed of inorganic hydrous oxides (largely Al) and adsorbed-coprecipitated organic matter (Table 1). Interactions of Pb with this matrix appear to be sensitive to pH fluctuations in a manner depicted in the conceptual model shown in Figure 6.

The middle panel of Figure 6 represents conditions of Pb at the substrate interface under pH 4.6 to 5.0. Surface species could include adsorbed-precipitated hydrous Al oxides (Al-OH) and organic matter (ORG), in addition to coprecipitated Al and organic matter (Al-ORG). Pb may interact with the substrate through adsorption-precipitation as organic complexes (ORG-Pb), or incorporated in a hydrous Al oxide matrix (Al-OH-Pb).

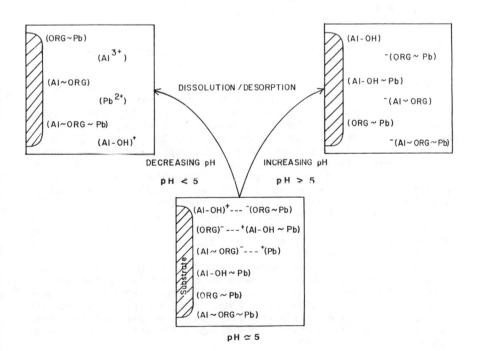

Figure 6.  A conceptual model of the surface reactions of Pb as a function of pH changes in dilute acidic waters.

The effect of decreasing pH on Pb interaction with the substrate is displayed in the left panel of Figure 6. Dissolution of hydrous oxides, and proton exchange reactions are likely to cause desorption-dissolution of inorganic forms of Al and Pb. For example, protonation of organic acid functional groups on particulate surfaces would decrease the binding capactity for Al and Pb, resulting in the desorption of Al and Pb to the bulk solution. Also, dissolution of hydrous oxides would cause release of coprecipitated Pb to solution, in addition to Al. With increasing pH, as shown in the right panel of Figure 6, release of organic complexes of Pb and Al are hypothesized to occur, leading to increases in DOC, organic Al, and Pb in the bulk solution. Dissolution of the inorganic hydrous oxide matrix by dissolved organic matter is also possible at elevated pH (Schwertmann, 1966; Perdu et al., 1976; Kodama and Schnitzer, 1977), and may account for observed increases in organic Al and Pb in the vernal hypolimnion of Darts Lake (Fig. 1) and in sediment titration experiments (Fig. 4). DOC that has been released from the sediment matrix may also complex trace-metal ions associated with surface exchange sites, resulting in desorption of metal ions (particularly Pb) from the surface into solution. Thus, DOC may either serve to dissolve the solid phase and/or desorb particulate-bound metal ions; either mechanism can account for mobilization of Pb and Al from particulates with increasing pH.

## A Summary of Parameters Depicting Conservative Behavior of Trace Metals in Acidic Systems

A number of parameters have been presented in this study which indicate the extent of particle-solution interaction of trace metals, and thus provide a method of comparing conservative behavior of elements. Table 2 presents mean values of a set of parameters for Pb, Zn, and Mn derived from Darts Lake, Big Moose Lake, and laboratory sediment titration experiments. The annual mean turnover rate (T; Equation 2) for Darts Lake represents the time (days) required to deplete the water column pool of a trace metal at a rate of deposition determined by sediment traps set at 14 m (whole water column rates). Note that the mean value of T for Pb is considerably faster (50-60 times) than Zn and Mn. Based on the mean values of $D_g$ for these trace metals normalized to the soluble pool in the water column, Pb behaves less conservatively in the system than Zn and Mn.

Mean $K_D$ values determined from the lake water column demonstrate that at pH 5 (the mean pH of Darts Lake) the average affinity of Pb for natural particulates is approximately 1.8 orders of magnitude greater than Zn and Mn. Sediment titration results for $K_D$ at pH 5 are consistent with these in-lake sediment trap results (1.3 and 1.7 orders of magnitude difference, respectively) (Table 2).

Table 2.   A Comparison of Parameters Measuring Conservative
Behavior of Trace Metals.

| Parameter | Pb | Zn | Mn |
|---|---|---|---|
| T (days) | 76 | 4,800 | 3,600 |
| log $K_D$-exp.[1] (pH 5.0) | 5.2 | 3.9 | 3.5 |
| log $K_D$-lake[2] (pH 5.0) | 5.5 ± 0.2 | 3.7 ± 0.2 | 3.7 ± 0.3 |
| log $K_D$-sed.[3] (pH 6.2) | 4.3 | 3.8 | 3.2 |
| Pool ratio | 0.05 | 1.9 | 13 |

[1]Distribution coefficients based on laboratory sediment
titration studies at pH 5.0.
[2]Mean (± S.D.) distribution coefficients determined from
sediment trap collections over the entire water column of
Darts Lake (mean pH = 5.0).
[3]Distribution coefficients determined from pore water samples
collected from the sediments of Big Moose Lake (pore water
pH = 6.2).

In comparison to particle-solution behavior of trace metals in
the water column of acid systems, $K_D$ data are also presented
(Table 2) from pore waters in Big Moose L (same watershed as Darts
Lake).   Since the pore waters are at higher pH (pH ~6.2) than the
water column in these dilute acidic lakes (Gubala and White,
Chapter 14), sediment titration results (Fig. 5) suggest a lower
affinity of Pb for particulates.   The observed mean value of log $K_D$
for Big Moose Lake pore water was lower than predicted by sediment
titration results, however DOC levels in the pore water were
significantly greater than were present in titration samples
(Gubala and White, Chapter 14).   Affinity of Zn and Mn for
particles were less impacted by the change in chemical conditions
from lake water to pore water as predicted by results from
titration experiments (Fig. 4).   Results of distribution
coefficient determinations from various environments (water column,
sediment pore water, sediment titrations) support the notion that
Zn and Mn are considerably more conservative in partitioning
between bulk solution and natural particulates than is Pb under
more acidic conditions (pH 3-6.2).   At pH >6, Zn and Mn exhibit a
greater affinity for natural particles than Pb (White and Driscoll,
1987b; White and Driscoll, 1987a).
The final parameter which addresses conservative behavior of
trace metals in lake systems is the pool ratio (Table 2; presented
as an annual mean).   The pool ratio parameter is the quotient of
the water column pool (mmols $m^{-2}$) and the surficial (top cm)
sediment pool (mmols $m^{-2}$).   Note that the more conservative Zn and
Mn species exhibited lower sediment concentrations as compared to
the pool present in the water column (pool ratios = 1.9 and 13,
respectively), while Pb concentrations in surficial sediments were

significantly greater in importance as a pool within the system than the soluble Pb component in the water column (pool ratio = 0.05). The large difference in pool ratio further supports the contention that Pb differs significantly in its particle-solution chemistry than Zn and Mn in dilute acidic systems.

## SIGNIFICANCE OF VARIATIONS IN THE PARTICLE-SOLUTION CHEMISTRY OF TRACE METALS

Differences in the partitioning of trace metals between particles and the bulk solution as influenced by pH may be a significant process in a large number of lakes. The likely population of lakes which might undergo large temporal fluctuations in pH similar to Darts Lake (pH 4.8-5.6) would include lakes which have alkalinities less than or equal to 10 $\mu$eq $1^{-1}$ (Darts Lake = 18 ± 12 $\mu$eq $1^{-1}$). Drainage lakes located in regions where melting snow (with large amounts of stored mineral acidity) is a significant hyrologic and chemical event, and which contain low alkalinities, may be particularly susceptible to large temporal fluctuations in pH. A recent survey of eastern lakes estimated the population of lakes with alkalinities at or below 10 $\mu$eq $1^{-1}$ during fall turnover to be 537 lakes in the northeast (a region of high acidic deposition), and 340 lakes in the upper midwest (Linthurst et al., 1986). Deposition of trace metals to sediments in these lakes may vary significantly with season. The seasonal variation is likely to depend upon changes in the rate of deposition of adsorbents (i.e., Al) and shifts in $K_D$. Seasonal changes in the rate of deposition of trace metals to sediments in lakes are important to studies of trace metal budgets in lakes. If a lake experiences large changes in the degree of retention of a particular element on a seasonal basis, chemical bugets derived from a portion of the annual cycle in that lake will be susceptible to significant error. Net retention of trace metals within lakes may actually be enhanced by seasonal fluctuations in pH. Periods of pH decline in spring serve to mobilize potential adsorbents (i.e., Al) and trace metals to the lake from the surrounding watershed, which is followed by an increase in pH and enhanced deposition of adsorbents and trace metals during summer stratification. During the course of watershed acidification (involving a number of decades, for example), the seasonal pattern of trace-metal deposition in a lake is likely to pass through different stages which depend on temporal and spatial patterns of pH, Al, and trace-metal chemistry. The pattern, on an annual basis, would be: (a) a continuous increase in annual deposition rates (of Pb, in particular) as the lake acidifies through a stage where Al is mobilized from the soil environment and subsequently retained in the lake during the summer low-flow period (a period of maximum pH and metal deposition); followed by (b) a continuous decrease in annual deposition rates as the retention of Al within the lake declines in response to a decrease in the summer-maximum pH, during advanced stages of acidification.

Over a large time frame (decades), changes in the efficiency of trace-metal scavenging within a lake may cause a consistent and significant shift in the rate of trace metal accumulation in sediments. Changes in the accumulation rates of trace metals within a lake that result from changes in lake chemistry have important implications for the interpretation of sediment stratigraphy. For example, Pb stratigraphy is used for dating sediments (Pb-210) and for the investigation of historical changes in atmospheric deposition of Pb (Heit et al., 1981; Hanson et al., 1982; Kahl et al., 1984; Baron et al., 1986). Current interpretation of sediment metal stratigraphy does not consider the potential for temporal variations in trace metal retention within lakes.

Models have been developed to date sediments using Pb-210 activity as a function of depth (Robbins, 1978). The Constant Initial Concentration model (CIC) assumes that the specific activity of Pb-210 in surficial sediment is constant over time. The flux of Pb-210 to sediments is assumed to be linearly dependent on the sedimentation rate, such that each unit mass of sedimenting material absorbs a specific amount of Pb-210 from the water column, which is constant over time. The fundamental assumption of the Constant Rate of Supply model (CRS) is that the flux of Pb-210 to sediments is constant and independent of sediment deposition rates. A change in the $K_D$ for Pb-210 over time as a lake acidified would violate the assumptions of the CIC dating model, since the uptake of Pb-210 activity from the water column by sedimenting particles would shift as the pH regime in the lake changed. If the rate of Pb-210 deposition were to change as a lake acidified, the CRS model would also be invalid. Although no data are available on temporal changes in Pb-210 deposition and partitioning in acidic lakes, our results for Pb are probably a good estimate of the response of Pb-210 in dilute acidic lakes.

To accurately predict the error in Pb-210 dating resulting from temporal changes in Pb deposition will require more data on Pb chemistry in acidic lakes, and sophisticated modeling and simulation studies. These studies are currently ongoing, and no concrete results are available; however, it is speculated that significant dating errors would occur in lakes that may have experienced a 20% change in Pb retention in the last few decades. The magnitude of the error in specific dates will change with depth in a core, and the model used to derive those dates.

Since trace metals exhibit different characteristic $K_D$-pH relationships (White and Driscoll, 1987a; Santschi et al., 1986; White and Driscoll, 1987b), trace metals are likely to differ in their response to historical changes in pH regime. Therefore, one can not generalize a change in scavenging efficiency exhibited by one trace metal to all trace metals. A knowledge of the relationship between pH and $K_D$ for different trace metals is critical to understantding biogeochemical cycling of trace metals. Whether the processes observed in Darts Lake are important in other lake systems or significantly influence the interpretation of sediment geochronology remains to be evaluated. However, consideration of the role of pH and DOC in effecting particle-solution chemistry in dilute acidic lakes over long time frames warrants consideration in future studies of sediment stratigraphy. Studies of trace-metal budgets in dilute lake systems also should involve an evaluation of short-term temporal variations in trace metal retention.

REFERENCES

Barnes, R. B. 1975.  The determination of specific forms of aluminum in natural water.  Chemical Geology. 15:177-191.

Baron, J., S. A. Norton, D. R. Beeson, and R. Hermann. 1986.  Sediment diatom and metal stratigraphy from Rocky Mountain lakes
   with special reference to atmospheric deposition.  Can. J.
   Fish. Aquat. Sci. 43:1350-1362.

Beamish, R., and J. VanLoon. 1977.  Precipitation loading of acid
   and heavy metals to a small acid lake near Sudbury, Ontario. J.
   Fish. Res. Bd. Can. 34:649-658.

Bloesch, J., and N. M. Burns. 1980.  A critical review of sediment
   trap technique.  Schweiz A. Hydrol. 42:15-55.

Charles, D. F., and S. A. Norton. 1986.  Paleolimnological evidence
   for trends in atmospheric deposition of acids and metals.  *In*
   Acid Deposition:  Long Term Trends. National Research Council,
   National Academy Press, Washington, D.C.

Cronan, C. S., J. Conlan, and S. Skibinski. 1987.  Forest
   vegetation in relation to surface water chemistry in the North
   Branch of the Moose River, Adirondack Park, N.Y.
   Biogeochemistry.  (In press)

Davis, J. A., and J. O. Leckie, 1978.  Surface ionization and
   complexation at the oxide/water interface, II.  Surface
   properties of amorphous iron oxyhydroxide and adsorption of
   metal ions.  J. Coll. Int. Sci. 67:90-107.

Davis, J. A., and R. Gloor. 1981.  Adsorption of dissolved organics
   in lake water by aluminum oxide.  Effects of molecular weight.
   Env. Sci. Technol. 15:1223-1229.

Davis, A. O., and J. N. Galloway. 1982.  Metal budgets for Al, Fe,
   Mn, and Zn during spring melt in an Adirondack watershed.
   Electric Power Research Institute, Project report No. RP1109-5,
   Palo Alto, CA.

Davis, J. A. 1982.  Adsorption of natural dissolved organic matter
   at the oxide/water interface.  Geochim. Cosmochim. Acta.
   46:2381-2393.

Dickson, W. 1978.  Some effects of acidification of Swedish lakes.
   Vern. Internat. Verein. Limnol.  20:851-856.

Driscoll, C. T., and G. C. Schafran. 1984.  Characterization of
   short-term changes in the base neutralizing capacity of an
   acidic Adirondack, N.Y., lake.  Nature. 310:308-310.

Driscoll, C. T. 1984. A procedure for the fractionation of aqueous aluminum in dilute acidic waters. Intern. J. Env. Anal. Chem. 16:267-283.

Driscoll, C. T., J. P. Baker, J. J. Bisogni, and C. L. Schofield. 1984. Aluminum speciation and equilibria in dilute acidic surface waters of the Adirondack region of New York State, pp. 55-75. In O.P. Bricker, ed., Geological aspects of acid rain. Ann Arbor Sci., Ann Arbor.

Driscoll, C. T., C. P. Yatsko, and F. J. Unangst. 1987. Longitudinal and temporal trends in the water chemistry of the North Branch Moose River. Biogeochemistry. (In press)

Galloway, J. N., and G. E. Likens. 1979. Atmospheric enhancement of metals deposition in Adirondack lake sediments. Limnology and Oceanogr. 24:427-433.

Gubala, C. P., and J. R. White. Processes at the sediment-water interface and their significance in aluminum and trace metal chemistry of an acidic lake. In The chemical quality of water and the hydrologic cycle. Lewis Publishers, Ann Arbor. (Chapter 14, this volume)

Hanson, D. W., S. A. Norton, and J. S. Williams. 1982. Modern and paleolimnological evidence for accelerated leaching and metal accumulation in soils in New England, caused by atmospheric deposition. Water, Air and Soil Poll. 18:227-239.

Heit, M., Y. Tan, C. Klusek, and J. Burke. 1981. Anthropogenic trace elements and polycyclic aromatic hydrocarbon levels in sediment cores from two lakes in the Adirondack lake region. Water, Air and Soil Poll. 15:441-464.

Henriksen, A., and R. F. Wright. 1978. Concentrations of metals in small Norwegian lakes. Water Research. 12:101-112.

Hesslein, R. H., W. S. Broecker, and D. W. Schindler. 1980. Fates of metal radiotracers added to a whole lake: sediment-water interactions. Can. J. Fish. Aquat. Sci. 37:378-386.

Hohl, H., and W. Stumm. 1976. Interactions of $Pb^{2+}$ with hydrous $Al_2O_3$. Journal of Colloid and Interface Science. 55:281-288.

Hooper, R. P., and C. A. Shoemaker. 1985. Aluminum mobilization in an acidic headwater stream: temporal variation and mineral dissolution disequilibria. Science. 229:463-465.

Hsu, P. H. 1977. Aluminum hydroxides and oxyhydroxides, pp. 114-115. In J. B. Dixon and S. B. Weed, eds., Minerals in soil environments. Soil Science Society of America, Madison.

Hunter, K. A., and P. S. Liss. 1979. The surface charge of suspended particles in estuaries and coastal waters. Nature. 282:823-825.

Hutchinson, N. J., and J. B. Sprague. 1986. Toxicity of trace metal mixtures to American Flagfish (Jordanella floridae) in soft, acidic water and implications for cultural acidification. Can. J. Fish. Aquat. Sci. 43:647-655.

Isachsen, Y. M., and D. W. Fisher. 1970. Geologic map of New York-Adirondack Sheet, New York State Museum of Science, Map and Chart Series No. 15.

Jackson, T. A., G. Kipphut, R. H. Hesslein, and D. W. Schindler. 1980. Experimental study of trace metal chemistry in soft-water lakes at different pH levels. Can. J. Fish. Aquat. Sci. 37:387-402.

James, R. O., and T. W. Healy. 1972. Adsorption of hydrolyzable metal ions at the oxide/water interface, III. A thermodynamic model of adsorption. J. Coll. Interf. Sci. 40:65-81.

Johnson, N. M., C. T. Driscoll, J. S. Eaton, G. E. Likens, and W. H. McDowell. 1981. Acid rain, dissolved aluminum and chemical weathering at the Hubbard Brook Experimental Forest, New Hampshire. Geochim. Cosmochim. Acta. 45:1421-1437.

Kahl, J. S., S. A. Norton, and J. S. Williams. 1984. Chronology, magnitude and paleolimnological record of changing metal fluxes related to atmospheric deposition of acids and metals in New England, pp. 23-35. *In* O. P. Bricker, ed., Geological aspects of acid deposition. Ann Arbor Sci., Ann Arbor.

Kinniburgh, D. G., and M. L. Jackson. 1981. Cation adsorption by hydrous metal oxides and clay, pp 91-160. *In* M. A. Anderson and A. J. Rubin, eds., Adsorption of inorganics at solid-liquid interfaces. Ann Arbor Sci., Ann Arbor.

Kodama, H., and M. Schnitzer. 1977. Effects of fulvic acid on the crystallization of Fe(III) oxides. Geoderma. 19:279-291.

Letterman, R. D., S. G. Vanderbrook, and P. Sricharoenchaikit. 1982. Electrophoretic mobility measurements in coagulation with aluminum salts. J. Am. Water Works Assoc. 74:44-51.

Linthurst, R. A., D. H. Landers, J. M. Eilers, D. F. Brakke, W. S. Overton, E. P. Meier, and R. E. Crowe. 1986. Characteristics of lakes in the eastern United States. Volume I. Population descriptions and physico-chemical relationships. EPA/600/4-86/007a, U.S. Environmental Protection Agency, Washington, D.C. 136 pp.

Menzel, D. W. and R. F. Vaccaro. 1964. The measurement of dissolved organic and particulate carbon in seawater. Limnology
and Oceanography. 9:138-142.

Millward, G. E. and R. M. Moore. 1982. The adsorption of Cu, Mn,
and Zn by iron oxyhydroxide in model estuarine solutions.
Water Res. 16:981-985.

National Academy Press. 1985. Acid deposition. Effects on
geochemical cycling and biological availability of trace
metals. Subgroup on metals of the tri-academy committee on
acid deposition. National Academy Press, Washington, D.C.
83 pp.

Newton, R. M, J. Weintraub, and R. April. 1986. The relationship
between surface water chemistry and geology in the North Branch
of the Moose River. Biogeochemistry. (In press)

Nordstrom, D. K., and J. W. Ball. 1986. The geochemical behavior
of aluminum in acidified surface waters. Science. 232:54-56.

Norton, S. A., and Hess, C. T. 1980. Atmospheric deposition in
Norway during the last 300 years as recorded in SNSF lake
sediments. I. Sediment dating and chemical stratigraphy,
pp. 268-269. *In* D. Drablos and A. Tollan, eds., Ecological
impact of acid precipitation. SNSF Project, Oslo, Norway.

Norton, S. A., C. T. Hess, and R. B. Davis. 1980. Rates of
accumulation of heavy metals in pre- and post-European
sediments in New England lakes, pp. 409-421. *In* S. J.
Eisenreich, ed., Atmospheric pollutants in natural waters.
Ann Arbor Sci., Ann Arbor.

Nriagu, J. O., H. K. T. Wong, and R. D. Coker. 1982. Deposition
and chemistry of pollutant metals in lakes around the smelters
at Sudbury, Ontario. Env. Sci. Technol. 16:551-560.

Perdu, E. M., K. Beck, and J. H. Reuter. 1976. Organic complexes
of iron and aluminum in natural waters. Nature. 260:418.

Robbins, J. A. 1978. Geochemical and geophysical applications of
radioactive lead, pp. 285-393. *In* J. O. Nriagu, ed., The
biogeochemistry of lead in the environment. Elsevier Press.

Salim, R., and B. G. Cooksey. 1981. The effect of centrifugation
on the suspended particles of river waters. Water Res.
15:835-839.

Santschi, P. H., U. P. Nyffeler, R. F. Anderson, S. L. Schiff,
P. O'Hara, and R. H. Hesslein. 1986. Response of radioactive
trace metals to acid-base titrations in controlled experimental
ecosystems: evaluation of transport parameters for application
to whole-lake radiotracer experiments. Can. J. Fish. Aquat.
Sci. 43:60-77.

Schafran, G. C., and C. T. Driscoll. 1987. Spatial and temporal variations in aluminum chemistry of a dilute, acidic lake. Biogeochemistry. (In press)

Schindler, D. W., R. H. Hesslein, R. Wageman, and W. S. Broecker. 1980. Effects of acidification on mobilization of heavy metals and radionuclides from the sediments of a freshwater lake. Can. J. Fish. Aquat. Sci. 37:373-377.

Schindler, D. W., and M. A. Turner. 1982. Biological, chemical and physical responses of lakes to artificial acidification. Water, Air and Soil Pollut. 18:259-271.

Schofield, C. L. 1976. Acid precipitation: effects on fish. Ambio. 5:228-230.

Schut, P. H., R. D. Evans, and W. A. Scheider. 1986. Variation in trace metal exports from small Canadian Sheild watersheds. Water, Air, and Soil Pollut. 28:225-237.

Schwertmann, U. 1966. Inhibitory effects of soil organic matter on the crystallization of amorphous ferric hydroxide. Nature. 212:645-646.

Standard Methods for the Examination of Water and Wastewater. 1976. American Public Health Association, 14th ed. New York, NY.

Stumm, W., and J. J. Morgan. 1981. Aquatic Chemistry, 2nd ed. Wiley Interscience, New York. 780 PP.

Tipping, E., C. Woof, and D. Cooke. 1981. Iron oxide from a seasonally anoxic lake. Geochim. Cosmochim. Acta. 45:1411-1419.

Troutman, D. E., and N. E. Peters. 1982. Deposition and transport of heavy metals in three lake basins affected by acid precipitation in the Adirondack Mountains, N.Y., pp. 33-61. *In* L. H. Keith, ed., Energy and environmental chemistry, vol. 2, Ann Arbor Sci., Ann Arbor.

White, J. R. 1984. Trace metal cycling in a dilute acidic lake system. Ph.D. thesis. Syracuse University, Syracuse, N.Y.

White, J. R., and C. T. Driscoll. 1985. Lead cycling in an acidic Adirondack lake. Environ. Sci. Technol. 19:1182-1187.

White, J. R., and C. T. Driscoll. 1987a. Manganese cycling in an acidic Adirondack lake. Biogeochemistry. (In press)

_____. 1987b. Zinc cycling in an acidic Adirondack lake. Environ. Sci. Technol. 21:211-216.

PROCESSES AT THE SEDIMENT-WATER INTERFACE AND THEIR
SIGNIFICANCE IN ALUMINUM AND TRACE METAL
CHEMISTRY OF AN ACIDIC LAKE

Chad P. Gubala and Jeffrey R. White, Indiana University,
Bloomington, Indiana

## ABSTRACT

Trace metal cycling across the sediment-water interface in the
hypolimnion of Big Moose Lake (Adirondack State Park, New York) was
studied during summer stratification.   Chemical analyses of pore
water, sediment solid phases, and water column samples indicate
significant chemical gradients in pH (5.0-6.4), dissolved organic
carbon (DOC) (250-1,250 $\mu$mols $L^{-1}$), soluble Fe(II) (0-600 $\mu$mols
$L^{-1}$), and Fe oxides (250-800 $\mu$mols $g^{-1}$ dry weight).   Results
support two mechanisms which may be important to the cycling of Al
and trace metals in sediments:  Organically complexed metals may be
released to the hypolimnion by diffusion of DOC through the pore
water, and mobilization and redeposition of Fe in an oxide layer in
sediments may concentrate metals through sorption-coprecipitation.
Post-depositional movement of metals within the sediment are of
concern  to  paleolimnologists  and  geochemists  attempting  to
reconstruct lake histories.

## INTRODUCTION

Natural  aquatic  ecosystems  respond  in  a  variety  of  ways  to
increased  loadings  of  hydrogen  ions.   The  effects  of  acidic
deposition  upon  lake  systems  include  increases  in  aluminum
concentrations (Haines and Alielaszek, 1983; Driscoll et al., 1984;
Schafran  and  Driscoll,  1987)  and  concentrations  of  other  trace
metals (Henriksen and Wright, 1978; Haines and Alielaszek, 1983;
White,  1984).   High  concentrations  of  aluminum,  lead,  and  other
trace metals are of concern due to their toxicological effects upon
aquatic biota (Driscoll et al., 1980; Dillon et al., 1984; Campbell
and Stokes, 1985).   Changes in trace metal chemistry within dilute

*Chemical Quality of Water and the Hydrologic Cycle*, Robert C. Averett and Diane M. McKnight (Eds.) © 1987 Lewis Publishers, Inc.,
Chelsea, Michigan. Printed in the United States of America.

acidic lake systems, as affected by acidic deposition, are also of concern to geochemists and paleolimnologists studying lake sediments (Schindler et al., 1980; Kahl et al., 1984; Campbell and Tessier, 1984; Carignan and Tessier, 1985).

In addition to geochemical studies of sediments as a sink for trace metals in lakes, sediments and sediment pore waters of acidic lake systems may be a potential source of trace metals to the water column (Schafran, 1985; Schindler et al., 1980; Carignan and Tessier, 1985; Sakata, 1985). The sediments of acidified lakes may, therefore, serve as both a source and a sink for trace metals, rather than a permanent resting place. However, while evidence exists to support the notion of sediments as a transient source of trace metals to the water column, the conditions and mechanisms controlling this phenomenon have not yet been clearly described. In this study, we investigated diagenetic transformations potentially affecting the transfer of Al and trace metals across the sediment-water interface within dilute acidic lake systems. Results are presented for Al and Pb as examples, since a treatment of the entire data set is beyond the scope of this chapter.

METHODS

Study Site Description

The study area centered around Big Moose Lake (74°51'W, 43°49'N) in the North Branch Moose River system within the west central region of the Adirondack Mountains of New York (Fig. 1). Approximately 95% of the watershed is forested by mixed varieties of secondary growth deciduous hardwoods (Cronan et al., 1987; Driscoll et al., 1987). Thin acidic soils overlie granitic gneisses and meta-sedimentary bedrock within the basin (Newton et al., 1987) offering little to no buffering capacity against the region's characteristic inputs of acidic deposition (Electric Power Research Institute, 1984).

Big Moose Lake (SA = 513 ha) is a large, dilute, acidic, drainage lake (mean pH = 5.1, mean ANC = 1.0 $\mu$eq $L^{-1}$) (Driscoll and Newton, 1985) located near the headwaters of the drainage system. Details on the water chemistry of the North Branch Moose River system are presented elsewhere (Driscoll et al., 1987).

Sampling and Analysis

Water column, sediment, and sediment pore water samples were collected at a 24-m pelagic station (Fig. 1) during a 2 week period in July of 1985. The water column was sampled at 4-m intervals with a polypropylene Kemmerer bottle. Dissolved oxygen was fixed in the field and measured by Winkler titration (Standard Methods, 1975). pH was determined potentiometrically on-site using a Ross electrode. Aliquots taken for DOC and dissolved inorganic carbon (DIC) determinations were immediately preserved with $H_3PO_4$ and $HgCl_2$ respectively, and measured by an automatic carbon analyzer.

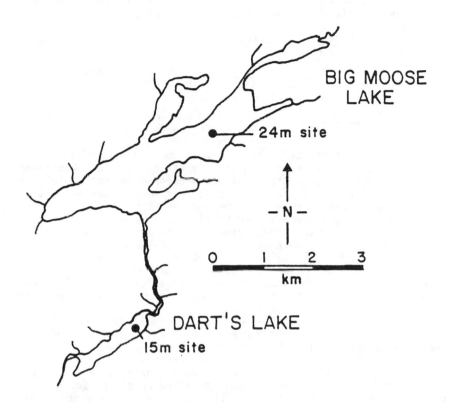

Figure 1.  Map of Big Moose Lake and Darts Lake showing locations
          of the sampling sites.

Unpreserved aliquots were analyzed by ion chromatography for
concentrations of $SO_4^-$, $NO_3^-$ and $Cl^-$. Acidified aliquots (2%
Ultrex $HNO_3$) were analyzed for trace metals (Zn, Pb, Al, Mn, Fe)
and major cations ($Ca^{2+}$, $Mg^{2+}$, $Na^+$, $K^+$) by graphite furnace and
flame atomic absorption spectrophotometry, respectively. A
temperature profile of the water column was also taken at the time
of sample collection (thermistor).

A sediment core (app. 40 cm$^2$ SA) was taken at the 24-m sampling
site with a gravity corer. The sediment was immediately sectioned
into 1-cm thick intervals to a depth of 20 cm. To avoid chemical
alteration of sediment caused by exposure to atmospheric oxygen
(Rapin et al., 1986), sectioning of the core was performed in a
specially designed glove box under a $N_2$ gas environment. Sediment
sections were then subjected to a sequential extraction procedure
for the speciation of trace metals (Tessier et al., 1979) within 24
hr of collection. Extraction reagents were deoxygenated with $N_2$
gas prior to use and all sediment processing was conducted within
the purged glove box. The results reported in this chapter are

for the first three fractions of the five-step series. They are operationally defined as follows (Tessier et al., 1979):

(1) <u>Exchangeable Metals</u> - Extracted for 1 hr with 1.0 M $MgCl_2$ initially at pH 7.0.

(2)  <u>Metals Bound to Carbonates or Specifically Adsorbed</u>   - The residue from (1) extracted for 6 hr with 1.0 M NaOAc adjusted to pH 5.0 with HOAc.

(3) <u>Metals Bound to Oxides</u> - The residue from (2) extracted for 6 hr at 96°C with 0.04-M $NH_2OH \cdot HCL$ in 25% (v/v) HOAc.

Concentrations of trace metals within each fraction for each sediment interval were determined by flame atomic absorption spectrophotometry. Concentrations are presented as mmols metal per gm dry wt. sediment. These first three fractions of the sequential extraction procedure represent the labile solid phases of trace metals most likely to be affected by diagenetic processes.

The pore water within the sediment at the 24-m site was sampled by in situ dialysis using a device similar to the sampler described by Hesslein (1976). A stabilizing platform was added to the basic design to provide clear designation of the sediment-water interface. The platform was positioned to allow 12 1-cm (8.5 $cm^3$ volume) dialysis cells to sample the water column and 45 1-cm cells to penetrate into the sediment.

The plexiglas sampler was assembled approximately 48 hr prior to installation using 0.2-μm Nuclepore dialysis membrane. The unit was then submersed in deionized water and purged with $N_2$ gas to remove dissolved oxygen from the compartments (Carignan, 1984a; Carignan et al., 1985) and remained in deoxygenated water until the time of installation. The unit was installed by divers and allowed to equilibrate for a period of 15 days. Once the sampler was retrieved, aliquots from each cell were quickly removed and preserved as described earlier for subsequent analyses. The pH for each cell was immediately measured potentiometrically using a needle-nose combination electrode. DIC, DOC, major anions and cations, and trace metals were determined for each interval as described for the water column analyses.

RESULTS AND DISCUSSION

pH as a Master Variable

The pH of the water column of Big Moose Lake in July of 1985 ranged from 5.2 near the surface of the lake to 5.0 near the sediment-water interface. The pH increased sharply across the sediment-water interface to a maximum of approximately 6.4 within the pore water (Fig. 2). This pH increase was likely due to reduction reactions generating alkalinity within the sediment column. Since pH is thought to be a master variable in controlling reactions involving trace metals (Stumm and Morgan, 1981), the differences in pH between the water column and pore waters are

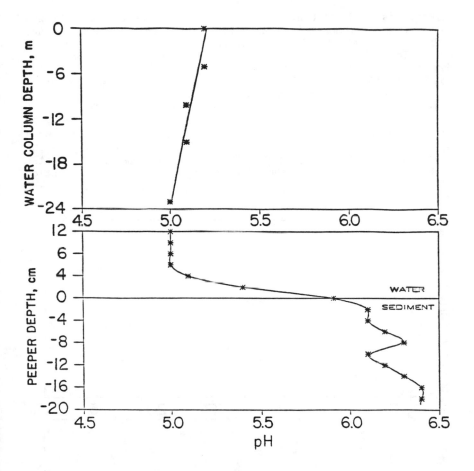

Figure 2.   pH profile of the water column (top panel) and pore
            water (bottom panel) from the 24-m station at Big Moose
            Lake.   The smooth curve drawn through the data points
            was generated by a cubic spline algorithm.

noteworthy.   Differences in the chemistry of trace metals between
the water column and pore water were expected as a result of
spatial changes in pH.

## The Role of DOC in Trace Metal Transport

Dissolved organic carbon represents a general class of
naturally occuring compounds which are important complexing ligands
for trace metals in dilute waters (Stumm and Morgan, 1981; Benes et
al., 1976).   Cycling of DOC within lakes may be important to the
fate and transport of trace metals (White and Driscoll, 1985).

Aluminum is a significant cation in the waters of the North Branch Moose River system ($8.9 \pm 2.7$ µmols $L^{-1}$) (Driscoll and Newton, 1986; Driscoll et al., 1986). The importance of DOC as an aluminum complexing ligand near the sediment-water interface has been demonstrated in previous work examining organic Al complexes in Big Moose Lake (Driscoll, 1980) and Darts Lake (Schafran and Driscoll, 1987). Water column results from Big Moose Lake during this study indicated enrichment of soluble Al in the hypolimnion (10-16 µmols $L^{-1}$), however, no concentration gradients in soluble Al or Pb within the sediment pore water were observed (Figs. 3, 4). Note that soluble Fe exhibited a large concentration gradient within the sediment (10 µmols $cm^{-1}$) (Fig. 5). Soluble lead within the water column of Big Moose Lake showed a roughly orthograde distribution (app. 85 nmol $L^{-1}$).

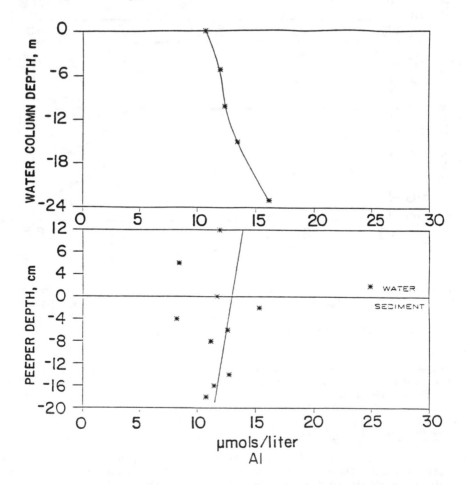

Figure 3.    Soluble aluminum profile from the 24-m station at Big Moose Lake.

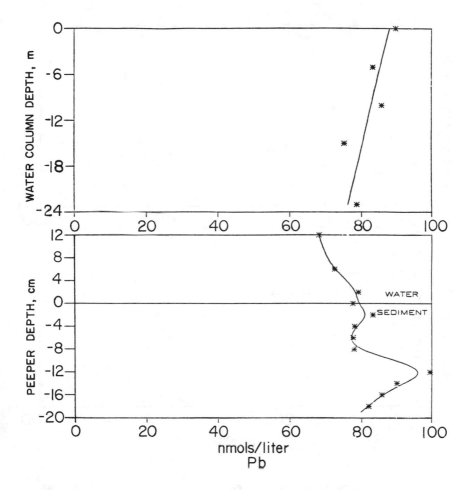

Figure 4.    Soluble lead profile from the 24-m station at Big Moose
             Lake.

Recent studies of Darts Lake reported significant enrichment in
concentrations of organic Al, DOC (Schafran and Driscoll, 1987), Fe
(White, 1984) and Pb (White and Driscoll, 1985) in the vernal
hypolimnion as compared to epilimnetic concentrations.    Hypolimnetic enrichment of DOC (50%) (Fig. 6), organic monomeric Al
(230%) (Fig. 7), Fe (1,300%) and Pb (200%) (Fig. 8) were attributed, in part, to release of solutes from the sediment,
concomitant with a decline in the volume of the hypolimnion during
late summer stratification.

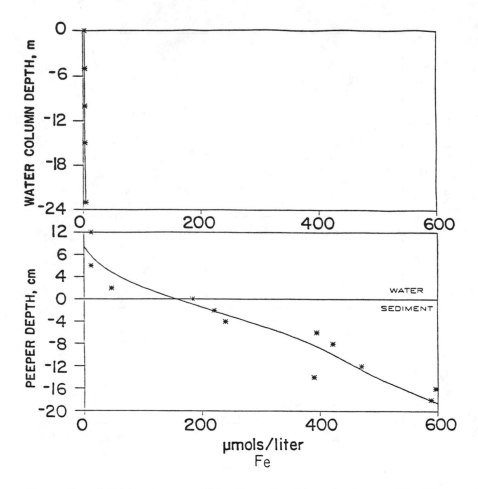

Figure 5.    Soluble iron profile from the 24-m station at Big Moose Lake.

The concurrent enrichment of DOC, Fe, Pb and organically complexed Al near the sediment-water interface is consistent with our hypothesis of DOC-related transport of metal ions from the sediment system to the water column. During our sampling period, concentrations of DOC were approximately 250 μmols $L^{-1}$ throughout the water column of Big Moose Lake. However, note the steep negative concentration gradient (app. 60 μmols $cm^{-1}$) across the sediment-water interface, approaching 1,250 μmols $L^{-1}$ at 20 cm into the sediment (Fig. 9). The significant concentration gradient of DOC from the pore water into the water column is likely to result in chemical diffusion of DOC and DOC-complexed solutes into the water column.

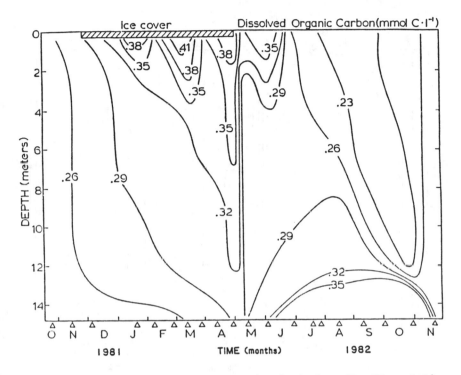

Figure 6. Dissolved organic carbon isopleth from the 15-m station at Darts Lake (Schafran, 1985).

Release of DOC-complexed solutes from sediments into the water column may occur in two ways. One mechanism might involve diffusion of soluble metal-organic complexes through the sediments and into the water column as controlled by DOC concentration gradients. Based upon chemical equilibrium calculations (MINEQL model; Westall et al., 1976), using eight model organic ligands to simulate the chemistry of DOC (Sposito, 1981), significant amounts of soluble Al (app. 100%) and Pb (app. 75%) within the pore water of Big Moose Lake were predicted to be organically complexed. While the metal-binding capacities of heterogeneous mixtures of natural organic ligands remain unknown, the model predictions are consistent with the observed relationships between concentrations of DOC and organically complexed Al for a number of lakes and streams within the Adirondack region (Bisogni and Driscoll, 1979). Theoretically, therefore, diffusion of DOC may provide a mechanism for movement of metal species (i.e. Al, Fe, Mn, trace metals) into the water column as organic complexes. If metal-organic complexes are diffusing in sediments along concentration gradients, then the observed pore water profiles of total soluble concentrations of metals should support this hypothesis. While profiles of Fe

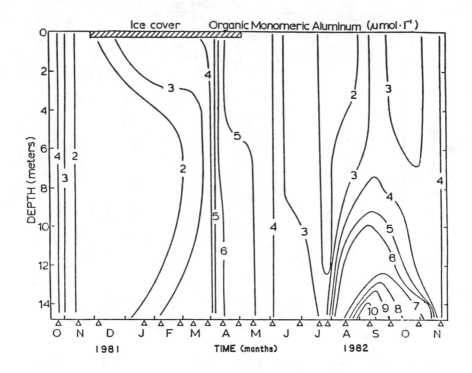

Figure 7.  Organic monomeric aluminum isopleth from the 15-m
           station at Darts Lake (Schafran, 1985).

(Fig. 5) and DOC (Fig. 9) are consistent with this mechanism,
results for Al (Fig. 3) and Pb (Fig. 4) do not directly support the
above mechanism.  However, the metal-binding capacity of DOC is
likely to be different for each metal.  Also, the overall
complexation capacity of DOC in pore waters of Big Moose Lake may
be largely consumed by soluble Fe(II).  The molar ratio of Fe to
DOC in the pore water was consistently 0.5 mol Fe  mol $C^{-1}$.

The study was conducted during mid-summer stratification, while
the volume of the hypolimnion was still large (depth to thermocline
10 m, maximum depth 24 m).  Solutes transported to the water column
from the sediments would be diluted into the hypolimnion, and not
exhibit significant increases in concentration near the sediment-
water interface during this time of year.  Concentration profiles
in lake sediments may be affected by the seasonal changes in water
chemistry in the overlying waters.  More intensive seasonal
sampling of the water column and pore water is necessary to
properly investigate the importance of these processes to trace
metal cycling in acidic lakes.

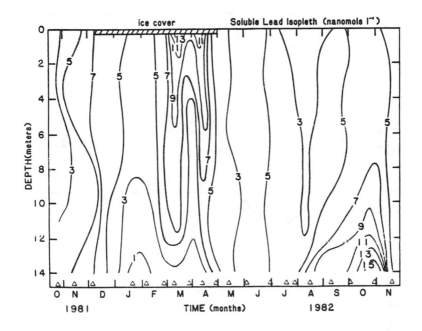

Figure 8.    Soluble lead isopleth from the 15-m station at Darts
Lake (White and Driscoll, 1985).


A second manner by which DOC may affect the transport of trace
metals into the water column is by direct desorption of metals from
sediment    exchange    sites    near    the    sediment-water    interface.
Exchangeable Al and Pb were enriched in the surface sediments of
Big Moose Lake (Fig. 10) and, although a small component of the
total mass of each element, represent important labile pools. The
exchangeable pool may be readily desorbed by DOC and released into
the water column.    Release of the pools of exchangeable Al and Pb
present in the first centimeter of sediment (4,000 $\mu$mols m$^{-2}$ Al,
75 $\mu$mols m$^{-2}$ Pb) would be more than sufficient to account for a
20-fold increase in Al and an 80-fold increase in Pb in the
hypolimnion of Big Moose Lake.    Thus, a gradient in metal
concentrations in pore waters may not be necessary to yield their
enrichment in the hypolimnion, if DOC is causing mobilization of
exchangeable metals from surface sediments.

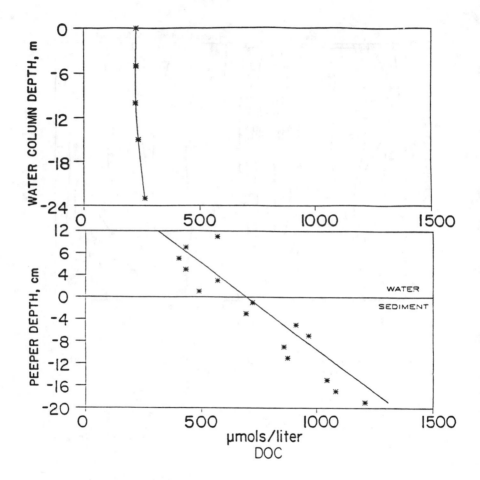

Figure 9.    Dissolved organic carbon (DOC) from the 24-m station at Big Moose Lake.


The Role of Iron Oxides in Sediments

It is well established that both aquo and organically complexed forms of trace metals may be strongly adsorped onto the surfaces of iron oxides (Carpenter et al., 1975; Davis, 1982; Lion et al., 1982). Within the sediments of Big Moose Lake, iron oxides ($NH_2OH \cdot HCL$ labile fraction) may account for between 5 and 25% of the total dry mass of sediment (Fig. 10). It was therefore hypothesized that solid phase iron oxides may affect the distribution of trace metals within the sediments of Big Moose Lake.

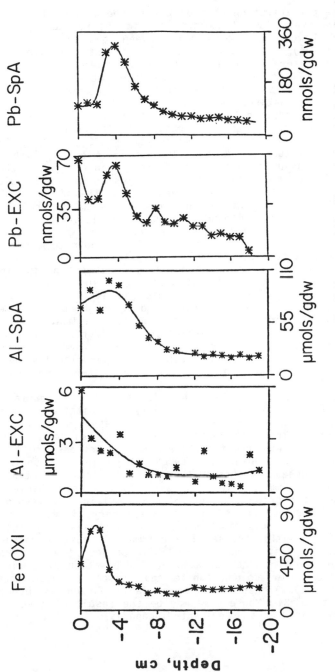

Figure 10.  Exchangeable aluminum and lead (Al-EXC and Pb-EXC), specifically adsorbed aluminum and lead (Al-SpA and Pb-SpA) and iron oxide (Fe-OXI) solid phase sediment profiles from the 24-m station at Big Moose Lake.

The concentration of iron within the water column of Big Moose Lake was very low (3 µmols $L^{-1}$) (Fig. 5) reflecting a high redox potential (minimum dissolved oxygen = 160 µmol $L^{-1}$ in the hypolimnion). The concentration gradient for iron across the sediment-water interface was large, resulting in the potential for upward diffusion of soluble iron into the water column. Similar profiles of iron in pore water have been observed in other lakes (Cook, 1984; Sakata, 1985).

MINEQL modeling was used to determine the degree of saturation of lake and pore water with respect to four mineral phases of iron which may be found in aquatic systems [$FeCO_3$, $Fe(OH)_2$, $Fe(OH)_3$ and $Fe_3(OH)_8$] (Stumm and Morgan, 1981). Soluble iron within the pore water was assumed to exist as ferrous iron [Fe (II)], while ferric iron [Fe(III)] was the dominant species in the water column (based upon redox potential as determined by the pH-DO couple). The logarithm of the saturation index (SI) was determined for each sample and each iron mineral phase (Fig. 11). An SI value of 0.0 indicates equilibrium, positive values of SI indicate super-saturation, and negative values of SI indicate undersaturation, with respect to a given mineral phase.

A zone of supersaturation with respect to the two ferric iron hydroxides [$Fe(OH)_3$ and $Fe_3(OH)_8$] exists just above the sediment-water interface (Fig. 11). This condition may be due to the diffusion of ferrous iron upward through the sediment pool, entering an oxidizing zone at or near the sediment-water interface. The potential for precipitation of iron in the region of supersaturation was corroborated by the sediment core extraction results. Elevated concentrations of operationally defined iron oxide were measured near the sediment-water interface (0-3 cm) (Fig. 10).

Iron oxide solid phases may act as effective surfaces for the adsorption of trace metals. Actively precipitating iron hydroxides may also coprecipitate Al, trace metals, or organometallic complexes, forming labile amorphous solids. Profiles of exchangeable and specifically adsorbed fractions of Al and Pb (as determined by sequential extraction) within the solid sediments support these hypotheses (Fig. 10). These fractions exhibit surface enrichment within the sediment and are closely correlated with the iron oxide profile. Stepwise multiple regression analysis confirm that iron oxide is a primary independent variable controlling the distributions of labile phases of Al (P<0.001) and Pb (P<0.05) within the sediment column. It must be noted, however, that the distributions of the labile phases of Al and Pb may also be controlled by the formation and deposition of aluminum oxides from the water column (White and Driscoll, 1985; Schafran and Driscoll, 1987). The profiles of labile Pb and Al observed in the sediment of Big Moose Lake may be the result of a combination of in-lake processes and sediment diagenesis.

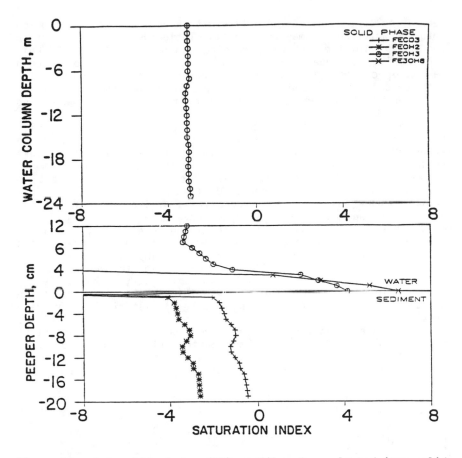

Figure 11.    Saturation index (SI) profiles for selected iron solid
phases from the 24-m station at Big Moose Lake.

## SUMMARY AND IMPLICATIONS

Two diagenetic mechanisms have been suggested as partially
controlling Al and trace metal chemistry near the sediment-water
interface of dilute acidic lakes.

DOC-mediated transport processes may contribute to the flux of
Al and trace metals from the pore water or surface sediments into
the water column (Fig. 12a).  The principal implication of this
process is the increase in concentration of potentially toxic
metals within hypolimnetic waters.  However, the toxicity of metals
may be mitigated by DOC complexation (Driscoll et al., 1980).  An
additional effect of DOC-mediated transport of metals is the
potential alteration of the sedimentary geochemical profile.

Decreasing concentrations of metals within pore waters by diffusion into the water column may force a dissolution/desorption of certain species within the sediment, causing misinterpretations of their depth (time) dependent trends (Carignan, 1984b). This phenomenon may be of critical importance to researchers employing $Pb^{210}$ as a dating tool, since lead may be susceptible to redistribution or mobilization from sediments through diffusion of DOC (White and Driscoll, 1985).

Sequestering of trace metals by coprecipitation/sorption with Fe oxides near the sediment-water interface may contribute to translocation of significant quantities of trace metals within lake sediments (Fig. 12b). Iron cycling and redeposition to sediments as oxides may also serve as a sink for trace metals present in hypolimnetic waters. DOC-mediated transport of metals within sediments, and transport and redeposition of iron in sediments may be interacting processes. DOC may serve as an adsorbing ligand which interacts with Fe oxides; combining a mode of trace metal transport (as an organic complex), with a means of redeposition (sorption to Fe oxides). Movement of important sedimentary markers, such as $Pb^{210}$ and Al, after deposition to deep sediments, is plausible based on the results of this study. The processes proposed in this chapter for dilute acidic lakes may contribute to the alteration of trace metal gechronology. Historical information derived from trace metal profiles in these systems may, therefore, be inaccurate. Although this research has highlighted important processes occurring at the sediment-water interface of acid lakes, the actual impact of such processes on the integrity of the sediment record remains to be evaluated.

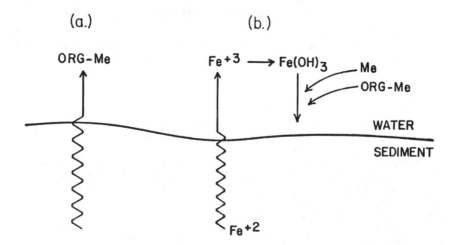

Figure 12.    Conceptual models of (a.) DOC related transport of trace metals, and (b.) iron cycling and sequestering of trace metals.

REFERENCES

Benes, P., E. Gjessing, and E. Steinnes. 1976. Interactions between humus and trace elements in fresh waters. Water Research. 10:711-716

Bisogni, J. J. Jr., and C. T. Driscoll. 1979. Characterization of the pH buffering systems in dilute Adirondack surface waters. Office of Water Research and Technology Proj. # A-073-NY.

Campbell, P. G. C., and A. Tessier. 1984. Paleolimnological approaches to the study of acidic deposition: Metal partitioning in lacustrian sediments. pp. 234-274. *In* Proceedings of a workshop on paleolimnological studies of the history and effects of acidic precipitation. USEPA # CR-881631-01-0.

Campbell, P. G. C., and P. M. Stokes. 1985. Acidification and toxicity of metals to aquatic biota. Canadian Journal of Fisheries and Aquatic Science. 42:2034-2049.

Carignan, R. 1984a. Interstitial water sampling by dialysis. Methodological notes. Limnology and Oceanography 29:667-670.

_____. 1984b. Trace metal diffusion across the sediment-water interface: Implications in the chronological interpretation of dated trace metal sediment profiles. pp. 198-199. *In* Proceedings of a workshop on paleolimnological studies of the history and effects of acidic precipitation. USEPA # CR-881631-01-0.

Carignan, R., F. Rapin, and A. Tessier. 1985. Sediment porewater sampling for metal analysis: A comparison of techniques. Geochimica et Cosmochimica Acta. 49:2493-2497.

Carignan, R., and A. Tessier. 1985. Zinc deposition in acid lakes: The role of diffusion. Science 228:1524-1526.

Carpenter, R. H., T. A. Pope, and R. L. Smith. 1975. Fe-Mn oxide coatings in stream sediment geochemical surveys. Journal of Geochemical Exploration. 4:349-363.

Cook, R. B. 1984. Distributions of ferrous iron and sulfide in an anoxic hypolimnion. Canadian Journal of Fisheries and Aquatic Science. 41:286-292.

Cronan, C. S., J. Conlan, and S. Skibinski. 1987. Forest vegetation in relation to surface water chemistry in the North Branch of the Moose River. Biogeochemistry, (in press).

Davis, J.A. 1982. Adsorption of natural dissolved organic matter at the oxide/water interface. Geochimica et Cosmochimica Acta. 46:2381-2393.

Dillon, P. J., N. D. Yan, and H. H. Harvey. 1984. Acidic deposition: Effects on aquatic ecosystems. CRC Critical Reviews in Environmental Control. 13:167-193.

Driscoll, C. T. 1980. Chemical characterization of some dilute acidified lakes and streams in the Adirondack region of New York State. Ph.D. Thesis. Cornell University, Ithaca, New York.

Driscoll, C. T., J. P. Baker, J. L. Bisogni, and C. L. Schofield. 1980. Aluminum speciation and its effects on fish in dilute acidified waters. Nature. 284:161-164.

Driscoll, C. T., J. P. Baker, J. J. Bisogni, and C. L. Schofield. 1984. Aluminum speciation and equilibria in dilute acidic surface waters of the Adirondack region of New York State. pp.55-75. *In* O. P. Bricker, ed., Geological aspects of acid rain. Ann Arbor Science, Ann Arbor, Michigan.

Driscoll, C. T., and R. M. Newton. 1986. Chemical characteristics of Adirondack lakes. Environmental Science and Technology. 19:1018-1024.

Driscoll, C. T., P. C. Yatsko, and F. J. Unaagst. 1987. Longitudinal and temporal trends in the water chemistry of the North Branch Moose River. Biogeochemistry, (in press).

Electric Power Research Institute. 1984. The integrated lake-watershed acidification study. EA-3221, v. 4 222 pp.

Haines, T. A., and J. Alielaszek. 1983. A regional survey of the chemistry of headwater lakes and streams in New England: Vulnerability to acidification. U.S. Fish and Wildlife Service, Eastern Energy and Land Use Team, FWS/OBS-80/40 v. 15 141 pp.

Henriksen, A., and R. F. Wright. 1978. Concentrations of metals in small Norwegian lakes. Water Research. 12:101-112.

Hesslein, R.H. 1976. An in situ sampler for close interval pore water studies. Limnology and Oceanography. 21:912-914.

Kahl, J. S., S. A. Norton, and J. S. Williams. 1984. Chronology, magnitude and paleological record of changing metal fluxes related to atmospheric deposition of acids and metals in New England. pp. 23-25. *In* O. P. Bricker, ed., Geological aspects of acid rain. Ann Arbor Science, Ann Arbor, Michigan.

Lion, L. W., R. S. Altmann, and J. O. Leckie. 1982. Trace-metal adsorption characteristics of estuarine particulate matter: Evaluation of contributions of Fe/Mn oxide and organic surface coatings. Environmental Science and Technology. 16:660-666.

Newton, R. M., J. Weintraub, and R. April. 1987. The relationship between surface water chemistry and geology in the North Branch of the Moose River. Biogeochemistry, (in press).

Rapin, F., A. Tessier, P. G. C. Campbell, and R. Carignan. 1986. Potential artifacts in the determination of metal partitioning in sediments by a sequential extraction procedure. Environmental Science and Technology. 20:836-840.

Sakata, M. 1985. Diagenetic remobilization of manganese, iron, copper and lead in anoxic sediment of a freshwater pond. Water Research. 19:1033-1038.

Schafran, G. C. 1985. Temporal and spatial variations in the acid/base and aluminum chemistry of a dilute acidic lake in the Adirondack region of New York State. Masters thesis. Department of Civil Engineering, Syracuse University. 139 pp.

Schafran, G. C., and C. T. Driscoll. 1987. Spatial and temporal variations in aluminum chemistry of a dilute acidic lake. Biogeochemistry, (in press).

Schindler, D. W., R. H. Hesslein, R. Wagemann, and W. S. Broecker. 1980. Effects of acidification on mobilization of heavy metals and radionuclides from the sediments of a freshwater lake. Canadian Journal of Fisheries and Aquatic Science. 37:373-377.

Sposito, G. 1981. Trace metals in contaminated waters. Environmental Science and Technology. 15:396-403

Standard Methods for the Examination of Water and Wastewater. 1975. American Public Health Association, 14th ed.

Stumm, W., and J. J. Morgan. 1981. Aquatic Chemistry. John Wiley and Sons, New York. 780 pp.

Tessier, A., P. G. C. Campbell, and M. Bisson. 1979. Sequential extraction procedure for the speciation of particulate trace metals. Analytical Chemistry. 51:844-851.

Westall, J.C., J.L. Zachary, and F.M.M. Morel. 1976. MINEQL a computer program for the calculation of chemical equilibrium composition of aqueous systems. Technical Note # 18, Department of Civil Engineering, Massachusetts Institute of Technology. Cambridge, Mass.

White, J. R. 1984. Trace metal cycling in a dilute acidic lake system. Ph.D. thesis, Department of Civil Engineering, Syracuse University. Syracuse, New York. 261 pp.

White, J. R., and C. T. Driscoll. 1985. Lead cycling in an acidic Adirondack lake. Environmental Science and Technology. 19:1182-1187.

# CHAPTER 15

IDENTIFYING IN-STREAM VARIABILITY:
SAMPLING IRON IN AN ACIDIC STREAM

Kenneth E. Bencala and Diane M. McKnight, U.S. Geological Survey,
Menlo Park, California and Denver, Colorado

## ABSTRACT

Small scale spatial and temporal variations of stream
chemistry were studied in the naturally acidic, metal-enriched,
Snake River near Montezuma, Colorado. Iron concentrations ranged
from 0.7 to 1.6 mg/L over a stream reach of 1 km and over a 24-hour
period at one stream site. The iron concentrations in the stream
also were higher than those in selected inflows to the stream
channel. Variations of sulfate and calcium concentrations were
smaller than the variations of iron, when based on comparisons
to values in the inflows. In the Snake River, iron concentrations
varied on characteristic spatial scales as short as tenths of a
kilometer and temporal scales as short as hours. These results
emphasize the importance of reconnaissance sampling over a range of
spatial and temporal scales in establishing monitoring programs for
studies of upland watersheds impacted by acid precipitation or
toxic mine drainage.

## INTRODUCTION

Transport of water and associated solutes within a small
(low-order), mountain stream is a segment of the hydrologic cycle
with a comparatively short residence time; i.e., the order of
hours, or perhaps days at the most. During this short residence
time, in-stream processes may occur rapidly and with significant
effect on the chemical composition of the stream water.
Concurrently, mixing of solutes from diverse sources may occur,
adjusting solute concentrations in proportion to the solute mass
flow.

*Chemical Quality of Water and the Hydrologic Cycle*, Robert C. Averett and Diane M. McKnight (Eds.) © 1987 Lewis Publishers, Inc.,
Chelsea, Michigan. Printed in the United States of America.

Weathering of different solid phases and the occurrence of daily cycles are examples of factors resulting in variation of instream solute concentrations. Spatial variation of solute concentrations along streams has been documented as a geochemical exploration tool (Rose et al., 1979). The study of temporal variations in small watershed streams was reviewed by Walling (1975). The fine scales of these variations become particularly significant in studies directed at providing quantitative interpretations of the dynamics and transport of reactive solutes. In spite of many decades of water-quality research, there is no *a priori* solution to the problem of knowing how many monitoring sites to establish and how frequently to sample within a given watershed.

For this work, samples were collected first on a spatial scale of tenths of a kilometer and on a temporal scale of hours. Results obtained from these sampling studies are presented to illustrate the extent of iron concentration variations present within a naturally acidic stream. As part of the spatial study the inflows sampled were visible groundwater seeps and small tributaries. These results are used to speculate on the relationship for iron between the variations in concentration within the stream channel and variations in concentration within the inflows. Comparisons are made to variations of sulfate and calcium concentrations and pH.

## CONCEPTUAL FRAMEWORK

The framework for this discussion is illustrated in Figure 1. If a change in solute concentration, either in space or in time, is observed within a stream channel, then several possible explanations can be considered. Figure 1 illustrates a 'tree' of increasingly complex considerations. At one level, the observed change could be considered as due to either a conservative mixing of inflows or a reaction occurring within the stream. Next, either of these processes could be occurring uniformly along the stream channel or they could be highly variable in location. Any of these four combinations could be steady or dynamic in time.

The two extremes of complexity would be to consider concentration differences as a result of (1) the conservative mixing of steady and spatially uniform inflows, or (2) the action of dynamic and spatially variable reactions. Variations in concentration alone are not necessarily indicative of whether a solute is reactive or conservative. For example, concentration may be highly variable for an ideally conservative solute in a physically complex watershed. On the other hand, concentration may be virtually unchanged for a highly reactive solute reaching steady-state on time scales faster than the movement of the water.

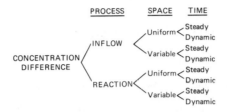

Figure 1. Combinations of process variability resulting in concentration differences.

Finally, it is crucial to recognize that designations such as 'conservative/reactive', 'uniform/variable', or 'steady/dynamic' are not absolute. These terms are all system dependent and related to the scales of hydrologic transport processes. At one point in time and space a particular solute may be conservative (that is, not undergo an observable chemical transformation), while in another system the same solute may be clearly reactive.

Routine monitoring programs are sometimes designed to establish 'typical' conditions either in the sense of the mode or the median. Thus, sampling can be over spatial and temporal scales which are long compared to the scales of chemical and physical processes. An intensive sampling effort works at the fine scales of system processes. It is the thesis of this chapter that an intensive sampling effort is an effective sampling scheme on which to base initial generalizations about (1) the reactive nature of a particular solute, and (2) the significance of instream processes. Such generalizations are then useful in the design of further research studies or long-term monitoring programs.

## ANALYSIS EQUATIONS

### Normalized Concentrations

The concept of 'variation' is not absolute, thus the assessment of significant versus insignificant variation is expressed relative to some defined measure. In the case of a conservative solute, and disallowing loss of water mass by such processes as evapotranspiration or diversion, the instream concentration will reach a value between the extremes of the values in the inflows to the stream. In this study one of the manners of data presentation will be in the form of a normalized concentration. By normalization with regard to the extremes of the sampled inflows, the variation of several solutes may be compared on a common basis. The normalized concentration is defined as follows:

$$C_{NORM} = \frac{(C - C^I_{MIN})}{(C_{MAX} - C^I_{MIN})} \tag{1}$$

where   $C_{NORM}$ = Normalized concentration

$C$       = Concentration

$C^I_{MAX}$ = Maximum inflow concentration

$C^I_{MIN}$ = Minimum inflow concentration.

Note, $C_{NORM}$ is a ratio of concentrations and is therefore itself dimensionless. $C_{NORM}$ = 0.0 if the concentration of interest is identical to the minimum inflow value. $C_{NORM}$ = 1.0 if the concentration of interest is identical to the maximum inflow value. Normalization of all concentrations in this manner allows for comparisons between solutes, while accounting for absolute differences in the values of concentrations of different solutes.

## Comparison with Inflow Concentrations

Variations of solute concentrations from inflows can be presented independently of corresponding information from stream samples. However, the concentrations in the stream are coupled to the concentrations of the inflows by the mass-balance relationship. In presenting the solute concentrations in the stream, it is possible to estimate the hypothetical concentration of inflows needed to produce observed stream concentrations. The derivation of the equations for this analysis assumes the solute to be conservative. Comparison of the results of the computation to actual field data allows for speculation as to whether or not observed variation in stream concentration is the result of variation of the inflows.

The equation for estimation of inflow concentration was developed as follows:

Conservation of flow requires

$$Q^S_A + Q^I = Q^S_B \tag{2}$$

where   $Q^S_A$ = Discharge of the stream above an inflow

$Q^I$ = Discharge of the inflow to the stream

$Q^S_B$ = Discharge of the stream below an inflow.

Conservation of mass of a conservative solute requires

$$c^S_A \, Q^S_A + c^I \, Q^I = c^S_B \, Q^S_B \qquad (3)$$

where  $c^S_A$  = Concentration of the stream above an inflow

$c^I$   = Concentration of the inflow

$c^S_B$  = Concentration of the stream below an inflow.

Substitution of (2) into (3) and rearrangement yields

$$c^I = \frac{(c^S_B \, Q^S_B - c^S_A \, Q^S_A)}{(Q^S - Q^S_A)} \qquad (4)$$

The results of computations with Equation (4) are estimates of the concentrations in an inflow. These estimates may then be compared to observed concentrations. Equation (4) should not be mis-construed as a 'model' for calculating actual concentrations. Rather it is used here as a test of the conservation of solute mass in the discussion of the relation between inflows and variation within the stream.

DATA COLLECTION

Site

The study area was in the upland reaches of the Snake River near Montezuma, Colorado. This system was previously studied by Theobald et al. (1963). The following excerpt from their abstract summarized water-quality conditions within the stream: "The oxidation of disseminated pyrite in relatively acid schists and gneisses of the Snake River drainage basin provides abundant iron sulfate and sulfuric acid to ground and surface water. This acid water dissolves large quantities of many elements, particularly aluminum and surprisingly large quantities of elements, such as magnesium and zinc, not expected to be abundant in the drainage basin. ... The principal precipitate on the bed of the Snake River is hydrated iron oxide with small quantities of the other metals." Details of the geologic setting and of ecological conditions in the stream were given by Theobald et al. (1963) and McKnight and Feder (1984), respectively. For the sampling described herein, the study area extended upstream 2,850 m from the confluence of the Snake River with Deer Creek (Fig. 2A).

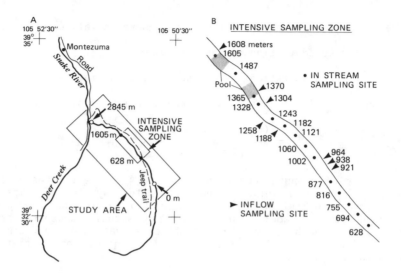

Figure 2.    A. Location of Snake River study area, near Montezuma,
              Colorado.  All  site  locations  are  designated  in
              meters below the site marked '0 m', which is ~2,850 m
              above  the  confluence  of  the  Snake  River  with  Deer
              Creek.
             B. Sketch of 'Intensive Sampling Zone' showing locations
              of 'Stream' and 'Inflow' sampling sites.

## Sampling Scheme

     Results  are  presented  from  (1)  a  study  of  spatial  variation,
and  (2)  a  study  of  temporal  variation.   For  the  spatial  study,
samples were obtained from 14 instream locations and eight inflows.
(Fig. 2B for locations.   All  site  locations  are  designated  in
meters  below  the  site  marked  '0 m'  on  Fig. 2A.   Site  '0 m'  is
~2,850 m  above  the  confluence  of  the  Snake  River  with  Deer  Creek.)
The  eight  inflows  represent  a  portion  of  the  visible,  surface
inflows  to  the  reach  of  the  Snake  River  within  the  intensive
sampling  zone.   These  samples  were  collected  between  the  hours  of
1050  and  1406  on  September  13,  1984.  For  the  temporal  study,
samples  were  collected  between  the  hours  of  1800  on  September  9,
1984 and 1700 on September 10, 1984.   Samples  were  collected  on
approximately  an  hourly  basis  at  the  site  designated  2,845  m
(Fig. 2A).   At  the  site  designated  628  m,  samples  were  collected  at
six arbitrarily spaced times.

## Sampling Methods

Samples were filtered immediately upon collection through 0.4-μm Nuclepore filters. Polyethylene sample bottles were rinsed in the field with stream water. Samples for sulfate were stored untreated. Samples for calcium and iron were acidified with an addition of 0.5 mL concentrated nitric acid to 250 mL of sample. pH was measured in the field with Beckman pHI-21 meters.

## Laboratory Procedures

Sulfate anaylses were performed on a Dionex 2000i ion chromatograph. Calcium and iron analyses were performed on a Jarrell-Ash 975 inductively coupled plasma spectrometer.

## RESULTS

### Data

The concentration ranges for sulfate, calcium, [H$^+$], and iron are presented in Table 1 for both the spatial and temporal sampling studies. The detailed data are graphically presented in Figures 3 and 4 in the form of the normalized concentration computed from Equation (1). This normalization is with respect to the maximum and minimum values of the sampled inflow concentrations. These concentrations are listed in the first column of Table 1. Within a 1,000-m subreach of the 2,845-m study reach, eight inflows were sampled. Although the concentrations of all solutes in unsampled inflows may not necessarily have been within the range defined by these eight samples, sulfate and iron concentrations in the eight sampled inflows spanned an order of magnitude and thus provide an appropriate range for performing the normalization.

Figure 3 presents the normalized concentration data from the spatial sampling for sulfate, calcium, [H$^+$], and iron. In this figure, the normalized concentrations of the inflows range, by definition, from 0.0 to 1.0. The sulfate and calcium concentrations showed patterns similar to each other. The most evident features are (1) the instream concentration values were between the extremes of the inflow values, and (2) the variation of instream values was small compared to that of the inflows. Furthermore, the spatial trends of the instream concentration were generally positive near the locations of the highest inflow concentration. The instream values of [H$^+$] were higher than most of the inflow values. The iron concentrations showed two patterns markedly different from the sulfate. The normalized iron concentrations of all but one of the inflows was below 0.7; however, the normalized concentrations of all instream samples was greater than 0.7. The total range of the instream sulfate concentrations was approximately 0.1 normalized units. In contrast, the sampling point-to-sampling point change in iron concentrations was as large as 0.15.

Table 1.   Concentration Ranges in Intensive Sampling of the Snake
           River September 1984.

| | Spatial sampling | | Temporal sampling | |
|---|---|---|---|---|
| | Inflows | Stream | 628 m | 2,845 m |
| $SO_4$ (mgL$^{-1}$) | | | | |
| Maximum | 132.6 | 54.8 | 42.8 | 49.6 |
| Median | 47.8 | 53.1 | 42.4 | 48.4 |
| Minimum | 12.9 | 44.3 | 41.0 | 47.3 |
| Ca (mgL$^{-1}$) | | | | |
| Maximum | 10.30 | 6.26 | 4.77 | 7.04 |
| Median | 6.06 | 5.56 | 4.72 | 6.44 |
| Minimum | 3.71 | 4.78 | 4.49 | 6.17 |
| pH | | | | |
| Maximum* | 3.75 | 3.86 | 3.88 | 3.98 |
| Median | 4.38 | 3.88 | 4.07 | 4.19 |
| Minimum* | 5.82 | 3.95 | 4.11 | 4.28 |
| Fe (mgL$^{-1}$) | | | | |
| Maximum | 1.34 | 1.55 | 1.52 | 1.02 |
| Median | .19 | 1.36 | 1.38 | 0.85 |
| Minimum | .07 | 1.00 | 1.13 | 0.67 |

*Maximum and Minimum with respect to [H$^+$], not numerical value
of pH.

Figure 4 presents the normalized concentration data from the
temporal sampling. The values from both the sites at 628 m and
2,845 m are shown. The sulfate values at both sites remained
almost constant throughout the sampling period. The calcium values
ranged over 0.14 normalized units. This range was, again by
definition, small compared to the range of the inflows. The values
of [H$^+$] vary by as much as 0.3 normalized units. As in the spatial
sampling, the iron concentrations showed patterns markedly
different from the sulfate. First, the normalized iron concentrations were higher than the sulfate values. Thus, relative to inflow
concentrations, the iron concentrations in the stream are higher
than the concentrations of the sulfate. Second, at both sites the
iron values ranged over 0.3 normalized units.
    As shown in Figures 3 and 4 there were similar degrees of
variation for [H$^+$] and iron. Figure 5 shows iron concentration
versus pH, with the iron represented as the log of molar iron
concentration. When all of the samples are taken into account, the
anticipated pattern of increased iron with decreased pH is evident.
However, for a given iron concentration the range for pH was 0.2 to
0.5 pH units. A clear one-to-one relationship between the
variation of iron concentrations and pH was not defined by this
data set.

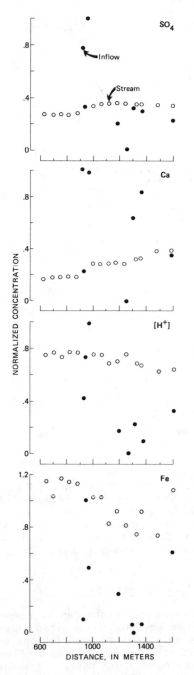

Figure 3.   Normalized sulfate, calcium, [H⁺], and iron concentrations along the Snake River and from selected inflows.
Samples obtained on September 13, 1984.

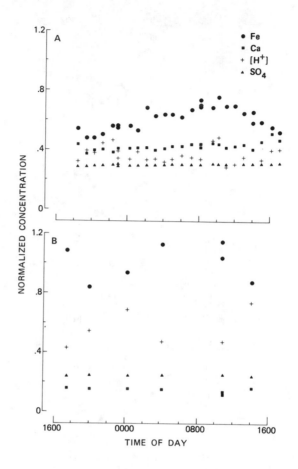

Figure 4.   Normalized sulfate, calcium, [H⁺], and iron concentra-
            tions in the Snake River between 1700 September 9, 1984
            and 1700 September 10, 1984.
                A.  Samples obtained at 'Site 2,845 m' (approximately
                    5 m above the confluence with Deer Creek).
                B.  Samples obtained at 'Site 628 m'.


## Comparison with Inflow Concentrations

     Measured inflow concentrations can be compared with concentra-
tions estimated from mass balance.  This comparison can then be
used to assess the role of inflow variations on the variations in
the stream.  The discharge of the Snake River was measured on
September 6, 1984 using a pygmy meter.  The discharge at the site
designated '628 m' was 0.16 $m^3s^{-1}$, and at '2,845 m' discharge was
0.28 $m^3s^{-1}$.  To perform the inflow comparison it was assumed that
discharge had not changed significantly in the week between the

time of discharge measurement and of chemical sampling, and that
discharge increased linearly with distance between the two measurement sites.    It is most likely that errors associated with
these assumptions were appreciably less than the order of magnitude
variations in measured inflow concentrations.

Figures 6A, 6B, and 6C present the estimated concentrations
from the inflows for sulfate, calcium, and iron, respectively.    The
estimated inflow concentrations were computed using Equation (4).
This analysis was approximate; i.e., better estimates could be
obtained using concentration values above, below, and in every
inflow, and discharge measurements at every sampling site.

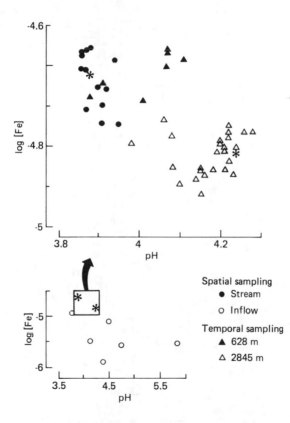

Figure 5.    Log [Fe] (molar concentration) versus pH for all samples
            collected in the spatial and temporal sampling studies
            at the Snake River study site.    The lower figure
            contains all of the data from the inflow sites. The
            boxed area is expanded in the upper figure and contains
            all of the data from the stream sites.

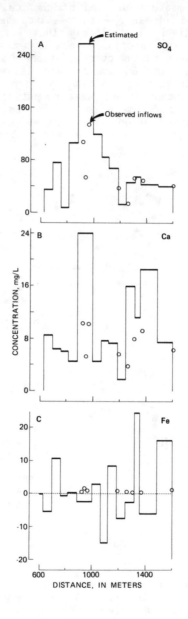

Figure 6. Comparison of estimated inflow concentrations to observed concentrations along the Intensive Sampling Zone in the Snake River.
    A. Sulfate.
    B. Calcium.
    C. Iron.

In Figure 6A there is a clear correspondence between the actual measured inflow concentrations and those estimated subject to several approximations. The maximum estimated sulfate inflow concentration was significantly greater than the highest value actually observed. However, the estimated inflows of highest and lowest concentration were located along the stream reach in the areas of the highest and lowest observed concentration. For calcium (Fig. 6B) the correspondence between the estimates and observations was not as consistent as for sulfate (Fig. 6A). However, over most of the distance along the stream, the estimated calcium values were within the range of the observations and the highest of the estimates are within a factor of three of the observations. In contrast to the sulfate and calcium, the estimated iron values were significantly different from the observations (Fig. 6C). The clearest deviation was in the estimation of several negative concentrations. This result does not signal an error in either the equation development or the field data. Rather, a loss of iron mass in the stream channel by such a process as precipitation of hydrous iron oxide would result in the estimated inflow concentrations being negative. The second deviation was in the estimation of several high inflow concentrations which could be interpreted as an instream iron source, from photoreduction for example. Several of these estimates were higher than the observations by an order of magnitude; not a factor of two to three as for sulfate and calcium.

## DISCUSSION AND SUMMARY

In Figure 7 the ranges and the medians of sulfate, calcium, [$H^+$], and iron are presented in terms of the normalized concentration. The variations of these solutes may be described as: (1) Sulfate concentrations in the stream stay approximately constant near the median value for the inflows; (2) Calcium concentrations vary spatially but also stay near the median value for the inflows; (3) [$H^+$] values were generally variable and high relative to either sulfate or calcium values; and (4) Iron concentrations vary significantly, both spatially and temporally. Iron concentrations in the stream are also significantly higher than the median value for the inflows and for a few cases exceeded the highest sampled inflow concentration.

The exercise of estimating the inflow concentrations demonstrated that instream changes in sulfate concentrations, and probably calcium concentrations, could be approximately accounted for by the inflows. Variation of the iron concentrations showed no consistency with the inflow concentrations.

In summary, significant variations in the concentration of iron were observed in the acidic Snake River. These variations were on characterisitic spatial scales as short as tenths of a kilometer and on time scales as short as hours. Variations of two other solutes, sulfate and calcium, were minor by comparison.

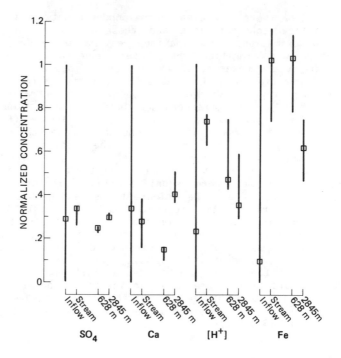

Figure 7.    Comparison of ranges and median concentrations of normalized concentrations for sulfate, calcium, [H⁺], and
iron.   Symbols along lines denote the median concentration values.   Ranges are shown for inflows and in stream
concentations from the intensive spatial sampling and at
Sites 628 m and 2,845 m from the intensive temporal
sampling.

The analysis presented in this paper has important implications
for verification of monitoring designs.    The impact of acidic
inputs to upland watersheds is an area of active process-oriented
research as well as continuing efforts to provide documentation of
actual trends.    In the design of monitoring schemes for a reactive
solute, the scale of the spatial and temporal variations of the
underlying processes should be taken into account.    Typically,
sampling on these scales will be relatively intensive compared to
the sampling demands of routine monitoring programs.    The situation
in the Snake River may not be similar to that of most streams, but
the documentation of the concentration variation within this
particular stream emphasizes the importance of intensive reconnaissance sampling as the basis for establishing monitoring programs.
Intensive sampling can indicate if a solute is reactive and thus
also can indicate the presence of active in-stream processes.
Identifying in-stream reactivity is a logical first step prior to
detailed studies of reaction processes in stream environments.

Furthermore, from a long-term monitoring perspective, it is critical to empirically ascertain the relative impacts of active in-stream processes and passive mixing inputs in adjusting solute concentrations. This study underscores Walling's conclusion that "the extent and rapidity of solute variations must not be underestimated," (Walling, 1975).

## ACKNOWLEDGMENT

Joe Blattner determined the stream discharge. Larry Barber, Peter Dunn, Judy McHugh, Christine Miller, Dorothy Pickney, Elaine Smith, Richard Smith, and Eric Stiles assisted with the sample collection. Eric Stiles performed the calcium and iron analyses. Gary Zellweger performed the sulfate analyses and has worked collaboratively with us on all aspects of the studies at the Snake River site on which this chapter is based.

## DISCLAIMER

The use of trade or product names in this chapter is for identification purposes only and does not constitute endorsement by the U.S. Geological Survey.

## REFERENCES

McKnight, D. M., and G. L. Feder. 1984. The ecological effect of acid conditions and precipitation of hydrous metal oxides in a Rocky Mountain stream in Colorado. Hydrobiologia. 119(2):129-138.

Rose, A. W., H. E. Hawkes, and J. S. Webb. 1979. Geochemistry in mineral exploration, Second Edition. Academic Press, New York.

Theobald, P. K., H. W. Lakin, and D. B. Hawkins. 1963. The precipitation of aluminum, iron, and manganese, at the junction of Deer Creek with the Snake River in Summit County, Colorado. Geochim. Cosmochim. Acta. 27(2):121-132.

Walling, D. E. 1975. Solute variations in small catchment streams: some comments. Transactions of the Institute of British Geographers. 64:141-147.

# CHAPTER 16

REACTION PRODUCTS OF MANGANESE-BEARING WATERS

Carol J. Lind, John D. Hem, and Charles E. Roberson,
U.S. Geological Survey, Menlo Park, California

ABSTRACT

In natural waters the mineral form of manganese oxides depends on slowly fluctuating temperatures and solute composition and concentration. Oxide deposition tends to occur on surfaces at redox fronts where water that contains dissolved manganese under reducing conditions encounters oxygen or possibly other oxidizing agents. One or more of several metastable manganese solid phases that initially precipitate will change to more stable products. Metastable $\gamma$MnOOH may remain for extended time periods. The final product has Mn(IV) as the major cation with lesser amounts of lower valent manganese and possibly other cations. Organic matter as well as changing solution conditions can interfere with the transformation of dissolved Mn(II) to Mn(IV) oxides.

Manganese oxide deposits commonly are associated with iron oxides and silicates in discrete but adjacent areas or layers. Trace metals are concentrated in manganese oxides by adsorption, cation exchange, and coprecipitation. Mechanisms of coprecipitation can include substitution of Zn(II) for Mn(II), Co oxidation at the surface of $\beta$MnOOH, and redox interaction of Mn with Cu, Ni, and Pb.

INTRODUCTION

Manganese oxides are widely dispersed in nature in contact with surface water and seawater in the form of suspended materials, sediments, coatings, and nodules and in contact with ground water and soils as micronodular deposits, as thin layers on mineral surfaces, or interspersed in clay minerals (Koljonen et al., 1976; Ross et al., 1976; Glasby, 1984). The oxides also can be found as dendrites coating rock surfaces where water has been leaching rocks or filling cracks in solid rock materials (Potter and Rossman,

*Chemical Quality of Water and the Hydrologic Cycle*, Robert C. Averett and Diane M. McKnight (Eds.) © 1987 Lewis Publishers, Inc., Chelsea, Michigan. Printed in the United States of America.

1979a).   The  form  and  location  of  these  reaction  products  are
important  as  a  sample  medium  for  geochemical  prospecting  and  as  a
key  to  the  availability  not  only  of  manganese  but  also  of  various
trace  metals  as  nutrients  and  as  toxic  substances  for  aquatic
organisms  (Nowlan,  1976;  Godo  and  Reisenauer,  1980;  Van  Der  Werff,
1984;  Roeming  and  Donovan,  1985).

This  chapter  describes  manganese  oxide  precipitation  in  natural
water  systems,  discusses  thermodynamic  predictions,  and  presents
some  of  the  more  common  manganese  oxide  interactions  with
associated  substances.    Much  of  the  subject  matter  was  drawn  from
published  results  of  work  done  in  the  aqueous  chemistry  research
program  of  the  U.S.  Geological  Survey.    Many  conclusions  by  other
workers  are  also  included  to  help  relate  these  more  theoretical
relationships  to  natural  settings  and  to  supplement  and  verify  our
conclusions.    Helpful  review  comments  were  furnished  by  O.  P.
Bricker,  U.S.  Geological  Survey,  and  by  Prof.  J.  W.  Murray  of  the
University  of  Washington.


MANGANESE OXIDE PRECIPITATION


## Conditions Favorable for Manganese Oxide Accumulation

Manganese  oxide  precipitates  formed  in  natural  water  are
generally  deposited  under  open-system  conditions  and  their  mineral
form  depends  on  slowly  fluctuating  temperatures  and  solute
composition  and  concentrations.    The  soluble  reactants  in  the
system  (aqueous  $Mn^{2+}$,  $OH^-$,  or  $H^+$)  and  the  electron  acceptor
(commonly  a  dissolved  oxygen  species)  are  transported  continuously
to  the  reaction  site  in  moving  water;  soluble  products  leave  the
site  in  the  same  way.    The  water  composition  is  influenced  by
associated  mineral  phases,  infiltrating  water  composition,  and
recharge  rate.    An  example  of  the  continuous  transport  of  reactants
and  reaction  products  to  and  from  the  reaction  site  is  described  by
Renard  et  al.    (1978)  who  studied  manganese  oxide  concretions
forming  on  granitic  rocks  at  the  level  where  a  fresh  water  spring,
high  in  manganese  and  low  in  iron,  is  flowing  out  and  mixing  with
seawater  in  the  intertidal  cove  of  Belmont  (Loire-Atlantique,
France).

Manganese  oxides  will  tend  to  be  deposited  at  redox  fronts
where  water  that  contains  dissolved  manganese,  under  reducing
conditions,  encounters  oxygen  or  other  oxidizing  agents.    Such
conditions  may  occur  in  surface  water  at  the  air-water  interface,
producing  coatings  on  rocks,  at  some  specific  level  within  the
water  column,  or  on  or  within  the  sediment  surface.    Examples  of
these  three  conditions  are  given  below.

(1) In  the  southeastern  United  States  these  coatings  were  found
to  be  confined  to  the  portion  of  the  rock  exposed  to  the
well-oxygenated,  fast-flowing  water  and  did  not  extend
below  the  water-sediment  interface  in  streams  with  muddy

bottom sediments. The coatings were less abundant or absent in areas of sluggish drainage, such as swamps and marshes (Carpenter and Hayes, 1980).

(2) In a study of two streams in Maine, concretionary oxides were shown to be generally formed at an interface between oxygenated stream waters and reduced manganese- and iron-charged sediment-pore waters (Nowlan et al., 1983).

(3) In Oneida Lake, New York, ferro-manganese nodules are concentrated mainly in the shallow, well-oxygenated central areas of the lake in water depths <10 m (Dean and Ghosh, 1978).

In ground waters oxidizing fronts may occur near the water table if water moving downward as recharge differs in redox properties from the main ground-water body. The oxidation fronts may also occur at places along the ground-water flow path owing to changes in vertical or horizontal permeability. Where flow occurs in channels, as in cavernous limestone and in the unsaturated zone above the water table, there may be intermittent or continuous replenishment of oxygen from the atmosphere. In such flow systems oxidation may occur intermittently or continuously at the water-solid interface.

The oxidation and precipitation reactions occur at favorable sites on solid surfaces and preferentially on preexisting manganese oxide surfaces. Wilson (1980) concluded that within the water column the oxidation process generally is not likely to be influenced by natural sediment concentrations and that only with unusually high sediment loads (50 or more ppm) does the water column provide opportunity for sufficient solute-surface interaction to produce measurable changes in the reaction rate. The importance of the solute-surface interaction is illustrated by Diem and Stumm (1984). They found no oxidation of Mn(II), even after 7 yr, in homogeneous Mn(II) solutions which were not oversaturated with respect to $MnCO_3$ and $Mn(OH)_2$. The solutions, stored in the dark at 20°C and occasionally stirred, did contain dissolved oxygen in excess of that required for the oxidation of Mn(II) but no surface catalysts or manganese bacteria were present. Even in the presence of small concentrations of $MnO_x$ (<50 $cm^2$ buserite surface per liter solution) no oxidation occurred for a period of 6 or 7 yr, but there was some catalytic effect in the presence of Co(II) and Ni(II) after 4 or 5 yr. Measurable oxidation rates were observable in solutions oversaturated with respect to $MnCO_3$ or in waters containing manganese bacteria or other surface catalysts. Evidently, in the absence of sufficient surface catalysts, chemical parameters (i.e. pH, $E_h$, and manganese and oxygen concentration) well above saturation are necessary to initiate precipitation.

Once a surface forms the solute-surface interaction proceeds. Other metal ions may be incorporated in the layered manganese oxide precipitate as it grows in thickness. A ground-water system can provide extensive opportunity for solute-surface interaction. Most mineral surfaces in ground-water systems have a significant cation exchange capacity (CEC) per unit area. Although clay minerals have a large CEC per unit weight, the poor water-transmitting properties of clay beds caused by the small size of the clay particles decreases the significance of these materials as CEC sites in ground-water systems and in sediments. In any event, manganese ions are attracted to the exchange sites and may be oxidized there if encountered by an oxidizing agent.

Once it has formed, a manganese oxide surface enhances further manganese oxide precipitation and creates conditions favorable for a multi-step precipitation process. The manganese oxidation rate is enhanced with increasing available surface area and is retarded by the complexation of oxidizable manganese with soluble organic matter (Wilson, 1980). The rates of the autocatalytic crystal growth of $Mn_3O_4$ and of $\beta MnOOH$ are first order with respect to the amount of unreacted manganese and are proportional to the oxide surface area available per unit volume of solution and to the degree of supersaturation with respect to $Mn^{2+}$ (Hem, 1981).

If the water circulation pattern and/or the chemical composition changes to give conditions favorable for manganese reduction, the precipitate may be redissolved. The changes in conditions generally take place slowly but may be hastened by man-induced changes in hydraulic head or by other anthropogenic effects and by dilution or influx of dissolved species from watersheds caused by a heavy downpour or rapid snow melt. Sunda et al. (1983) showed that light enhanced the dissolution of manganese oxides by marine humic substances and Stone and Morgan (1984) not only confirmed these findings but also found that several natural organic compounds reduced and dissolved manganese oxides at pH 7.2 and that at pH 6.35 photoreductions of manganese oxides by marine fulvic acid was much faster than at pH 7.40 to 7.65. These chemical substances are examples of materials that, if added in sufficient concentration, could change conditions in favor of manganese reduction.

## Precipitation Process

Initially, one or more of several metastable manganese solid phases precipitate. If a sufficiently oxidizing medium remains available, these precipitates alter to more stable products by structural rearrangement, protonation, or disproportionation.

The particular reaction that predominates will depend on solution conditions and composition, i.e. temperature, hydrogen ion and manganese concentrations, and the identity of the major anion present. At one atmosphere and at 25°C or higher where nitrate, chloride, or sulfate are the major anions present, the initial precipitate will be primarily $Mn_3O_4$ (hausmannite). Where temperatures are near 5°C and chloride or nitrate are the major anions present, the initial precipitate will be $\beta MnOOH$ (feitknechtite). If sulfate is the major anion present, $\gamma MnOOH$ (manganite) will precipitate. At intermediate temperatures the nitrate and chloride-containing waters will develop mixtures of $\beta MnOOH$ and $Mn_3O_4$ as their initial precipitates and those containing sulfate will develop mixtures of $\gamma MnOOH$ and $Mn_3O_4$. After aging, if a sufficient oxygen supply is maintained and no large excess of manganese is added to the system, these initial products will alter to form more stable oxides. The $\beta MnOOH$ will rearrange to $\gamma MnOOH$; $Mn_3O_4$ will protonate to $\gamma MnOOH$; and $\beta MnOOH$, $\gamma MnOOH$, and $Mn_3O_4$ will disproportionate to $MnO_2$. The $\gamma MnOOH$, although metastable, is sufficiently stable to remain for extended time periods. The reaction affinity (discussed later in the section "Thermodynamics-Predictions") calculated from the $P_{O_2}$, pH, and manganese activity, can predict the potential for a particular solid to be formed (Hem and Lind, 1983).

The formation of $\gamma MnOOH$ in the sulfate medium is attributed to the presence of the $MnSO_4°$ ion pair which may inhibit formation of the $[Mn(OH)_2]_n$ hexagonal structure (the precursor of $\beta MnOOH$) formation. Other ion pairs may possibly exist in natural waters that can have this effect.

Oxalate, a strong manganese chelator, has been shown to inhibit conversion of $Mn_3O_4$ to $\gamma MnOOH$, at pH 7.4 ± 0.2 for a period of a year. The inhibition capacity increases with oxalate concentration (Lind, 1986). Under near natural conditions and in the absence of other solute or suspended matter, it has been suggested that the rate-determining step in $\gamma MnOOH$ formation is the transformation of the rapidly formed initial products hausmannite ($Mn_3O_4$) and feitknechtite ($\beta MnOOH$) to manganite ($\gamma MnOOH$) (Giovanoli, 1980b). Because $\gamma MnOOH$ persists metastably, conditions regulating $\gamma MnOOH$ formation (i.e. concentrations of reactive organic constituents) would influence the fate of aqueous manganese. In addition to inhibiting $\gamma MnOOH$ formation the oxalate complexing capacity also increases the total dissolved manganese concentration and decreases the free $Mn^{2+}$ concentration. Oxalate models the action of many dissolved organic compounds and in specific environments (e.g. some algal blooms, sediments, and root zones) oxalate concentrations alone may be sufficiently high to cause these effects (Lind, 1986).

## Biological Mediation of Manganese Oxidation

Under certain circumstances, biological mediation of manganese oxidation may occur in natural water by providing additional surfaces for precipitation and by contributing to the chemical activity in terms of $E_h$, pH, chemical species (especially organic materials), and by catalytic action. This chapter does not thoroughly describe this type of manganese oxidation mediation but the following discussion gives some examples and illustrates that biological mediation is a factor to be considered in some settings.

Biogenic controls may be dominant in many lacustrine deposits (Crerar et al., 1980). Although the soil zone is small in comparison to the other water-mineral interaction zones in aqueous systems important controlling reactions may occur there because of intensive soil biological activity. In soils microbial activity is largely responsible for oxidation and reduction of manganese compounds as well as for the formation of manganese concretions and is one of the most important factors governing the solubility and thus the bioavailability of manganese (Kabata-Pendias and Pendias,

1984). In addition, physiochemical relationships between bacteria and mineral surfaces may lead to diverse effects of dissolution and secondary precipitation of trace metal ions. These effects include trace metal valence changes and/or organometallic compound formation.

As for the biological mediation mechanisms in natural waters, the previous discussion concerning oxalate illustrates one possible contribution due to biologically produced organic matter. Kabata-Pendias and Pendias (1984), in discussing soil bacteria, conclude that the two main types of metal uptake by microorganisms are nonspecific binding of the cation to cell surfaces, slime layers, extracellular matrices, etc., and metabolic-dependent intracellular uptake. Polygalacturonic acid, a common constituent of the outer slime layer of bacterial cells, can complex several trace metals. Hastings and Emerson (1986) describe the dormant spores of marine *Bacillus* as being able to bind and oxidize manganese extracellularly at pH values below which Mn(II) autooxidizes. The catalytic mechanism is believed to be the complexing of divalent manganese by a spore coat protein. The manganese is then rapidly oxidized. Biological manganese oxidation by living organisms was shown in Saanich Inlet and in Framvaren Fjord, by Tebo et al. (1984). They showed that manganese and cobalt binding by biological catalysts was inhibited by bacterial poisons and that manganese was being oxidized and not simply bound.

The possible magnitude of biological mediation is demonstrated by marine *Bacillus* used in the laboratory work of Hastings and Emerson (1986). In seawater at pH 7.8 dormant spores of marine *Bacillus* facilitated divalent manganese oxidation so that initially it was four orders of magnitude faster than would be expected by abiotic autocatalysis on a colloidal $MnO_2$ surface. As the spores became coated with manganese oxide, the rate became slower and approached that for abiotic surface catalysis. At pH 6.5 these spores did not exhibit this catalytic property.

Results obtained in the laboratory in pure cultures of microorganisms may differ from those naturally occurring in soils because more than 80% of soil microorganisms are believed to be adsorbed to organic matter and clay minerals (Kabata-Pendias and Pendias, 1984).

## Reaction Mechanisms

The protonation and disproportionation processes are cyclic. First the metastable oxides of $Mn_3O_4$ or $MnOOH$ form and release hydrogen ions and then the higher oxides form by taking up hydrogen ions and releasing divalent manganese. The released manganese reprecipitates again as metastable manganese oxide. Equations 1 through 7 describe these processes.

<u>The $Mn^{2+}$ - $Mn_3O_4$ - $\gamma MnOOH$ Cycle</u>

$$Mn^{2+} + 1/6\ O_2aq + H_2O \rightarrow 1/3\ Mn_3O_4 + 2\ H^+ \tag{1}$$

$$Mn_3O_4 + 2\ H^+ \rightarrow 2\ \gamma MnOOH + Mn^+ \quad \text{(protonation)} \quad (2)$$

### The $Mn^{2+}$ - $Mn_3O_4$ - $MnO_2$ Cycle

$$2\ Mn^{2+} + 1/3\ O_2aq + 2\ H_2O \rightarrow 2/3\ Mn_3O_4 + 4\ H^+ \quad (3)$$

$$Mn_3O_4 + 4\ H^+ \rightarrow MnO_2 + 2\ Mn^{2+} + 2\ H_2O \quad \text{(disproportionation)} \quad (4)$$

### The MnOOH Phase Change

$$\beta MnOOH \rightarrow \gamma MnOOH \quad \text{(rearrangement)} \quad (5)$$

### The $Mn^{2+}$ - MnOOH - $MnO_2$ Cycle

$$Mn^{2+} + 1/4\ O_2aq + 3/2\ H_2O \rightarrow MnOOH + 2\ H^+ \quad (6)$$

$$2\ MnOOH + 2\ H^+ \rightarrow MnO_2 + Mn^{2+} + 2\ H_2O \quad \text{(disproportionation)} \quad (7)$$

Figure 1 illustrates the various oxidation pathways and Figures 2A, 2B, and 2C illustrate the crystal forms of $Mn_3O_4$, $\beta MnOOH$, and $\gamma MnOOH$. Figure 2C also shows how the $Mn_3O_4$ crystals reprecipitate on the $\gamma MnOOH$ needles.

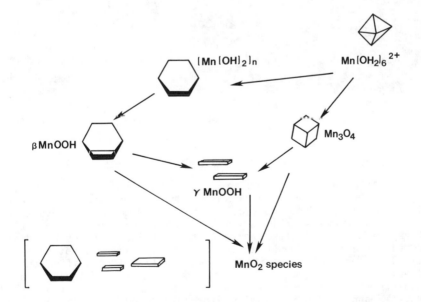

Figure 1.  Some manganese oxidation pathways possible in natural water settings.

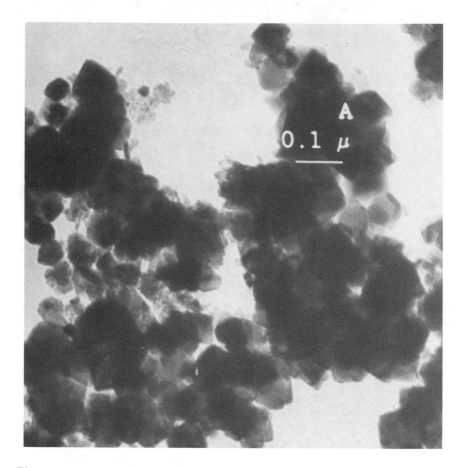

Figure 2A.  Electron micrograph showing the crystal form of $Mn_3O_4$.

## Commonly Identified Manganese Oxide Minerals in Natural Settings

Naturally occurring manganese oxide, generally expressed as $MnO_2$, has $Mn^{4+}$ as the major cation with lesser amounts of lower valent manganese and possibly other cations.  Some of the naturally occurring manganese oxide minerals frequently mentioned in the literature are pyrolusite, ramsdellite, nsutite, birnessite, disordered birnessite, vernadite, buserite, todorokite, cryptomelane, hollandite, and lithiophorite.  Giovanoli (1969) classified these oxides as true $MnO_2$ modifications--pyrolusite ($\beta MnO_2$) and ramsdellite, nsutite family ($\gamma MnO_2$), birnessite family (formerly "$\delta MnO_2$"), and hollandite family (formerly "$\alpha MnO_2$").  The composition of pyrolusite most closely approximates the simplified $MnO_2$

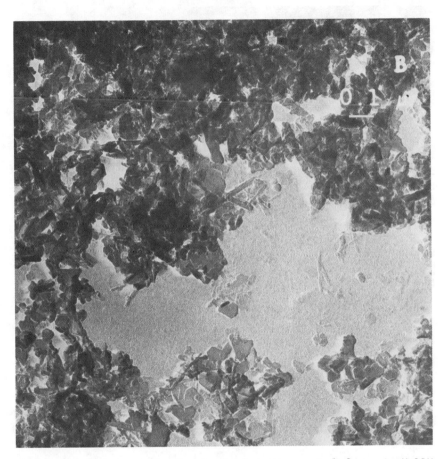

Figure 2B.   Electron micrograph showing the crystal form of βMnOOH.

formula (Frenzel, 1980) but even this mineral generally does not attain the 63.2% Mn content of $MnO_2$.  At the other extreme is birnessite ($\delta MnO_2$) which really is not a true $MnO_2$ form (Frenzel, 1980).  Usually the birnessite phases are so poorly crystallized that their forms can only be recognized under the electron microscope and their X-ray diffraction (XRD) patterns have few lines, sometimes having only two indistinct lines, d≅2.4 and 1.4 Å, and occasionally d≅7.3 Å.

Figure 2C.    Electron micrograph showing the crystal form of γMnOOH needles with tiny $Mn_3O_4$ crystals precipitated on the needles.

Giovanoli and Brütsch (1979) classified the less pure manganese oxides as follows:

Me-Buserite (=10Å-Me-manganate) as $[Mn_5^{4+}(^{Me^{2+}}_{Mn^{3+}})O_{12}]^+ \cdot [OH^-(H_2O)_{11}]^-$
and

Me-Birnessite (=7Å-Me-manganate) as $[Mn_5^{4+}(^{Me^{2+}}_{Mn^{3+}})O_{12}]^+ \cdot [OH^-(H_2O)_5]^-$.

They defined

Na-Buserite (Na-10Å manganite) as $Na_4Mn_{14}O_{27} \cdot 21H_2O$ and

Na-Birnessite (Na-7Å manganite) as $Na_4Mn_{14}O_{25} \cdot 9H_2O$.

Giovanoli et al. (1975) classified

Mn-birnessite ($Mn_7O_{13} \cdot 5H_2O$) as a 7Å manganite.

Over the years there has been extensive discussion concerning the names and structural identities of naturally occurring manganese oxides. Giovanoli (1980a) and Chukhrov and Gorshkov (1980) discuss the terms vernadite and random stacked birnessite. Giovanoli (1985) and Burns et al. (1983 and 1985) discuss the terms todorokite, buserite, and romanechite (psilomelane). The authors of the last two papers agree that todorokite is a valid mineral form but otherwise have some differences of opinions regarding these minerals.

Birnessite was the major manganese oxide coating identified on gravel and sand, and in soil, stream deposits and desert varnish, but X-ray amorphous material was also frequently observed (Koljonen et al., 1976; Ross et al., 1976; McKenzie and Osmund, 1977; Potter and Rossman, 1979a and b; Taylor et al., 1964; Taylor, 1968). Birnessite usually occurs in alkaline soils (Taylor et al., 1964) but has been observed in some acid soils (Ross et al., 1976). Todorokite has been identified in calcareous soil horizons (Ross et al., 1976) and in dendrites collected in underground mine workings while romanechite or a hollandite-group mineral has been identified in dendrites collected in surface exposures (Potter and Rossman, 1979a). Lithiophorite and hollandite are among common crystalline manganese oxide minerals in soils and todorokite and pyrolusite are less common (McKenzie and Osmund, 1977). Giovanoli (1980b) concluded that there is strong evidence that buserite can precipitate in the presence of transition metals and that cryptomelane and hollandite can precipitate in the presence of $K^+$, $Ba^{2+}$, and some other ions.

These natural manganese particulates are difficult to identify because of their extreme microcrystallinity and because they coexist with and coat a multitude of other minerals, biological forms, and organic materials. Consequently, the initial, freshly formed oxidation products described in the above equations are often unidentifiable by XRD and may comprise some of the unidentified X-ray amorphous natural material.


THERMODYNAMICS

Equilibrium controls influence diagenetic behavior of manganese during marine and lacustrine sedimentation (Crerar et al., 1980).

Predictions

The thermodynamic feasibility for a chemical reaction to occur under a specific set of conditions may be evaluated by comparing the computed activity quotient ($Q$) with the thermodynamic equilibrium constant ($K$) for the reaction. If the value of $Q$ is less than that of $K$ the reaction will be thermodynamically favored to proceed to the right as written. The net change in standard free energy in a reaction, $\Delta G_R^o$, is related to the equilibrium constant by the relationship

$$-\Delta G_R^o = RT \ \ln K \tag{8}$$

where R is the gas constant, T is the temperature in degrees Kelvin and $K$ is the equilibrium constant. Thermodynamic feasibility can be evaluated in energy units by comparing observed or computed free energies of reaction.

Because these terms have not been uniformly applied in the literature the thermodynamic feasibility of reaction is expressed in this chapter in terms of the reaction affinity, $A$. This quantity is defined (Rysselberghe, 1963; Prigogine, 1978) as

$$A = -RT \ (\ln Q - \ln K). \tag{9}$$

At standard conditions (temperature = 25°C and pressure = 1 atmosphere) and using base-10 logarithms to give $A$ values in kilocalories:

$$A = -1.364 \ (\log Q - \log K) \text{ at } 25°C. \tag{10}$$

A positive reaction affinity value indicates a favorable potential for the reaction to proceed to the right, a zero reaction affinity indicates equilibrium conditions, and a negative reaction affinity indicates no potential for the reaction to occur. The rate at which the reaction actually proceeds will also depend on the reaction kinetics of the processes involved. Thus, if the first reaction in the process is much more rapid than the second, the main products present will be the products of the first reaction.

Manganite and metastable $MnO_2$ phases are examples of intermediate mineral forms that may persist in natural-water settings. There is a range of solution parameters that favor the formation of metastable manganite ($\gamma MnOOH$) as an intermediate along the oxidation pathway from Mn(II) to $MnO_2$. Under these conditions the mechanism governing transformation to $MnO_2$ may become more difficult to maintain, and because $\gamma MnOOH$ is oxidized slowly, $\gamma MnOOH$ may persist metastably. In regard to the $MnO_2$ phases, although pyrolusite has the lowest standard free energy of formation, all four $MnO_2$ crystalline forms (pyrolusite, ramsdellite, nsutite, and birnessite) occur naturally. This is possible because the free energies of the $MnO_2$ crystal forms differ by less than 2%. Because of these small differences there may be little thermodynamic incentive for equilibration once a kinetic path has produced one of the three metastable oxides (Crerar et al., 1980).

Equilibrium conditions in aerated solutions for the reactions described in Equations 1 through 4, 6 and 7 are plotted as lines in Figures 3 through 6 in terms of manganese ion activity and pH. Birnessite, as defined by Bricker (1965), is the $MnO_2$ form referred to in this discussion and the birnessite free energy determined by him has been applied to calculations and diagrams. Solutions with manganese activities and pH values that plot to the right of the lower lines or that plot to the left of the upper lines have positive reaction affinities for the formation of the solid product specified in the equation for the line in question. Thus, the area between the two lines in each graph defines conditions favorable for both reactions to occur and the cyclic process may proceed as described in the section "Reaction Mechanisms." The lines in Figures 3 through 5 represent activities at standard conditions of temperature and pressure and Figure 6 represents activities at 5°C. When Figures 3, 4, and 5 are overlaid, the overlapping cyclic areas indicate the $Mn^{2+}$ activities and pH ranges over which more than one of the described cycles may occur. However, barring kinetic factors the reaction with the most positive reaction affinity would have the greatest potential to predominate.


## Laboratory and Field Data Confirmation

Laboratory experiments were carried out in which the manganese was initially precipitated from 0.01 M manganese solutions by titrating at the specified pH with microadditions of NaOH and simultaneously flushing with $CO_2$-free air while maintaining constant temperature (Hem and Lind, 1983). The first set was held at pH 8.5 and 25 to 35°C during preparation and had chloride, nitrate, or sulfate as the anion. The second set was held at pH 9.0 and near 0°C during preparation and had sulfate as the anion. The third was held at pH 8.5 and near 5°C during preparation and had chloride or nitrate as the anion present. The temperatures, pH values, and the major anions used in these laboratory experiments and the experiments mentioned later in this chapter may be found in many natural-water settings. According to Hem (1985), most United States ground waters have a pH range of 6.0 to 8.5 and unpolluted river water generally has a pH range of 6.5 to 8.5 but, where there is photosynthesis, the pH value may go as high as 9.0 during diurnal fluctuations.

After the initial preparation, and from time to time during aging, the solution pH values, the manganese concentrations, the oxidation number of the solids, and the crystal form of the solids were determined. The crystal form was determined by XRD (where possible), electron diffraction, and electron micrographs. In each case the identified mineral form(s) agreed with the oxidation numbers determined and with the calculated reaction affinities for the cyclic process predicting the identified minerals. Also, the solution parameters were plotted in the areas on the graphs that would predict the identified solids and the accompanying cyclic process (Hem and Lind, 1983). Data from these experiments for aged suspensions are plotted in Figures 3 and 6. Somewhat similar

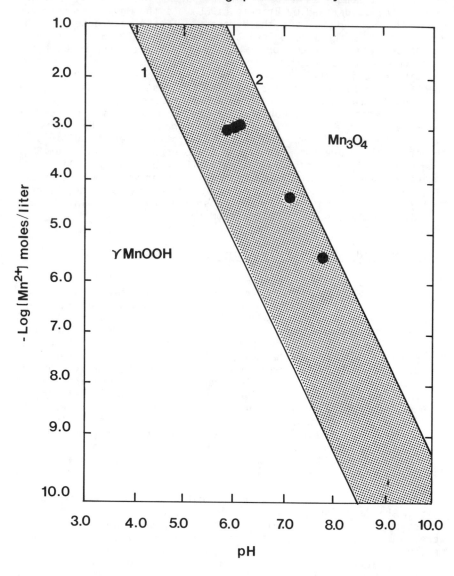

Figure 3.    Lines denoting equilibrium conditions for reactions 1
and 2--the $Mn^{2+}$ - $Mn_3O_4$ - $\gamma MnOOH$ cycle. Both reactions may occur in the areas between the lines. The
dots represent solution concentrations for the aged
laboratory suspensions.

## The $Mn^{2+}$-$Mn_3O_4$-$MnO_2$ cycle

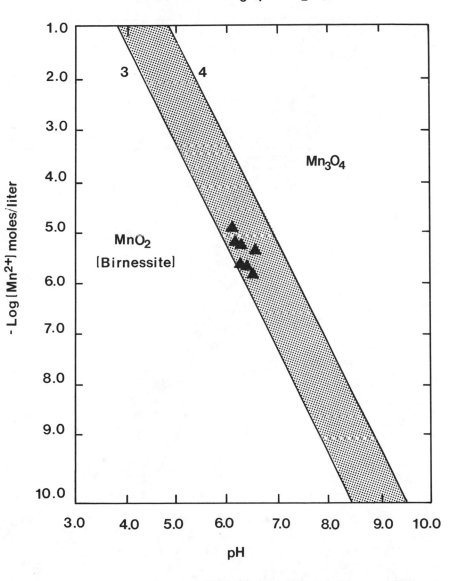

Figure 4.   Lines denoting equilibrium conditions for reactions 3 and 4--the $Mn^{2+}$ - $Mn_3O_4$ - $MnO_2$ cycle.   Both reactions may occur in the areas between the lines.   Triangles represent solution concentrations for seven California springs.

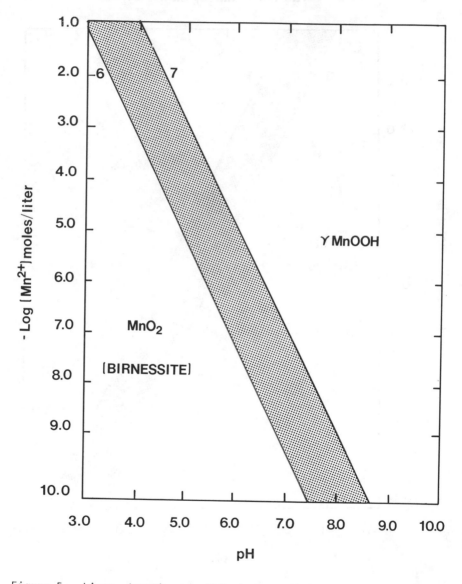

## The $Mn^{2+}$- $\gamma MnOOH$ - $MnO_2$ cycle

Figure 5.    Lines denoting equilibrium conditions for reactions 6 and 7--the $Mn^{2+}$ - $\gamma MnOOH$ - $MnO_2$ cycle.  Both reactions may occur in the areas between the lines.

## The Mn$^{2+}$ - βMnOOH - MnO$_2$ cycle

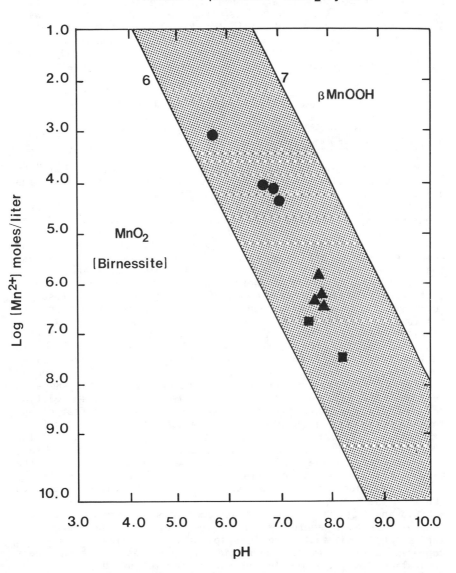

Figure 6.  Lines denoting equilibrium conditions for reactions 6 and 7--the Mn$^{2+}$ - βMnOOH - MnO$_2$ cycle. Both reactions may occur in the areas between the lines. Dots represent conditions for the precipitates in the aged laboratory suspensions; triangles, four manganese bog springs; and squares, drippings from the ceiling and from the ceiling and a stream in a cave containing birnessite deposits.

experiments by Murray et al. (1985), at pH 9.0, 25°C and in $NH_4Cl$, also confirm the suggested reaction pathways. Analytical data from seven California springs are consistent with the reaction cycle, $Mn^{2+}$ - $Mn_3O_4$ - $MnO_2$, and data from four springs from Dawson Settlement Bog manganese deposit, New Brunswick and from drippings and a stream in Matts Black Cave, West Virginia are consistent with the described reaction mechanism cycle, $Mn^{2+}$ - $\beta MnOOH$ - $MnO_2$, (Hem et al., 1982) and are plotted in Figures 4 and 6 respectively.

## MANGANESE OXIDE INTERACTION WITH ASSOCIATED SUBSTANCES

### Manganese-Iron and -Silica Associations

Manganese oxide deposits frequently are associated with iron oxides and silicates in discrete but adjacent areas or layers. The adjacent areas or layers occur in desert varnish, dendrites, particulate matter, stream deposits, soil deposits or wherever manganese oxides are found as nodules (Ross et al., 1976; Potter and Rossman, 1979b; Giovanoli, 1980b). Scanning electron microscopic examination shows that uniformly thick laminations completely encase marine manganese nodules with layers from less than 0.25 μm to greater than 10 μm in thickness (Margolis and Glasby, 1973). Electron microprobe profiles prove beyond any doubt that there is a kind of iron-manganese intergrowth on a microscopic scale and that iron and manganese never coincide in marine nodules (Giovanoli, 1980b). Nowlan et al. (1983) observed a steep $E_h$ gradient reflected by coatings and stains on rocks in stream sediments ranging from no-oxide deposition in the bottom zone, to mostly iron oxides in the middle zone, to manganese oxides containing iron oxides in the upper zone. Postma (1985) proposed some kinetic factors concerning the concentration of manganese in sediments as: (1) The reaction kinetics between Fe(II) in solution and birnessite and probably other manganese oxides are fast enough to be important in natural environments; (2) The reaction is extremely fast below pH 4, releasing $Fe^{3+}$ and $Mn^{2+}$; (3) The reaction is much slower above pH 4, probably because of surface blocking of reactive sites by FeOOH being precipitated on the birnessite; and (4) The reaction rate above pH 4 is controlled by surface reactions rather than by transport of dissolved species to and from the surface.

Giovanoli (1980c) found that amorphous silica can entirely suppress 10 Å-manganate (Na buserite) nucleation and thus allow z-disordered manganate to precipitate. He also found that coprecipitating ferric hydroxide suppresses 10 Å-manganate nucleation but previously prepared ferric hydroxide has no influence. He concluded that the oscillating reaction controlled by depletion and enrichment of silica and ferric hydroxide in the aquatic phase would explain the layered texture found in deep-sea nodules.

Hem (1977) states that the oxidation of ferrous iron by oxygen is substantially faster than the oxidation of manganese at pH 7 to 9 and that iron could be coupled to manganese oxidation as shown in Equations 11, 12, and 13.

$$4\ Fe^{+2} + O_2aq + 10\ H_2O = 4\ Fe(OH)_3 + 8\ H^+ \qquad (11)$$

$$Fe(OH)_3 + H^+ = Fe(OH)_2^+ + H_2O \qquad (12)$$

$$2\ Fe(OH)_2^+ + 3\ Mn^{+2} = Mn_3O_4 + 2\ Fe^{+2} + 4\ H^+ \qquad (13)$$

Thus, when there is an influx of divalent iron, the iron rapidly precipitates and provides a surface for the iron-manganese interaction. Continued interaction will produce a layer of manganese oxide on which the cyclic manganese oxidation processes can occur. Seasonal or other changes in iron and manganese concentrations and in other reaction conditions could explain layering of manganese and iron in such deposits.


## Manganese-Trace Metal Associations

The relatively high Co, Ni, and Cu concentrations in deep-sea manganese nodules indicate that marine sediments are potential ore deposits (Burns and Burns, 1977). Manganese wads also often contain significant amounts of Ni, Co and Cu and may represent a future commercial source of these metals (Canterford, 1984). Hydrous ferromanganese nodules, concretions and coatings on rock and mineral particles are common features of many surficial environments and may contain high concentrations of Ag, Co, Cu, Cr, Hg, Mo, Pb, Ti, Zn, as well as Mn and Fe (Robinson, 1983). Hydrous oxides, including those of manganese, are the principal controls on the fixation of Mn, Fe, Co, Ni, Cu, and Zn in soils and fresh-water sediments (Jenne, 1968). Nowlan (1976) found that manganese-iron oxides weakly scavenged Cu, Mo, Pb and Sr and manganese oxides strongly scavenged Ba, Cd, Co, Ni, Tl, and Zn in 400 sieved stream-sediment samples and 325 fluvial concretionary manganese-iron oxide samples. He found that Cu, Mo, and Pb are strongly scavenged where rocks are mineralized. Cobalt was related to manganese in all marine and soil nodules tested by McKenzie (1975) and Ni was related to manganese in all marine and some soil nodules. Strong Co enrichments occur in marine and soil manganese nodules, soils high in manganese, wads, and natural and synthetic minerals such as hollandite, cryptomelane, psilomelane, lithiophorite, birnessite, and $\delta MnO_2$ (Burns, 1976; and McKenzie, 1979). Significant proportions of the cobalt in fresh water and marine manganese nodules commonly consists of $Co^{+3}$ (Dillard et al., 1982). Coupling of manganese oxidation cycles to redox processes involving other adsorbed metals may be important (Hem, 1978).

The zero point of charge, $pH_{Z.P.C.}$ , may be a factor in trace metal association with manganese and iron oxides. For example, Glasby (1984) summarizes the work of many researchers by stating that manganese oxides, negatively charged in seawater at pH 7.8, strongly adsorb cations but by contrast iron oxides, being near their $pH_{Z.P.C}$ , do not do so under these conditions. The $pH_{Z.P.C.}$ of a commonly identified natural manganese oxide, $\delta$-$MnO_2$, is 2.25 (Murray, 1974) while that for one of the more likely initial manganese oxidation products, $Mn_3O_4$, is 5.4 in 0.01 M NaCl and 5.7 in 0.001 M $NaClO_4$ (Tipping and Heaton, 1983). Based on these $pH_{Z.P.C.}$ values, and considering the normal pH values for natural waters mentioned earlier, both $Mn_3O_4$ and $\delta MnO_2$ should strongly adsorb cations in natural waters but $\delta MnO_2$ would likely do so over a greater pH range.

Adsorption kinetics and catalytic behavior become most important in the deposition of marine ferromanganese crusts and nodules. Freshly precipitated Mn oxides have an unusually high adsorption capacity and are important scavengers of trace metals in soils and in marine and freshwater sedimentary environments (Crerar et al., 1980). Experimental evidence has shown that $\delta MnO_2$ adsorbs Co and Zn (Loganathan and Burau, 1973; Loganathan et al., 1977; Murray, 1975a); birnessite adsorbs Pb, Cu, Mn, and Zn (McKenzie, 1979); a manganese oxide adsorbed metals in the order Co>Mn>Zn>Ni >Ba>Sr>Ca>Mg (Murray, 1975b); and nine synthetic manganese oxides and three synthetic iron oxides adsorbed Co, Cu, Mn, Ni, Pb, and Zn (McKenzie, 1980). In fact McKenzie (1980) found that manganese oxides adsorbed up to 40 times more lead than the iron oxides and that all of the oxides except goethite adsorbed lead more strongly than any other ions studied.

Analogs of hausmannite ($Mn_3O_4$) should be produced when divalent metal ions having compatible dimensions and electron configurations are available to substitute for $Mn^{2+}$ in the $Mn_3O_4$ crystal lattice. Hausmannite and several other manganese oxides have spinel-type crystal lattices. The normal spinel crystal structure consists of $O^{2-}$ ions arranged in cubic close-packing with their charges balanced by divalent cations occupying 1/8 of the available tetrahedral coordination sites and trivalent cations occupying half of the octahedral coordination sites. The kind of site a par- ticular cation might be expected to occupy depends on its size, ionic charge, and bonding configuration. Ionic radii suitable for tetrahedral coordination with oxygen are characteristic of a considerable number of divalent metal ions such as $Mn^{2+}$, $Co^{2+}$, $Fe^{2+}$, $Ni^{2+}$, and $Zn^{2+}$. However, at the normal pH and respective metal concentration ranges of natural waters, only zinc coprecip- itates with well-defined manganese oxide ($ZnMn_2O_4$) due to its spinel structure. This oxide, hetaerolite, was produced in laboratory experiments when the entering fluxes of $Mn^{2+}$ and $Zn^{2+}$ were at a ratio of 2:1. Hausmannite and hetaerolite appear to be end members of a solid solution series. Thus, with lower fluxes of zinc, the precipitate is a mixture of hetaerolite and hausmannite. The $Zn^{2+}$ ion is not affected by redox processes involving $Mn^{3+}$ and the similarity in crystal field stabilization energies for tetrahedral coordination of $Mn^{2+}$ and $Zn^{2+}$ would predict this result (Hem and Roberson, 1985). Figure 7 shows hetaerolite crystals produced under similar conditions to that of the hausmannite shown in Figure 2A with the exception that the ratio of Mn:Zn added during preparation was 2:1 while only Mn was added during the hausmannite preparation.

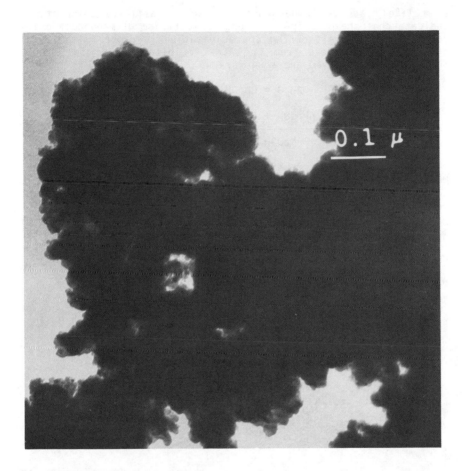

Figure 7. Hetaerolite produced under the same conditions as hausmannite shown in Figure 2A except the ratio of Mn:Zn added during precipitation was 2:1 whereas in Figure 2A only Mn was added.

$Co^{3+}$ may substitute for $Mn^{4+}$ in disordered birnessite, which has an hexagonal crystal structure (Burns, 1976). Cobalt is oxidized by $Mn^{4+}$ at the precipitate surface and appears to form CoOOH at solution values of pH 7 or less (Crowther et al., 1983). In natural, open aqueous systems the adsorbed layers of cobalt can be expected to be covered by continued deposition of manganese oxide as long as a flux of reactants continues to reach the precipitate surface.

Precipitations made in the laboratory at 0 to 5°C generally produce feitknechtite ($\beta$MnOOH) (Hem et al., 1982). This oxyhydroxide has an hexagonal crystal structure similar to that of CoOOH. Coprecipitation reactions involving these species would seem likely at low temperatures. Reaction affinity calculations indicate that the principal scavenging mechanism of manganese oxide for uptake of cobalt during coprecipitation is probably mediation of electron transfers from $Co^{2+}$ to precipitated $Mn^{3+}$ in $\beta MnOOH$. The reactions involved are shown below.

$$Co^{2+} + 2H_2O = Co(OH)_2(c) + 2H^+ \tag{14}$$

$$Co(OH)_2(c) + \beta MnOOH + 2H^+ \rightarrow CoOOH + Mn^{2+} + 2H_2O$$
$$\overline{\phantom{Co(OH)_2(c) + \beta MnOOH + 2H^+ \rightarrow CoOOH + Mn^{2+} + 2H_2O}} \tag{15}$$
$$Co^{2+} + \beta MnOOH = CoOOH + Mn^{2+}$$

The ultimate electron sink (oxidizing agent) is aqueous oxygen. However, the mediation of electron transfers by manganese is capable of intensifying oxidation effects and driving oxidation reactions more toward completion than where manganese oxidation processes are not involved (as in the direct oxidation of $Co^{2+}$ by oxygen). Similar effects can be expected in other geochemical redox processes where manganese is present (Hem et al., 1985). Figure 8 shows a mixture of $Co_3O_4$ and $CoOOH$, aged 23 months, and Figure 9 shows a mixture of $CoOOH$ and $\beta MnOOH$, aged 188 days. The data for both of these are from Hem et al. (1985).

Thermodynamic data predict that cobalt, lead, and nickel can be incorporated into the manganese oxide disproportionation reaction to release higher oxides in the form of $Co_3O_4$, $Ni_3O_4$, and $PbO_2$ (Hem 1978). Equations 16 through 18 describe these reactions.

$$\underline{Co^{2+} \text{ to } Co_3O_4}$$

$$\begin{aligned} 2\,Mn_3O_4(c) + 3\,Co^{2+} + 4\,H^+ \rightarrow \\ MnO_2(c) + Co_3O_4(c) + 5\,Mn^{2+} + 2\,H_2O \end{aligned} \tag{16}$$

$$\underline{Ni^{2+} \text{ to } Ni_3O_4}$$

$$\begin{aligned} 2\,Mn_3O_4(c) + 3\,Ni^{2+} + 4\,H^+ \rightarrow \\ MnO_2(c) + Ni_3O_4(c) + 5\,Mn^{2+} + 2\,H_2O \end{aligned} \tag{17}$$

$$\underline{Pb^{2+} \text{ to } PbO_2}$$

$$\begin{aligned} 2\,Mn_3O_4(c) + Pb^{2+} + 8\,H^+ \rightarrow \\ MnO_2(c) + PbO_2(c) + 5\,Mn^{2+} + 4\,H_2O \end{aligned} \tag{18}$$

By this combination of disproportionation and coprecipitation mechanisms manganese oxides may control minor trace metal concentrations.

The disproportionation coprecipitation process appears to be a possible mechanism for the incorporation of Co and probably of Ni and Pb into marine nodules. The concentrations of Mn and of Co and Ni in seawater at pH 8.2 are compatible with such a disproportionation mechanism. Saturation index values calculated from groups of

Figure 8.   A mixture of $Co_3O_4$ and CoOOH, aged 23 months, initially
            precipitated at pH 9.0, as described in Hem et al.
            (1985).

ten analyses indicated this coprecipitation mechanism controlled
the Co concentration in some stream waters containing metal mine
drainage (Hem, 1978).

   In summation, there are at least three modes of trace metal
coprecipitation for manganese oxides:   (1) Zn substitution for $Mn^{2+}$
in the hausmannite lattice to produce a solid solution series
ranging in composition from $Mn_3O_4$ to $ZnMn_2O_4$; (2) Co oxidation at
the surface of MnOOH to produce microenvironments of Co enrichment
in the manganese oxide structure; and (3) Mn versus Ni, Co, and Pb
redox interactions that create mixtures of distinct mineral forms.

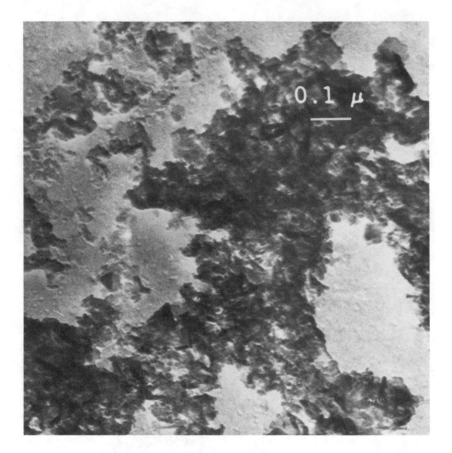

Figure 9.    A mixed precipitate of manganese and cobalt oxynyaroxides, aged 188 days, initially precipitated at pH 9.0, as described in Hem et al. (1985).

## Manganese Carbonate-Manganese Oxide Competition

Manganese carbonates can be found as impurities in calcium and magnesium carbonate deposits, but the potential for the oxidation of manganese over that for the formation of $MnCO_3$ is great enough that even in limestone caves manganese oxide precipitates are found. When conditions are favorable, manganese can quickly equilibrate to form $MnCO_3$. However, for the reaction affinity for the formation of $MnCO_3$ to be as positive as that for the formation of $\beta MnOOH$ in aerated solutions, a $P_{CO_2}$ of $10^{-1}$ atm (atmosphere) would be required (Hem and Lind, 1983). At $P_{CO_2}$ of air, $10^{-3.5}$ atm, only if the oxygen activity is reduced to a value of $10^{-16}$ M, would conditions be favorable for $MnCO_3$ precipitation in preference to manganese oxides.

## CONCLUSIONS

Manganese-bearing natural waters precipitate manganese oxides at solid-solution interfaces located in oxidizing fronts. The $Mn^{2+}$ is attracted to the exchange sites on the solid surfaces and oxidized to a metastable phase when there is an influx of oxidizing agent. With continued oxidizing conditions the metastable oxide alters to a more stable, and sometimes more highly oxidized, form. Slowly fluctuating changes in the water composition and redox conditions may redissolve the oxides and later reprecipitate them. The precipitation rate is positively correlated with the available surface area and along with the $Mn^{2+}$ concentration, may be retarded by the complexing action of organic matter and is influenced by permeability controls on the replenishment of reactants.

Barring the interference of other cations the initial metastable oxide is $\gamma MnOOH$ when precipitated near 5°C in the presence of sulfate and is $\beta MnOOH$ in the presence of chloride or nitrate. When the precipitation temperature is increased, the proportions of these oxides decreases and that of $Mn_3O_4$ increases until at 25°C and above almost all of the initial precipitate is $Mn_3O_4$. Depending on the pH and $Mn^{2+}$ activity, $\beta MnOOH$ will alter to $\gamma MnOOH$; $Mn_3O_4$ will protonate to $\gamma MnOOH$; and $\beta MnOOH$, $\gamma MnOOH$, and $Mn_3O_4$ will disproportionate to a form of $MnO_2$.

The most common forms of manganese oxide identifiable in natural-water systems are $\gamma MnOOH$ and birnessite. There is often much X-ray amorphous material present which may include $\gamma MnOOH$ because of its microcrystallinity. Buserite can precipitate if transition metals are present and cryptomelane and hollandite can precipitate if $K^+$, $Ba^{2+}$ and some other ions are present. Some other $MnO_2$ species that have been identified in soil minerals are todorokite, lithiophorite, romanechite or hollandite-group minerals and pyrolusite. $MnCO_3$ is not likely to form, except where $P_{O_2}$ is extremely low or $P_{CO_2}$ is abnormally high.

Silica, iron oxides, and manganese oxides associate in adjoining but distinctly separate microdeposits. Silica and coprecipitating ferric hydroxide suppress nucleation of certain manganese oxides. The more rapidly precipitating iron oxide can provide a substrate for iron-manganese interactions, and if a sufficient manganese oxide layer is formed the cyclic manganese oxidation process can proceed until a change in solution conditions encourages other reaction mechanisms.

Manganese oxides are known for their content of trace metals. Adsorption, cation exchange, and coprecipitation are mechanisms for uptake of trace metals by these oxides. Coprecipitation can occur through at least three mechanisms: (1) Substitution of $Zn^{2+}$ for $Mn^{2+}$ in the $Mn_3O_4$ structure to produce a solid solution series ranging in composition from $Mn_3O_4$ to $ZnMn_2O_4$; (2) Cobalt oxidation at the surface of $\beta MnOOH$ to produce cobalt enrichment in the manganese oxide structure; and (3) Manganese versus nickel, cobalt, and lead redox interactions that control and create mixtures of distinct mineral forms of oxides of these metals.

## REFERENCES

Bricker, O. P. 1965. Some stability relations in the system $Mn-O_2$ $-H_2O$ at $25°$ and one atmosphere total pressure. Amer. Mineral. 50:1296-1354.

Burns, R. G. 1976. The uptake of cobalt into ferromanganese nodules, soils, and synthetic manganese (IV) oxides. Geochim. Cosmochim. Acta. 40:95-102.

Burns, R. G., and V. M. Burns. 1977. The mineralogy and crystal chemistry of deep-sea manganese nodules, a polymetallic resource of the twenty-first century. Phil. Trans. R. Soc. London. A286:283-301.

Burns, R. G., V. M. Burns, and H. W. Stockman. 1983. A review of the todorokite-buserite problem: Implications to the mineralogy of marine manganese nodules. Amer. Mineral. 68:972-980.

Burns, R. G., V. M. Burns, and H. W. Stockman. 1985. The todorokite-buserite problem: Further considerations. Amer. Mineral. 70:205-208.

Canterford, J. H. 1984. Cobalt extraction and concentration from manganese wad by leaching and precipitation. Hydrometal. 12:335-354.

Carpenter, R. H., and W. B. Hayes. 1980. Annual accretion of Fe-Mn-oxides and certain associated metals in a stream environment. Chem. Geol. 29:249-259.

Chukhrov, F. V., and A. I. Gorshkov. 1980. Reply to R. Giovanoli's comment. Mineral. Deposita. Berlin. 15:255-257.

Crerar D. A., R. K. Cormick and H. L. Barnes. 1980. Geochemistry of manganese: An overview, p. 293-334. *In* I. M. Varentsov and G. Grasselly, eds., Geology and geochemistsry of manganese, v. 1. Akademiai Kiado, Budapest.

Crowther, D. L., J. G. Dillard and J. W. Murray. 1983. The mechanism of Co(II) oxidation on synthetic birnessite. Geochim. Cosmochim. Acta. 47:1399-1403.

Dean, W. E., and S. K. Ghosh. 1978. Factors contributing to the formation of ferromanganese nodules in Oneida Lake, New York. J. Res. U.S. Geological Survey. 6:231-240.

Diem, D., and W. Stumm. 1984. Is dissolved $Mn^{2+}$ being oxidized by $O_2$ in absence of Mn-bacteria or surface catalysts? Geochim. Cosmochim. Acta. 48:1571-1573.

Dillard, J. G., D. L. Crowther, and J. W. Murray. 1982. The oxidation states of cobalt and selected metals in Pacific ferromanganese nodules. Geochim. Cosmochim. Acta. 46:755-759.

Frenzel, G. 1980. The manganese ore minerals. p. 25-158. In I. M. Varentsov and G. Grasselly, eds., Geology and geochemistry of manganese, v. 1. Akademiai Kiado, Budapest.

Giovanoli, R. 1969. A simplified scheme for polymorphism in the manganese dioxides. Separat. Chimia. 23:470-472.

Giovanoli, R., P. Bürki, M. Giuffredi, and W. Stumm. 1975. Layer structured manganese oxide hydroxides. IV: The buserite group; structure stabilization by transition elements. Chimia. 29:517-520.

Giovanoli, R., and R. Brütsch. 1979. Über dieoxidhydroxide des Mn(IV) mit schichtengitter. 5. mitteilung: Stöchiometrie, Austauschverhalten und die Rolle bei der Bildung von Tiefsee-Mangankonkretionen. Chimia. 33:372-376.

Giovanoli, R. 1980a. Vernadite is random-stacked birnessite. Mineral. Deposita. 15:251-253.

_____. 1980b. On natural and synthetic manganese nodules, p. 159-202. In I. M. Varentsov and G. Grasselly, eds., Geology and geochemistry of manganese, v. 1. Akademiai Kiado, Budapest.

_____. 1980c. Layer structured manganese oxide hydroxides VI: Recrystallization of synthetic buserite and the influence of amorphous silica and ferric hydroxide on its nucleation. Chimia. 34:308-310.

Giovanoli, R. 1985. A review of the todorokite-buserite problem: Implications to the mineralogy of marine manganese nodules: Discussion. Amer. Mineral. 70:202-204.

Glasby, G. P. 1984. Manganese in the marine environment. Oceanogr. Mar. Biol. Ann. Rev. 22:169-194.

Godo, G. H. and H. M. Reisenauer. 1980. Plant effects on soil manganese availability. Soil Sci. Soc. Am. J. 44:993-995.

Hastings, D., and S. Emerson. 1986. Oxidation of manganese by spores of a marine bacillus: Kinetic and thermodynamic considerations. Geochim. Cosmochim. Acta. 50:1819-1824.

Hem, J. D. 1977. Surface chemical processes in ground-water systems, p. IV 76-IV 85. *In* Proc. of The 2nd Intern'l. Symp. on Water-Rock Interactions, Strasbourg, France.

Hem, J. D. 1978. Redox processes at surfaces of manganese oxide and their effects on aqueous metal ions. Chem. Geol. 21:199-218.

Hem, J. D. 1981. Rates of manganese oxidation in aqueous systems. Geochim. Cosmochim. Acta. 45:1369-1374.

Hem, J. D., and C. E. Roberson, and R. B. Fournier. 1982. Stability of $\beta$MnOOH and manganese oxide deposition from spring water. Water Resour. Res. 18:563-570.

Hem, J. D., and C. J. Lind. 1983. Nonequilibrium models for predicting forms of precipitated manganese oxides. Geochim. Cosmochim. Acta. 47:2037-2046.

Hem, J. D. 1985. Study and interpretation of the chemical characteristics of natural water. U.S. Geological Survey Water-Supply Paper 2254, Third edition. 263 p.

Hem, J. D., and C. E. Roberson. 1985. Synthesis and stability of divalent-metal manganites, p. 333-335. *In* Preprint of Div. Environ. Chem., Am. Chem. Soc. 190th Nat'l. Meet., Chicago.

Hem, J. D., and C. E. Roberson, and C. J. Lind. 1985. Thermodynamic stability of CoOOH and its coprecipitation with manganese. Geochim. Cosmochim. Acta. 49:801-810.

Jenne, E. A. 1968. Controls on Mn, Fe, Co, Ni, Cu, and Zn concentrations in soils and water: The significant role of hydrous Mn and Fe oxides, p. 337-387. *In* R. F. Gould, ed., Trace inorganics in water (Advances in chemistry series 73), Am. Chem. Soc., Washington, D.C.

Kabata-Pendias, A., and H. Pendias. 1984. Trace elements in soils and plants. CRC Press, Inc., Boca Raton, Fla. 315 p.

Koljonen, T., P. Lahermo, and L. Carlson. 1976. Origin, mineralogy, and chemistry of manganiferous and ferruginous precipitates found in sand and gravel deposits in Finland. Bull. Geo. Soc. Finland. 48:111-135.

Lind C. J. 1986. Alteration of hausmannite oxidation process in dilute oxalate solution, p. A-17. *In* Program and Abstracts of 8th Rocky Mountain Regional Meet., Am. Chem. Soc., Denver.

Loganathan, P., and R. G. Burau. 1973. Sorption of heavy metal ions by hydrous manganese oxide. Geochim Cosmochim Acta. 37:1277-1293.

Loganathan, P., R. G. Burau, and D. W. Fuerstenau. 1977. Influence of pH on the sorption of $Co^{2+}$, $Zn^{2+}$ and $Ca^{2+}$ by a hydrous manganese oxide. Soil Sci. Soc. Am. J. 41:57-62.

Margolis, S. V., and G. P. Glasby. 1973. Microlaminations in marine manganese nodules as revealed by scanning electron microscopy. Geol. Soc. Am. Bull. 84:3601-3610.

McKenzie, R. M. 1975. An electron microprobe study of the relationships between heavy metals and manganese and iron in soils and ocean floor nodules. Aus. J. Soil Res. 13:177-188.

McKenzie, R. M., and G. Osmond. 1977. Manganese oxides and hydroxides, p. 181-193. *In* J. B. Dixon and S. B. Weed, eds., Minerals in Soil Environments, Soil Sci. Soc. Am. Madison.

McKenzie, R. M. 1979. Proton release during adsorption of heavy metal ions by a hydrous manganese dioxide. Geochim. Cosmochim. Acta. 43:1855-1857.

McKenzie, R. M. 1980. The adsorption of lead and other heavy metals on oxides of manganese and iron. Aust. J. Soil Res. 18:61-73.

Murray, J. W. 1974. The surface chemistry of hydrous manganese dioxide. Jour. Coll. Interface Sci. 46:357-371.

Murray, J. W. 1975a. The interaction of cobalt with hydrous manganese dioxide. Geochim. Cosmochim. Acta. 39:635-647.

_____. 1975b. The interaction of metal ions at the manganese dioxide-solution interface. Geochim. Cosmochim. Acta. 39:505-519.

Murray, J. W., J. G. Dillard, R. Giovanoli, H. Moers, and W. Stumm. 1985. Oxidation of Mn(II): Initial mineralogy, oxidation state and ageing. Geochim. Cosmochim. Acta. 49:463-470.

Nowlan, G. A. 1976. Concretionary manganese-iron oxides in streams and their usefulness as a sample medium for geochemical prospecting. Jour. Geochem. Explor. 6:193-210.

Nowlan, G. A., J. B. McHugh, and T. D. Hessin. 1983. Origin of concretionary Mn-Fe oxides in stream sediments of Maine, U.S.A. Chem. Geol. 38:141-156.

Postma, D. 1985. Concentration of Mn and separation from Fe in sediments--I. Kinetics and stoichiometry of the reaction between birnessite and dissolved Fe(II) at 10° C. Geochim. Cosmochim. Acta. 49:1023-1033.

Potter, R. M., and G. R. Rossman. 1979a. Mineralogy of manganese dendrites and coatings. Amer. Mineral. 64:1219-1226.

____. 1979b. The manganese- and iron-oxide mineralogy of desert varnish. Chem. Geol. 25:79-94.

Prigogine, I. 1978. Time, structure, and fluctuations. Science. 201:777-785.

Renard, D., J. Boulégue, E. A. Perseil, and F. Chantret 1978. Concretions de bioxyde de manganese sur des affleurements granitiques dans une zone de balancement des marees. (Anse de Belmont, France)-1-Analyse et mineralogie des encroutements, Mineral. Deposita. Berlin. 13:65-81.

Robinson, G. D. 1983. Heavy-metal adsorption by ferromanganese coatings on stream alluvium: Natural controls and implications for exploration. Chem. Geol. 38:157-174.

Roeming, S. S., and T. J. Donovan 1985. Correlations among hydrocarbon microseepage, soil chemistry, and uptake of micronutrients by plants, Bell Creek Oil Field, Montana. Jour. Geochem. Explor. 23:139-162.

Ross, Jr., S. J., D. P. Franzmeier, and C. B. Roth. 1976. Mineralogy and chemistry of manganese oxides in some Indiana soils. Soil Sci. Soc. Am. J. 40:137-143.

Rysselberghe, P. 1963. Thermodynamics of irreversible processes. Blaisdell Pub. Co., New York. 165 p.

Stone, A. T., and J. J. Morgan 1984. Reduction and dissolution of manganese (III) and manganese (IV) oxides by organics: 2. Survey of the reactivity of organics. Environ. Sci. Technol. 18:617-624.

Sunda, W. G., S. A. Huntsman, and G. R. Harvey 1983. Photoreduction of manganese oxides in seawater and its geochemical and biological implications. Nature. London. 301:234-236.

Taylor, R. M., R. M. McKenzie, and K. Norrish 1964. The mineralogy and chemistry of manganese in some Australian soils. Aust. J. Soil Res. 2:235-248.

Taylor, R. M. 1968. The association of manganese and cobalt in soils-further observations. J. Soil Sci. 19:77-80.

Tebo, B. M., K. H. Nealson, S. Emerson, and L. Jacobs. 1984. Microbial mediation of Mn(II) and Co(II) precipitation at the $O_2/H_2S$ interfaces in two anoxic fjords. Limnol. Oceanogr. 29:1247-1258.

Tipping, E. and M. J. Heaton. 1983. The adsorption of aquatic humic substances by two oxides of manganese. Geochim. Cosmochim. Acta. 47:1393-1397.

Van Der Werff, M. 1984. The effect of natural complexing agents on heavy metal toxicity in aquatic plants, p. 441-444. *In* C. J. M. Kramer, and J. C. Duinker, eds., Complexation of trace metals in natural waters, v 1. Martinus Nijhoff/Dr. W. Junk, Boston.

Wilson, D. E. 1980. Surface and complexation effects on the rate of Mn(II) oxidation in natural waters. Geochim. Cosmochim. Acta. 44:1311-1317.

# CHAPTER 17

CADMIUM SPECIATION IN AQUATIC-LIFE
FLOW-THROUGH BIOASSAY DILUTERS

John E. Poldoski, Duane A. Benoit, Anthony R. Carlson, and
Vincent R. Mattson, U.S. Environmental Protection Agency,
Duluth, Minnesota

## ABSTRACT

A framework of cadmium speciation studies is described for
flow-through minidiluter bioassay systems which employs Lake
Superior water containing added concentrations of calcium carbonate
alkalinity, Aldrich humic acid, and/or <2 µm suspended Lake
Superior red clay. Aspects of the flow-through diluter function
were qualitatively related to major changes in the observed $Cd^{2+}$
concentration as a function of time and water-quality chemical
variables. Red clay caused the greatest change, since it markedly
accelerated the rate of loss of $Cd^{2+}$ from Lake Superior water
containing added $CaCO_3$. This acceleration factor was highly
sensitive to other variables such as pH and humic acid. In
addition, ion selective electrode (ISE) potentiometry, dialysis,
and membrane filtration were employed to formulate an ionic
chemical equilibria model designed to describe dissolved cadmium
specie concentrations. A linear regression of $Cd^{2+}$ concentrations,
measured in situ with the ISE, with the corresponding calculated
$Cd^{2+}$ concentrations (n = 84), gave the relationship (µg/L) of,

$$MODEL = 0.90 \cdot ISE + 0.6$$

with a correlation coefficient of 0.988.

## INTRODUCTION

In recent years the Environmental Protection Agency's (EPA)
water-quality program in Duluth has directed substantial effort at
deriving water-quality criteria for freshwater aquatic life. Much

*Chemical Quality of Water and the Hydrologic Cycle*, Robert C. Averett and Diane M. McKnight (Eds.) © 1987 Lewis Publishers, Inc.,
Chelsea, Michigan. Printed in the United States of America.

of the input data has resulted from toxicity bioassays on heavy metals, such as cadmium, that presently exist in the scientific literature (U.S. Environmental Protection Agency, 1980). It is also well known that the nation's surface waters contain concentrations of natural organic and inorganic ligands, clays, algae, and other substances which can affect the chemical speciation of cadmium. These factors give cadmium speciation in bioassays a special importance since the aquated divalent form of cadmium, sometimes referred to as "free cadmium" or $Cd^{2+}$, is currently recognized as one of the most common toxic forms to aquatic life (Canton and Sloof, 1982). The primary analytical tools found to be particularly useful for speciation were ISE potentiometry, dialysis, and 0.45-$\mu$m membrane filtration. These will be discussed later in this chapter.

One of the initial objectives of this water-quality study was to determine if a chemical characterization of bioassays employing minidiluters (Benoit et al., 1982), which use a variety of water quality types, could be successfully conducted under relatively controlled laboratory conditions. The primary purpose of this chapter is to describe some of the successes and the difficulties that were encountered in this endeavor. In addition, new preliminary findings regarding the rate of precipitation of cadmium in Lake Superior water is reported relative to its possible effect on bioassays. The chemical conditions used here were not necessarily meant to duplicate typical or average situations representing the environment, but to demonstrate the need and utility of a chemical-speciation framework for aquatic life bioassays.

A note of thanks is extended to Scott Heinritz for helping to maintain the minidiluter systems, to Barbara Halligan for drawing the figures, to Dr. Rosemarie Russo, who was responsible for the general water-quality types used in the bioassays, and to Robert Spehar, Dr. Philip Cook, and the U.S. Geological Survey for providing review comments on the manuscript.


## METHODS


### Chemicals

All chemicals were of reagent-grade quality or better, unless otherwise indicated. Lake Superior water was employed in all bioassay experiments and was obtained from the laboratory's private pipeline into the lake. Calcium bicarbonate from the dissolution of limestone (Lemke, 1969), suspended <2 $\mu$m Lake Superior red clay, a clay with a very low organic content (Marklund et al., 1981) obtained from the lake's south shore near Cornucopia, Wis., and/or humic acid (Aldrich) were added in various combinations to Lake Superior water to form a total of eight different water qualities. Concentrations were chosen near the upper limits that have been reported for natural waters. Table 1 shows overall values for some of the most significant water-quality parameters, specifically pH in the range of 7.9 to 8.4 and total hardness and alkalinity in the

Table 1.    Major Water Quality Parameters for Flow-Through
            Bioassays.

| Parameter | Value |
|---|---|
| Added total organic carbon (mg/L) | 20 |
| Total suspended solids added as | 0.07 |
| <2 μm suspended Lake Superior red clay (g/L) | |
| Turbidity with added clay (NTU) | 90 |
| pH | 7.9-8.4 |
| Total hardness (mg/L as $CaCO_3$) | |
| softwater | 48 |
| hardwater | 205 |
| Total alkalinity (mg/L as $CaCO_3$) | |
| softwater | 48 |
| hardwater | 212 |
| Calcium (mg/L) | |
| softwater | 14 |
| hardwater | 72 |
| Magnesium (mg/L) | 4 |
| Range of added cadmium (mg/L) | 0.01-1 |

range of either ~48 mg/L for Lake Superior water (softwater), or
205 mg/L hardness and 212 mg/L alkalinity for Lake Superior water
passed through the limestone water hardener (hardwater).  Added
suspended Lake Superior red clay (suspended solids, SS) and humic
acid (HA) were approximately 0.07 g/L and 0.05 g/L, respectively.
Humic acid, as dissolved organic carbon (DOC) in hardwater, was
measured to be approximately 12 mg/L.  In softwater, nearly all the
humic acid passed through a 0.45-μm membrane filter.  The values of
SS and HA were elevated by at least a factor of 50 and 20 respectively, relative to observed background values.  Preliminary
details on organic carbon determinations and the mineralogy of the
suspended solids may be found elsewhere (Marklund et al., 1981).
In terms of decreasing abundance, the major minerals they reported
to be present were quartz, calcite, kaolinite, illite, montmorillonite, plagioclase, a minor unknown clay, minor amphibole, and
very minor talc.  In addition, a surfacial hydrous iron oxide
component was inherent to the clay particles.


## Apparatus and Procedure

Measurements of total cadmium and other metals were made using
EPA standard procedures (U.S. Environmental Protection Agency,
1979) and a Perkin Elmer model 5000 atomic absorption spectro-
photometer equipped with deuterium arc background correction, a
model HGA-500 graphite furnace, a model AS-40 autosampler, a model
AS-50 autosampler, and a model 56 recorder.  Free ion and pH
measurements (in situ) were made according to previous work
(Poldoski, 1984) and equipment included an Orion model 801 meter

equipped with an Orion model 605 electrode switch, a 500-mL FEP teflon cell, an Orion model 9448A cadmium ion selective electrode, and either an Orion-Ross pH-double junction reference electrode pair or a Beckman pH (#39004) Orion double junction reference (#900200) electrode pair. Calibration curves for $Cd^{2+}$ via ISE tended to average at 95% of Nernstian response for standard solutions containing 0.03 mg/L and greater of total cadmium. Solutions were membrane filtered (Millipore, type HA, 0.45-μm membrane filters) in a pressurized filtration apparatus. Dialysis experiments were conducted as described elsewhere (Poldoski 1984, 1986) using 1000 molecular weight cut-off (MWCO) dialysis bags (#132634, Spectrum Medical Industries). Kinetics experiments were conducted at 20°C using 50 mL of a stirred solution in a FEP teflon cell to which $Cd^{2+}$ was added to initiate the reaction. Ionic equilibrium calculations employed the method of successive approximations (Nordstrom et al., 1979) on mass balance equations (0.1% convergence criteria). Calculations were performed with a Hewlett Packard model 3354 computer (BASIC) and a Texas Instruments model TI59 programmable calculator. The diluter tanks generally contained either fathead minnow larvae or *Daphnia magna* and the bioassay procedures and apparatus have been previously described (Benoit et al., 1982; Carlson et al., 1986).

## Analytical Precision and Accuracy

Standards addition was routinely made to samples to check their percent recovery in the sample matrix relative to standards based on deionized distilled water media. In addition, EPA reference water samples (EMSL, Cincinnati) were also routinely measured to compare their values to normal laboratory standards (as percent recovery). The means for the former ranged from 99% to 104% and for the latter they ranged from 95% to 106%, over the range of 14 elements (including cadmium) that were measured to characterize and provide background information on bioassay waters. The number of trials varied from 134 for cadmium to five for undetectable elements such as cobalt.

All water samples (50 mL) taken for analysis by AAS were withdrawn from the center of each tank at 1-cm depth using an acid-cleaned polypropylene disposable syringe (Nalgene), with care taken to minimize any disturbance that may have resulted to the tank's contents. Since all tanks normally contained an air bubbler (0.8 mL/min), any spatial variation of dissolved components within the tank was assumed to be negligible. After withdrawal of the sample, it was immediately subjected to the desired sample processing procedure. To check the overall repeatability of sampling and analysis, replicate samples were also taken immediately after the initial sampling, at about a 5% to 10% frequency. In addition, tanks which represented a replicate concentration level prepared by the diluter (replicates A,B,C, and D) were also sampled to check on this aspect of repeatability. The overall mean relative precision of replicate sampling and determination of cadmium in water samples representing a bioassay tank was in the range of 4% to 6% with n varying from 19 to 41, depending on which cadmium parameter was being determined. This precision was generally similar for samples of other elements that were determined when their concentrations were significantly above instrumental detection limits.

RESULTS AND DISCUSSION

Effect of Water Quality

Table 2 lists the water-quality types employed for flow-through bioassays and corresponding average values for observed total and percent dissolved (0.45 μm) cadmium. The relative standard deviation of various cadmium parameter concentrations with time for most bioassay water types was generally less than 10% (Table 3); however, as an apparent result of some water-quality modifications, marked instability of concentrations with time was observed in some cases. This could conceivably affect the ability of the diluter to maintain constant concentrations of cadmium species. The most unstable combination was the hardwater and suspended solids system. An example of cadmium parameter stability as a function of time, for a high concentration tank employing this combination, is displayed in Figure 1. It can be readily seen that total cadmium varies at least two-fold over the bioassay period. The ISE, dialysis, and filtration values, which are seen to be approximately an order of magnitude lower, also vary similarly, but not necessarily in accordance with the total cadmium values.

Table 2.   Percentage of Dissolved Cadmium as a Function of Water Quality for Flow-Through Bioassays.

| Water quality type | Total Cd high tank (μg/L) | % Dissolved high tank (0.45 μm) |
|---|---|---|
| SW | 112 | 98 |
| SW + HA | 251 | 78 |
| SW + SS | 128 | 83 |
| SW + HA + SS | 182 | 63 |
| HW | 129 | 87 |
| HW + HA | 435 | 41 |
| HW + SS | 668 | 16 |
| HW + HA + SS | 459 | 48 |

SW = Softwater (Lake Superior water).
HW = Hardwater ($CaCO_3$ added to Lake Superior water).
SS = <2 μm suspended solids (Lake Superior south shore red clay).
HA = Humic acid (commercial grade by Aldrich).

Table 3.    Percent Relative Standard Deviation of Total and 0.45 μm
            Dissolved Cadmium Concentrations in Flow-Through
            Bioassays over Time.

| Water Type[1] | SW + HA | SW + SS | HW + HA | HW + SS | HW |
|---|---|---|---|---|---|
| 4-day bioassay | | | | | |
| Total | 2 | 3 | 8 | 4 | 6 |
| Dissolved | 1 | 2 | - | 1 | 0 |
| | | | | | |
| 28-day bioassay | | | | | |
| Total | 8 | 5 | 38 | 26 | 26 |
| Dissolved | 8 | 6 | 9 | 35 | 12 |

[1]See Table 2 definitions.
- Not determined.


The theoretical solubility of cadmium, as limited by carbonate, was related to the highest observed mean total cadmium concentration for each corresponding bioassay system (Fig. 2). Precipitation of $CdCO_3$ was overwhelmingly favored for hardwater and hardwater and humic acid bioassays in the range of pH 8.0 to 8.4. In contrast, there appeared to be only a slight tendency for $CdCO_3$ to precipitate in softwater bioassays, with no tendency for this to occur when humic acid was present.

The latter was experimentally supported by comparing measurements of total cadmium to dissolved (0.45 μm) cadmium in softwater for which the linear regression equation (μg/L) was

$$DISSOLVED = 0.97 \cdot TOTAL + 0.6$$

with a correlation coefficient of 1.000. Even in softwater + suspended solids, there appeared to be only a relatively minor tendency for cadmium to rapidly associate with clay particles, since an identical regression produced a slope, intercept, and correlation coefficient of 0.81, -0.8, and 1.000, respectively. Furthermore, equilibrium calculations indicated that most dissolved cadmium should exist primarily as $Cd^{2+}$ in softwater.

For the systems containing hardwater, measurements of 0.45-μm filtrates of cadmium indicated that the loss of dissolved cadmium from solution could be a very important time-dependent process, (i.e., $CdCO_3$ precipitation continually occurring as related to the rate of test media replacement in the tanks). Since it was quite possible that the rate of precipitation could proceed with a half-life of similar magnitude relative to the time (2.5 h) for 95% replacement of the water in each tank (Carlson et al., 1986), it followed that the cadmium forms in some test waters likely were not at equilibrium, but probably existed in a steady state which was obviously sensitive to chemical and physical variables. It can be surmised that how well a steady state can be maintained by a flow-

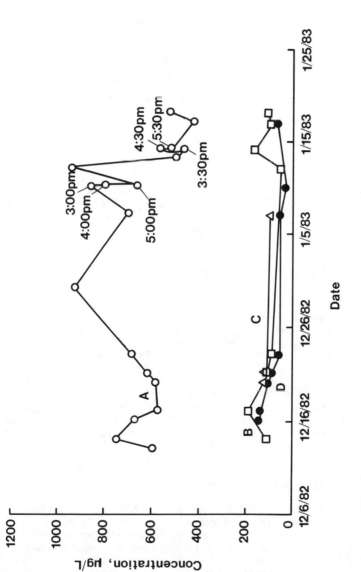

Figure 1.  Stability of cadmium parameters with time for a hardwater + suspended solids diluter.
A - Total cadmium,
B - Dissolved cadmium (0.45 µm),
C - Dialyzable cadmium (1000 MWCO),
D - Cd²⁺ via ISE.

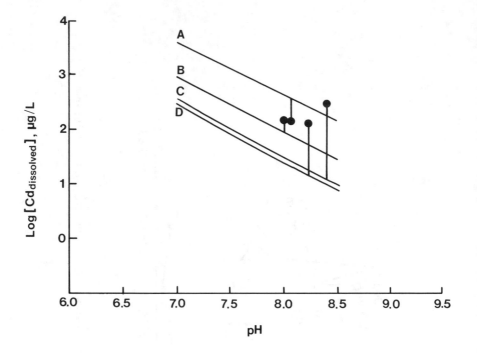

Figure 2.   Calculated solubility of cadmium.
            A - Softwater + humic acid,
            B - Softwater,
            C - Hardwater + humic acid,
            D - Hardwater,
            ● - Indicates the maximum mean total cadmium concentra-
                tion measured for a particular bioassay water type.


through diluter system could be dependent on (1) the magnitude and
stability of flow to each test chamber, (2) the water volume in
each test chamber, (3) the stability of values for relevant water-
quality variables, and (4) the magnitude of the rate for each
reaction pathway.


## Kinetics of Cadmium Loss

The loss of free cadmium ion as a function of time was
investigated in exploratory kinetics experiments by employing a
rapidly responding cadmium ion selective electrode in the following
media:   hardwater, hardwater + suspended solids, and hardwater +
suspended solids + humic acid.
Figure 3 shows a typical example of the variation of the cell
potential with time when 10 mg/L of $Cd^{2+}$ was rapidly added to a
stirred sample of hardwater + suspended solids at pH 7.73.   It can

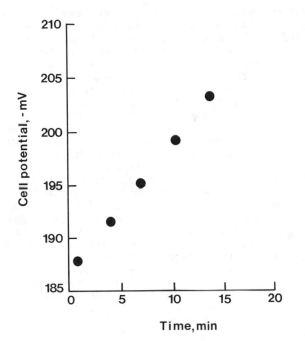

Figure 3.   Dependence of cell potential with time of reaction for hardwater + suspended solids.   Added $Cd^{2+}$ = 10 mg/L, pH = 7.73.

be seen that a well-defined linear change occurred with time.   In contrast, decreasing the $Cd^{2+}$ from 10 mg/L to 1 mg/L in deionized distilled water standardization media produced a stable cell potential within a matter of seconds, indicating the rapid response capability of the electrode.   Therefore, the characteristic linear change observed during approximately the first half of the reaction was attributed solely to the loss of free cadmium from solution. Later stages of the reaction were not investigated to any major extent.

Stumm and Morgan (1981) have summarized the general role that various surfaces (such as clays) could play in an interface-controlled crystal-growth process.   Based on this model, pseudo first-order kinetic behavior can be assumed in order to characterize these time-dependent potential changes.   Hence, it follows that the expression for the rate constant (k), with all associated variables held constant, can be given by the following:

$$k = \frac{E_2 - E_1}{t_2 - t_1} \cdot \frac{nF}{2.3\ RT}$$

Since terms in the expression are derived from the Nernst equation, they have their usual meaning, except for t which is time.

Preliminary values of the slope for potential change versus time ($k_o$) were obtained as a function of pH for the various media and presented in Figure 4. Since these data are still preliminary, absolute values of the rate constant are not given. As one can see, the variation of these values covers a wide range, with the most notable case being the marked effect of pH in the example of hardwater + suspended solids. The values increased sharply, beginning around pH 7.9 and changed by more than an order of magnitude in the pH range of approximately 7.7 to nearly 8.4. Moreover, as the initial added $Cd^{2+}$ was decreased from 10 mg/L to 3 mg/L and finally down to 1 mg/L, the curve shape remained similar, but it was shifted toward higher pH. The reason for this high rate of $Cd^{2+}$ loss cannot be attributed only to simple sorption of $Cd^{2+}$ to the surface of the clay, since this rate of loss of cadmium was not observed in the case of softwater + suspended solids. In addition, the effect of suspended solids can be seen by comparing these data to data taken similarly in hardwater without any suspended solids addition (Fig. 4). While it can be clearly observed that there is only a gradual increase in $k_o$ with pH, the striking feature is that there is no indication of a similar sharp breakpoint as when red clay is present. Thus, it is apparent that at least three ingredients need to be present to observe this previously unreported accelerated reduction of $Cd^{2+}$ concentrations from Lake Superior water, namely (1) suspended Lake Superior red clay, (2) sufficiently high pH, and (3) adequate total carbonate alkalinity. In brief, it is believed that this precipitation is markedly promoted by the greater ability for nucleation sites to form on the surface of red clay particles.

Further investigations of changes in $k_o$ with humic acid present displayed an equally pronounced effect. Since cadmium precipitation by humic acid was largely complete seconds after mixing, this factor did not substantially interfere with observing longer-lived $CdCO_3$ precipitation. The accelerating phenomenon of the clay could be virtually eliminated by the addition of 50 mg/L humic acid. Values of 14 or possibly greater could be reduced to less than 0.1 (Fig. 4). Because it has been reported that humic acid sorbs onto clay surfaces (Manos and Chia-ei, 1980; Nriagu, 1979) and prevents metal carbonate precipitation (Rashid and Leonard, 1973), the reason for its suppression of $k_o$ is likely its sorption onto active surfaces, thus preventing growth of nucleation sites.

It is apparent that these studies translate into some plausible explanations regarding stability of cadmium specie concentrations in bioassay tanks. Specifically, it can be said that hardwater systems containing humic acid are not even close to equilibrium nor are they making any significant progress toward reaching that state; however, they should be in a steady state unaffected by this precipitation. In contrast, dissolved cadmium concentrations in hardwater systems without added humic acid are subject to fluctuations due to high sensitivity to the presence of surfaces promoting the formation of nucleation sites, and in particular, small changes in pH. Table 3 illustrates this point by showing that the overall variabilities of dissolved cadmium concentrations in bioassay tanks were generally less than 10% for both 4-day and 28-day bioassays, except in the case of hardwater and hardwater + suspended solids.

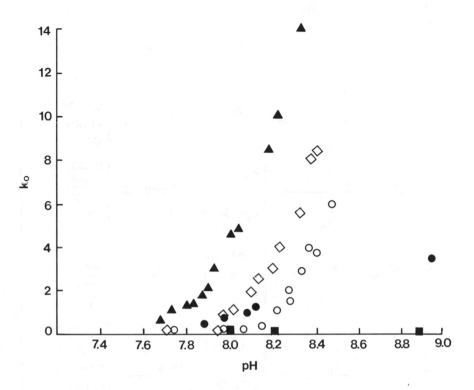

Figure 4.   Water quality variables affecting the kinetics of $Cd^{2+}$
            loss from solution.
            ▲ - Hardwater + clay + 10 mg/L $Cd^{2+}$,
            ◇ Hardwater + clay + 5 mg/L $Cd^{2+}$,
            o - Hardwater + clay + 1 mg/L $Cd^{2+}$,
            ● - Hardwater + 10 mg/L $Cd^{2+}$,
            ■ - Hardwater + clay + humic acid + 10 mg/L $Cd^{2+}$.

## Chemical Modeling

     Figure 5 summarizes, in a qualitative sense, the likely
transient or equilibrium states existing for each bioassay system
utilizing hardwater.  The relative rates assigned to each reaction
indicate that all transient states are rapidly attained.  Although
progress toward attaining the equilibrium state is substantial for
the hardwater and hardwater + suspended solids systems, at best it
is only partial; therefore, the major species likely prevailing in
the tanks are representative of both the equilibrium and transient
states.  For the latter two systems with humic acid present,
species in the transient state tend to be in much greater abundance
relative to those in the equilibrium state.  Moreover, it should
have the characteristics of a steady state, which may lend
stability to its chemical species composition.

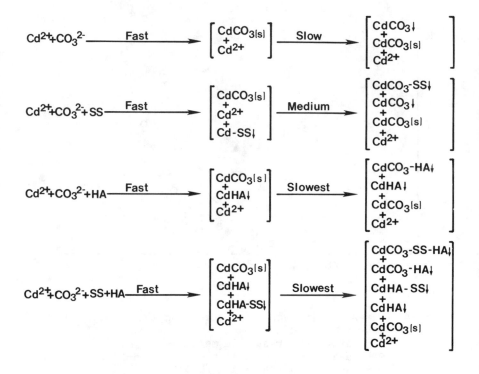

Figure 5.    Possible major reaction pathways and relative reaction
             rates for cadmium in bioassays.
             (s) - Soluble form,
             ↓  - Particulate and/or colloidal,
             SS - Suspended solids,
             HA - Humic acid.


     The ultimate goal of this work was to demonstrate that a
reasonable quantitation of the speciation of dissolved cadmium
could be acquired so that any possible correlations to observed
toxicity could be made in future work.  In achieving this goal, it
was necessary to make as many in situ measurements as possible
using cadmium ion selective electrode potentiometry, 1000 MWCO
dialysis, and 0.45 μm membrane filtration.  In order to assure
optimum comparability of the analytical methods, data were
collected for each tank utilizing the comparative methods at about
the same point in time, minimizing the possibility for fluctuations

of in-tank concentrations that could degrade comparison of the methods. This resulted in attaining a relatively high level of internal consistency between methods, as will be considered later. In turn, this confirmatory experimental analytical data prompted testing of an empirical ionic chemical equilibrium model to quantitate the various dissolved cadmium species present in the bioassay test waters.

The model was based on the total concentration of cadmium in $0.45-\mu m$ filtrates, the concentrations representing other appropriate water quality variables, and the stability constants for various species (Sillen and Martell, 1964, 1971; Martell and Smith, 1974, 1977; Gardiner, 1974) as given in Table 4. An empirical representation of the Aldrich humics used in this study was provided by two model ligands, namely citrate (L) and benzyliminodiacetate (X). The criteria used in this choice of ligands was that a complete set of stability constants be available and that they reasonably emulate the binding and some of the possible functional group structural features of the natural organics, as used under the specific conditions of these bioassay systems. It is also noted that these ligands are not necessarily a unique choice and that other combinations may be possible. Furthermore, the experimentally observed binding by humic acid was found to be consistent with models proposed by earlier workers investigating natural organics (Bresnahan et al., 1978; Buffle et al., 1977; McKnight et al., 1983). These basically consisted of a low concentration of a strongly binding (X) ligand and a higher concentration of a weaker binding ligand (L). In one case, data was interpreted as the formation of a 2:1 complex (Buffle et al., 1977). The concentration of benzyliminodiacetate was based on the molar equivalent endpoint ($\sim 10^{-6.15}$ M) in a titration of humic acid with $Cd^{2+}$. The citrate concentration ($\sim 10^{-3.57}$ M) was empirically adjusted to account for observed binding, but was limited to a corresponding dissolved organic carbon concentration less than that estimated for the bioassay waters. The value of the stability constant for the 2:1 complex of this ligand was adjusted to a value that would further help mimic the experimentally observed binding. All other stability constants were used as they were reported.

Calculations using this model were made over the range of all hardwater and softwater bioassays, comprising a total of eight different water-quality types. Input data was the mean 0.45 µ dissolved cadmium concentration observed at each dilution level for each bioassay. This produced a corresponding calculated $Cd^{2+}$ concentration, in addition to concentrations for several other species. Some examples are given (Table 5) for the calculated percent abundance of each species for both hardwater and softwater, with and without added humic acid. According to the model, it is seen that the general decreasing order of percent specie contribution is $Cd^{2+}$, $CdCO_3$, $CdX$, $CdL^-$, $CdL_2^{4-}$, $CdOH^+$, $CdCl^+$, and $CdSO_4$. In terms of the percentage of total cadmium bound to the added humics, it can be estimated that the value is 85% for softwater and 69% for hardwater. This compares favorably with corresponding experimentally observed values of 85% and 63%, respectively, via dialysis.

Table 4.  Ionic Equilibrium Constants.

| Metal | Ligand | Equilibrium constant (K) | Log $K^a$ Softwater | Hardwater |
|-------|--------|--------------------------|-----------|-----------|
| H  | $OH^-$        | $K_1$    | -13.96      | -13.92      |
| H  | $CO_3^{2-}$   | $K_1$    | 10.24       | 10.16       |
| H  | $CO_3^{2-}$   | $K_{12}$ | 6.31        | 6.27        |
| H  | $^b X^{2-}$   | $K_1$    | 9.27        | 9.19        |
| H  | $X^{2-}$      | $K_{12}$ | 2.36        | 2.25        |
| H  | $^c L^{3-}$   | $K_1$    | 6.24        | 6.13        |
| H  | $L^{3-}$      | $K_{12}$ | 4.72        | 4.64        |
| H  | $L^{3-}$      | $K_{13}$ | 3.05        | 3.02        |
| Cd | $X^{2-}$      | $K_1$    | 7.61        | 7.47        |
| Cd | $L^{3-}$      | $K_1$    | 4.85        | 4.63        |
| Cd | $L^{3-}$      | $K_2$    | $4.54^d$    | $4.65^d$    |
| Cd | $OH^-$        | $K_1$    | 4.77        | 4.69        |
| Cd | $CO_3^{2-}$   | $K_1$    | 4.22        | 4.07        |
| Cd | $CO_3^{2-}$   | $K_{so}$ | -11.40      | -11.25      |
| Cd | $Cl^-$        | $K_1$    | 1.92        | 1.84        |
| Cd | $Cl^-$        | $K_2$    | 0.66        | 0.62        |
| Cd | $SO_4^{2-}$   | $K_1$    | 2.13        | 1.98        |
| Ca | $X^{2-}$      | $K_1$    | 3.90        | 3.76        |
| Ca | $L^{3-}$      | $K_1$    | 4.60        | 4.38        |
| Mg | $X^{2-}$      | $K_1$    | 3.38        | 3.24        |
| Mg | $L^{3-}$      | $K_1$    | 4.47        | 4.25        |

[a] Davies equation used to adjust constants to proper ionic
strength; equilibrium constant definitions may be found in
Sillen and Martell (1964).

[b] N-Benzyliminodiacetate.

[c] Citrate.

[d] Value empirically determined.

Table 5.  Speciation of 0.45 μm Dissolved Cadmium in Flow-Through Bioassays via Ionic Chemical Equilibrium Modeling.

| Water type | Mean total Cd high tank (μg/L) | 0.45 μm Dissolved Cd % of total | Calculated individual specie concentration as % of total dissolved Cd | | | | | | | |
|---|---|---|---|---|---|---|---|---|---|---|
| | | | $Cd^{2+}$ | $CdCO_3$ | CdX | $CdL^-$ | $CdL_2^{4-}$ | $CdOH^+$ | $CdCl^+$ | $CdSO_4$ |
| Softwater | 112 | 98 | 84[1] | 9 | nd | nd | nd | 7 | 0.4 | 0.4 |
| Softwater[2] + humic acid | 251 | 78 | 16 | 1 | 14 | 33 | 34 | 1 | <0.1 | <0.1 |
| Hardwater | 129 | 87 | 61 | 33 | nd | nd | nd | 5 | 0.2 | 0.2 |
| Hardwater + humic acid | 435 | 41 | 33 | 36 | 20 | 4 | 0.6 | 5 | 0.1 | <0.1 |

[1]Values are underlined which contribute 10% or more.

[2]Estimated % of Cd-organics relative to total:  via model – 85% (softwater), 69% (hardwater); via dialysis – 85% (softwater), 63% (hardwater).

nd – Not detected.

Figure 6 provides a visual perception of the degree of agreement of each calculated $Cd^{2+}$ value with each corresponding $Cd^{2+}$ value, determined via the cadmium ISE. It is also noted that this comparison reflects both analytical errors and those associated with any possible fluctuations of concentrations in the tanks. It can readily be seen that a highly significant linear regression correlation was achieved (R = 0.988). Moreover, the slope and intercept of the corresponding regression equation,

$$MODEL = 0.90 \cdot ISE + 0.6$$

indicates a reasonable 1:1 correspondence and a near-zero intercept. Consequently, this type of information lends confidence to its application. However, it must also be noted that this model is applicable only to the specific conditions of this study and not necessarily in the sense of general applicability to the environment. Because of its empirical nature, this work also does not preclude the possibility that other chemical models may emulate the binding of cadmium equally well or better.

## Comparison of ISE and 1000 MWCO Dialysis

Since the validity of any prediction using this model is also based on the validity of the $Cd^{2+}$ determination via ISE, the necessity of comparing this determination to an independent estimation of $Cd^{2+}$ is obvious. Dialysis proved to be one of the most suitable alternative in situ techniques available for this specific task. Prior research (Poldoski, 1984, 1986) indicated that a 5-h dialysis period with adequate stirring was sufficient to attain at least 95% equilibrium in these waters using the smallest available diameter 1000 MWCO dialysis bags. Moreover, other researchers (Truitt and Weber, 1981) have reported that the 1000 MWCO membrane was adequate to retain the majority of natural organics. Tests indicated that Aldrich humic acid behaved similarly. Total cadmium was measured in the dialyzates; however, other dissolved water-quality parameters determined for the bulk tank solutions were assumed to also be valid for the dialyzate. In addition, the humics were considered to not appreciably permeate the membrane. Using these data, assumptions, and ionic equilibrium calculations, the corresponding $Cd^{2+}$ concentration was calculated for each dialyzate, including those for hardwater using a modified procedure (Poldoski, 1986). Each value was directly compared to its corresponding ISE $Cd^{2+}$ value, taken during the same time period. Values of $Cd^{2+}$ varied from 2 µg/L to 132 µg/L (n = 49) for hardwater (Fig. 7) and from 2 µg/L to 158 µg/L (n = 43) for softwater (Fig. 8). The linear regression equation for hardwater was,

$$DIALYSIS = 0.84 \cdot ISE + 4$$

with a correlation coefficient of 0.934. Similarly, values of the slope, intercept, and correlation coefficient were 0.81, 0.4, and 0.989, respectively, for bioassays containing softwater.

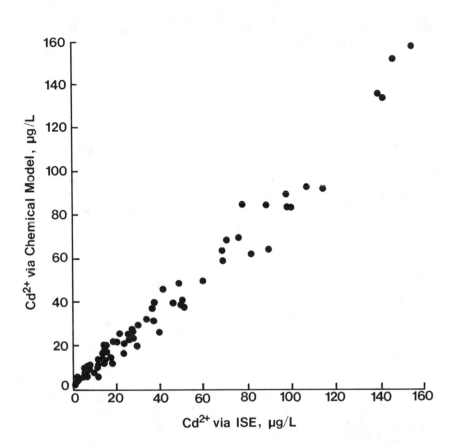

Figure 6. Plot of Cd$^{2+}$ values calculated via the chemical model versus measured Cd$^{2+}$ values via ISE, for all bioassays. Model = 0.90 · ISE + 0.6 (R=0.988).

There appears to be an adequate correlation between these methods, although dialysis tended to produce significantly lower values than the ISE method (95% confidence level). The precise reason for this is not clear at present, but it is suspected that it might be partly related to some degree of slow adsorption and/or chemical instability such as precipitation associated with the dialysis bags. If such effects gradually increased to a greater extent after the time when the ISE measurement was taken (midway during the dialysis period), the specie concentrations could be temporarily depleted in the bulk tank solutions by several percent, thus causing a relatively lower value in the dialyzate. Unfortunately, ISE measurements of free cadmium were not routinely taken on a single tank as a function of time throughout the day, so it cannot be concluded that this did indeed occur. However, it is noteworthy to point out the more variable nature of the hardwater data (Fig. 7) compared to softwater data (Fig. 8). This is in support of chemical instability being a major contributing factor to observed variability.

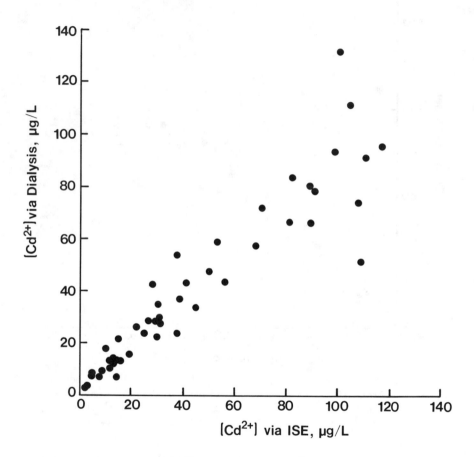

Figure 7.    Plot of Cd$^{2+}$ via dialysis versus Cd$^{2+}$ via ISE for bioassays containing hardwater.
Dialysis = 0.84 · ISE + 4 (R = 0.934).

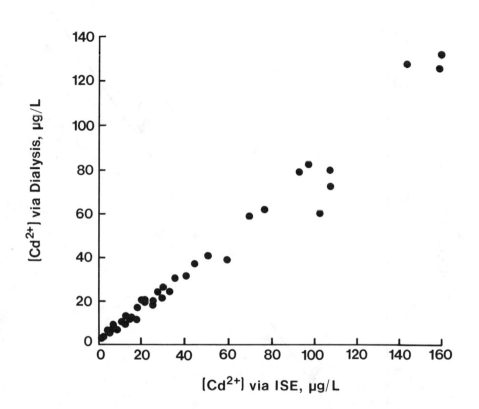

Figure 8.   Plot of Cd²⁺ via dialysis versus Cd²⁺ via ISE for
bioassays containing softwater.
Dialysis = 0.81 · ISE + 0.4 (R = 0.989).

## Comparison of ISE and 0.45-μm Filtration

Several Cd²⁺ ISE measurements were taken precisely at the same
point in time during which 0.45-μm filtrates were taken from
several hardwater + suspended solids bioassay tanks.  Since the
0.45-μ membrane was known to filter out the bulk of the particles
in this medium, the filtrate was thought to be a good representation of the true dissolved phase.  As can be seen (Fig. 9), an
excellent relationship  was  obtained  between  these  two  methods

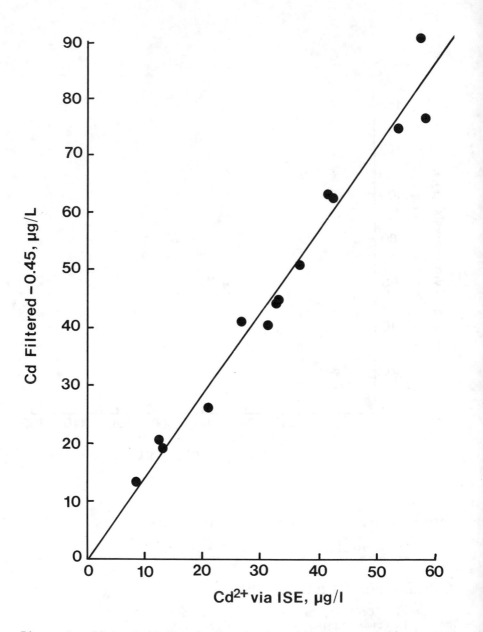

Figure 9.   Plot of (0.45 μm) dissolved cadmium versus Cd²⁺ via ISE
            for hardwater + suspended solids bioassays.
            Dissolved = 1.44 · ISE - 0.7 (R = 0.988).

in this unstable medium, with a linear regression equation of

$$DISSOLVED = 1.44 \cdot ISE - 0.7$$

and a correlation coefficient of 0.988. The observed slope of 1.44 is in good agreement with the expected theoretical value of 1.5, based on ionic equilibrium calculations for the dissolved phase.


SUMMARY

Cadmium ISE potentiometry, dialysis, and filtration were capable of providing useful in situ speciation results for flow-through minidiluters. These results allowed formulation of a chemical model which has potential use in correlation to bioassay data. Although ionic chemical equilibria modeling was the major framework of consideration, kinetic factors were also shown to have a profound influence, an important point which has received open attention in the past publications of relatively few researchers, such as Canton and Sloof (1982).


DISCLAIMER

This chapter has been reviewed in accordance with U.S. Environmental Protection Agency policy and approved for publication. Mention of trade names or commercial products does not constitute endorsement or recommendation for use.


REFERENCES

Benoit, D. A., V. R. Mattson, and D. L. Olson. 1982. A continuous-flow minidiluter system for toxicity testing. Water Res. 16:457-464.

Bresnahan, W. T., C. L. Grant, and J. H. Weber. 1978. Stability constants for the complexation of copper(II) ions with water and soil fulvic acids measured by an ion selective electrode. Anal. Chem. 50:1675-1679.

Buffle, J., F. Greter, and W. Haerdi. 1977. Measurement of complexing properties of humic and fulvic acids in natural waters with lead and copper ion selective electrodes. Anal. Chem. 49:216-222.

Canton, J. H., and W. Sloof. 1982. Toxicity and accumulation studies of cadmium ($Cd^{2+}$) with freshwater organisms of different trophic levels. Ecotoxicol. Environ. Saf. 6:113-128.

Carlson, A. R., D. A. Benoit, and V. R. Mattson. 1986. Manuscript in preparation, U.S. Environmental Protection Agency, Environmental Research Laboratory-Duluth. Duluth, Minn.

Gardiner, J. 1974. The chemistry of cadmium in natural water - I. A study of cadmium complex formation using the cadmium specific-ion electrode. Water Res. 8:23-30.

Lemke, A. E. 1969. A water hardener for experimental use. J. Am. Water Works Assoc. 61:415-416.

Manos, G. P., and T. Chia-ei. 1980. Mechanisms of humic material adsorption on kaolin clay and activated carbon. Water Air Soil Pollut. 14:419-427.

Marklund, D., P. Morten, S. Kohlbry, and E. Ruenger. 1981. Results of analyses performed under EPA task plans NAARD 0309 and NAARD 0308. Lake Superior Basin Studies Center. University of Minnesota-Duluth. Duluth, Minn.

Martell, A. E., and R. M. Smith. 1974. Critical stability constants. Vol. 1. Plenum Press. N.Y. and London. 469 p.

Martell, A. E., and R. M. Smith. 1977. Critical stability constants, Vol. 3. Plenum Press. N.Y. and London. 493 p.

McKnight, D. M., G. L. Feder, M. Thurman, R. L. Wershaw, and J. C. Westall. 1983. Complexation of copper by aquatic humic substances from different environments. Sci. Tot. Environ. 28:65-76.

Nordstrom, D. K., L. N. Plummer, T. M. L. Wigley, T. J. Wolery, J. W. Ball, E. A. Jenne, R. L. Bassett, D. A. Crerar, T. M. Florence, B. Fritz, M. Hoffman, G. R. Holdren, Jr., G. M. Lafon, S. V. Mattigod, R. E. McDuff, F. Morel, M. M. Reddy, G. Sposito, and J. Thrailkill. 1979. A comparison of computerized chemical models for equilibrium calculations in aqueous systems. pp. 857-892. In E. A. Jenne, ed., ACS Symposium 93. American Chemical Society. Washington, D.C.

Nriagu, J. O., ed. 1979. Cadmium in the environment. Part I: Ecological cycling. Wiley Interscience, New York. 682 p.

Poldoski, J. E. 1984-1985. Unpublished work. U.S. Environmental Protection Agency, Environmental Research Laboratory-Duluth. Duluth, Minn.

Poldoski, J. E. March 1986. Control of cadmium carbonate precipitation interferences during the dialysis of cadmium in high bicarbonate alkalinity aquatic-life bioassay waters. EPA-600/53-86-001. U.S. Environmental Protection Agency, Environmental Research Laboratory-Duluth. Duluth, Minn.

Rashid, M. A., and J. D. Leonard. 1973. Modifications in the solubility and precipitation behavior of various metals as a result of their interaction with sedimentary humic acid. Chem. Geol. 11:89-97.

Sillen, L. G., and A. E. Martell. 1964. Stability constants of metal-ion complexes. Special publication No. 17. Chemical Society. London.

Sillen, L. G., and A. E. Martell. 1971. Stability constants of metal-ion complexes. Supplement No. 1. Special publication No. 25. Chemical Society. London.

Stumm, W., and J. J. Morgan. 1981. Aquatic chemistry. Wiley Interscience. New York. 780 p.

Truitt, R. E., and J. H. Weber. 1981. Determination of complexing capacity of fulvic acid for copper(II) and cadmium(II) by dialysis titration. Anal. Chem. 53:337-342.

U.S. Environmental Protection Agency. March 1979. Methods for chemical analysis of water and wastes. EPA-600/4-79-020. Environmental Monitoring and Support Laboratory. Cincinnati, Ohio.

U.S. Environmental Protection Agency. 1980. Water quality criteria documents availability. Fed. Reg. 45:79318-79379.

SECTION IV

EXTENDED ABSTRACTS

# BULK ATMOSPHERIC HEAVY-METAL DEPOSITION AS IT AFFECTS WATER QUALITY IN EASTERN OKLAHOMA

M. H. Bates, A. K. Huggins, J. Karleskint, M. Tate, T. Ung, and Y. Yang, Oklahoma State University, Stillwater, Oklahoma

It has long been recognized that heavy-metal pollution of surface water is resulting in serious water quality degradation. One source of metals to waters is bulk atmospheric deposition. The significance of this source varies with geographical location and the proximity of the water body to urban and industrialized areas. Once in an aquatic system, the metals may be taken up by organisms, chemically precipitated, or adsorbed onto sediments and algae, bacteria, or other suspended matter.

The city of Tulsa, Oklahoma, in conjunction with the Tulsa River Parks Authority, has improved the east and west banks of the Arkansas River as it flows through an area adjacent to downtown Tulsa. Greenbelt areas, parks, and a pool below a low-water dam have been established to improve and enhance the aesthetic and recreational potential of the area and the Arkansas River. Water-quality monitoring of the Arkansas River in the River Parks area, as well as at other points, indicates the existence of several important water-quality problems that could limit the recreational potential of the river. One of the major problems is the presence of heavy metals.

Realizing that the atmosphere is a potential source of metals for the Arkansas River, bulk atmospheric fallout was collected and examined at six sites along the river. The metals studied included Cd, Cr, Ni, Pb, and Zn. Samplers, which consisted of a poly-ethylene funnel placed in a 1-L polyethylene bottle, were placed at sites A-E. Sites A and B were located in an industrialized area and samplers were attached to poles at a height of 7 to 8 feet above ground level. Site C was located in a commercial-residential area on top of a five-story building, and site D, established in a residential area, was located on top of a two-story building. Samplers at sites E and F were attached to poles 8 feet above ground level. Site E was in a residential district, and site F was in a largely residential area with a few small surrounding

*Chemical Quality of Water and the Hydrologic Cycle*, Robert C. Averett and Diane M. McKnight (Eds.) © 1987 Lewis Publishers, Inc., Chelsea, Michigan. Printed in the United States of America.

businesses.   Sites A, B, D, and F were in heavily traveled regions of the metropolitan area.

During the study period, 23 precipitation events occurred. Samples were collected as soon as possible after each event.  If no form of precipitation occurred during any month, samplers were washed down with distilled deionized water and the collected sample was analyzed for the metals in question.  All samples were acid digested using reagent grade nitric acid.  Metal determinations were made using a Perkin-Elmer 5000 atomic absorption equipped with a HGA 400 graphite furnace.

Loadings of cadmium, nickel, lead, and zinc to the Arkansas River varied with sampler location, sampler height, and the amount of rainfall.  During the 10-month period of sample collection, zinc was found to be the metal in greatest abundance.  Average loadings of zinc were approximately 10 times higher than the average loading of lead, 25 times higher than chromium, 80 times higher than nickel, and 582 times higher than cadmium.

The total river area over which samples were collected was 766.7 acres and the total average loading for all six sites was 8,295.47 µg.  The equivalent loading per acre was 4.18 lbs/acre. Therefore, the total loading over the entire area for the 10-month period was 3,207.45 lbs or 1.6 tons.

It is obvious that the atmosphere is a source of heavy metals to the Arkansas River.  The fate of these metals, and others, in aquatic systems were subsequently examined in a series of laboratory studies designed to investigate the uptake of metals by various sediments and the interactions between the metals and the aqueous phase.

To study metal uptake, sediments and water were obtained from four sources in Oklahoma--Lake Carl Blackwell, Sooner Lake, the Cimarron River, and the Arkansas River.  Soil and water samples were then characterized in the laboratory.  The river sediments were composed primarily of sand, while the lake sediments were composed of silt, clay, and organic matter.  Lake Carl Blackwell, an older lake, had the highest organic matter content.

Batch-model aquatic systems were set up for each sediment and each metal in one-gallon glass containers.  All systems were kept aerobic through the addition of air.  Soluble metal concentrations were monitored daily using atomic absorption.  Initial systems for each sediment contained only sediment and a distilled water solution of the metal being studied.  This allowed a study of the capacity of each soil to fix the metals in the absence of any competition from the chemical components of water.  Additional systems were set up for each metal and each natural water to study metal removal from solution as a function of aqueous water chemistry.  Finally, individual metal removal by the sediments and the natural waters was studied in systems containing both phases. The metals studied included Cu, Cr, Cd, and Zn.

For all metals, substantially higher quantities were fixed by the lake sediments as compared to the river sediments indicating that soil type has a pronounced effect on heavy-metal uptake.  The largest amounts of all four metals were fixed by Lake Carl Blackwell sediments.  The higher fixation capacity of these sediments was attributed to the higher organic matter content.

Water chemistry also played an important role in the removal of soluble metal from the systems. Copper was readily precipitated in all natural waters used. However, the heaviest copper precipitation occurred in the water obtained from the Cimarron River. The alkalinity, chloride, and sulfate concentrations in the river water were substantially higher than those found in the two lake waters. The amount of soluble copper in solution in the various natural waters decreased with increase in the alkalinity, chloride, and sulfate concentrations.

Although some of the added chromium was removed by interaction with the natural waters, the amount was very small. In addition, the greatest removal occurred with the lake waters rather than the river waters. Based on the data available, no correlation between chromium removal and water chemistry could be made.

Zinc and cadmium both reacted with the water components and were removed from solution. The pH and the alkalinity of the natural waters used appeared to be the most important water-quality parameters affecting the solubility of both metals. The amounts of both cadmium and zinc removed increased with an increase in pH and alkalinity of the waters used for the study.

Metals can be removed from aqueous systems by adsorption or fixation by the sediments or interaction with other ions in solution. The degree to which they remain precipitated, adsorbed, or complexed is a function of the pH and the oxidation-reduction potential. The release of copper, cadmium, chromium, lead, and zinc from sediments obtained from the Arkansas River was examined in laboratory systems at various pH and oxidation-reduction potential values. pH was artificially altered in duplicate systems through the addition of HCl or $NH_4OH$. The altered pH values of the overlying water were 1, 2, 3, 4, 5, 6, 7, 8, 9, 10, 11, 12. The flasks were shaken for 3 days, the final pH recorded, and soluble-metal content determined by atomic absorption. Maximum release of Cr, Cd, and Cu occurred in the strongly acidic pH range; whereas, maximum release of lead and zinc was found above a pH of 9.96. The μg/g of all the metals in solution was essentially constant and at a minimum in the pH range of 7.36 to 8.36. This pH range of minimum solubility is also within the pH range that might be expected in a natural water system. In addition, it should be pointed out that the amount of metal released, even under strongly acidic or basic conditions, was small compared to the amount of metal bound in the sediments. At the pH values of maximum release, only 5.5% of the bound chromium, 2.4% of the bound cadmium, 11.4% of the bound copper, 4.6% of the bound lead, and 4.9% of the bound zinc were released to the aqueous phase.

The effects of redox potential on metal release from Arkansas River sediments were studied using 27 different redox potentials and Cu, Pb, and Zn. Studies were done in duplicate in 250-mL Erlenmeyer flasks. Eh was altered through the addition of nitrogen or air. Flasks were sealed and shaken for 3 days. At the end of the equilibration period, the final redox potential was determined and the aqueous phase was separated by filtration. The concentrations of Cu, Pb, and Zn were determined by atomic absorption spectrocopy.

The solubility of Cu, Pb, and Zn decreased under oxidizing conditions. The maximum amounts of metal released over the Eh range of -300 to +400 mv was small for all the metals except for lead when compared to the amounts of metals bound in the sediments. At the redox value of maximum solubility, 5% of the copper, 42% of the lead, and 17% of the zinc were released.

## GEOCHEMICAL PROCESSES IN CLASTIC AQUIFERS OF THE SOUTHEASTERN COASTAL PLAIN[1]

Roger W. Lee, U.S. Geological Survey, Atlanta, Georgia

Cretaceous and Tertiary clastic aquifers of the southeastern Coastal Plain regional aquifer system contain sodium bicarbonate dominated water with local concentrations of dissolved bicarbonate greater than 1,200 mg/L in the deep, downgradient parts of the aquifers. Geochemical evolution of sodium bicarbonate waters in coastal sand aquifers was described by Foster (1950). Application of mass balance calculations (Parkhurst et al., 1982) and mass transfer modeling (Parkhurst et al., 1980) to aqueous geochemical data from southeastern Coastal Plain aquifers have shown that five subsurface geochemical reaction zones can be identified between recharge areas and deeper, downgradient parts of the aquifers (Lee, 1985). Chemical processes occurring in the zones may be characterized as follows:

RECHARGE ZONE:  Rainfall-input chemistry; aerobic system open
                to atmospheric  $CO_2$ + soil $CO_2$ + $O_2$.
ZONE A:  Aerobic system partially open to atmosphere,
         calcite dissolution; sodium feldspar hydrolysis
         to kaolinite; oxidation of carbonaceous matter
         ($CH_2O$) by dissolved $O_2$.
ZONE B:  Closed to atmosphere, anaerobic system; calcite
         dissolution; sodium feldspar hydrolysis to
         kaolinite; ferric iron reduction to ferrous
         iron, to siderite saturation, with oxidation of
         carbonaceous matter.
ZONE C:  Closed to atmosphere, anaerobic system; calcite
         dissolution; sodium feldspar hydrolysis to
         kaolinite; ferric iron reduction to ferrous iron,
         at siderite saturation.

[1]Investigation part of U.S. Geological Survey Regional Aquifer Systems Analysis (RASA).

*Chemical Quality of Water and the Hydrologic Cycle*, Robert C. Averett and Diane M. McKnight (Eds.) © 1987 Lewis Publishers, Inc., Chelsea, Michigan. Printed in the United States of America.

ZONE D:   Closed to atmosphere, anaerobic system; calcite
          dissolution to near saturation; sodium-for-
          calcium ion exchange; ferric iron reduction to
          ferrous iron; gypsum dissolution; sulfate
          reduction, pyrite precipitation with oxidation
          of carbonaceous matter; sodium beidellite
          precipitation.

ZONE E:   Mixing of meteorically-derived freshwaters from
          Zone D with downgradient brines or seawater.

Approximate distributions of the geochemical zones A through E in
Cretaceous and Tertiary Age aquifers for flow path-1 (Fig. 1)
are shown in Figure 2.

In the upgradient zones A, B, and C, feldspar hydrolysis and
oxidation-reduction are largely responsible for the sodium
bicarbonate character of ground water. In zone D, sodium and
bicarbonate ions are derived principally from calcite dissolution with some dolomite dissolution coupled with sodium-for-calcium
cation exchange, which is driven by evolution of $CO_2$ from organic
matter oxidized in the metabolic cycles of sulfate-reducing and
methanogenic bacteria. Geochemistry in zone E is not well
established owing to uncertainties and variability of the chemistry
of the end member saline waters that mix with the meteorically-
derived fresh ground water (Plummer, 1984).

In clastic aquifers low in calcite content, such as the Black
Creek, Middendorf, and Tuscaloosa-Cape Fear aquifers (Fig. 2), the
transition from an oxidizing environment to a reducing one near the
recharge areas is marked by significant increases in dissolved
ferrous iron (greater than 15 mg/L). In clastic aquifers
containing proportionately more calcite, such as the Claiborne
aquifer (Fig. 2), extensive calcite dissolution limits dissolved
ferrous iron concentrations in the upgradient areas due to siderite
($FeCO_3$) precipitation. In the deeper, downgradient parts of the
aquifers, reduction of dissolved sulfate to sulfide further lowers
dissolved ferrous iron, owing to the low solubility of pyrite
($FeS_2$). Ferrous iron increases in solution as dissolved chloride
and salinity increase in zone E.

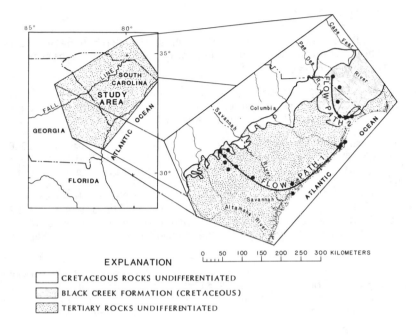

EXPLANATION

☐ CRETACEOUS ROCKS UNDIFFERENTIATED

☐ BLACK CREEK FORMATION (CRETACEOUS)

▦ TERTIARY ROCKS UNDIFFERENTIATED

Figure 1.    Study area with two ground-water flow paths and sample sites.

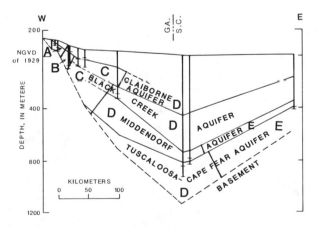

Figure 2.    Generalized geologic section for flow path 1 showing the approximate locations of geochemical zones in the aquifers.

REFERENCES CITED

Foster, M. D. 1950. The origin of high sodium bicarbonate waters in the Atlantic and Gulf Coastal Plains. Geochimica et Cosmochimica Acta. 1:33-48.

Lee, R. W. 1985. Geochemistry of ground water in Cretaceous sediments of the southeastern Coastal Plain, eastern Mississippi--western Alabama. Water Resour. Res. 21:1545-1556.

Parkhurst, D. L., D. C. Thorstenson, and L. N. Plummer. 1980. PHREEQE-A computer program for geochemical calculations. U.S. Geological Survey, Water-Resources Investigations Report, 80-96.

Parkhurst, D. L., L. N. Plummer, and D. C. Thorstenson. 1982. BALANCE-A computer program for calculation of mass balance. U.S. Geological Survey, Water-Resources Investigations Report, 82-14.

Plummer, L. N. 1984. Geochemical modeling--A comparison of forward and reverse methods, pp. 149-177. *In* B. Hitchon and E. I. Wallick, eds., Proceedings of the first Canadian/American conference on hydrogeology, Banff, Alberta, Canada.

## AMPEROMETRIC MEASUREMENT OF CHLORINE AT A PULSED SILVER IODIDE ELECTRODE

T. N. Morrison and C. O. Huber, University of Wisconsin-Milwaukee, Milwaukee, Wisconsin

Chlorination of water represents response to a dilemma in water management.  Protection from infectious diseases dictates chlorination, yet the toxic and mutagenic effects of chlorinated organics produced in natural waters dictates that minimal concentrations of active chlorine remain.  There is need for new, sensitive, selective, and rapid analytical devices for measurement of chlorine species.  The practical need for such monitoring indicates that analytical devices which can be operated with minimal training are desirable.

When chlorine is added to water it is rapidly hydrolized to form hypochlorous acid ($K_a = 3.0 \times 10^{-8}$).

$$Cl_2 + H_2O \rightarrow HOCl + H^+ + Cl^-$$

In natural waters which have been chlorinated, the chlorine is usually found reacted with amines or ammonia.

$$HOCl + NH_3 \rightarrow NH_2Cl + H_2O$$

$$HOCl + NH_2Cl \rightarrow NHCl_2 + H_2O$$

Each of these chlorine species has a different activity as a disinfectant and a different toxicity to other forms of life (Brooks, 1977).  In most potable, surface, or waste-water samples, the two most important chlorine species are hypochlorous acid and monochloramine because they are relatively stable at neutral pH.

Classical methods for measurement of hypochlorous acid and monochloramine make use of very rapid reactions with iodide as a reducing agent.

$$HOCl + 2I^- + H^+ \rightarrow Cl^- + I_2 + H_2O$$

$$NH_2Cl + 2I^- + H^+ \rightarrow Cl^- + I_2 + NH_3$$

Silver iodide is a semiconductor material which, at cathodic potentials, is reduced to silver and iodide. By coupling this reduction with the reaction between iodide and chlorine, a signal is obtained which is sensitive to the chlorine concentration.

$$AgI + e^- \rightarrow Ag + I^-$$

$$2I^- + H^+ + HOCl \rightarrow I_2 + Cl^- + H_2O$$
$$\text{or} \qquad\qquad \text{or}$$
$$NH_2Cl \qquad\qquad NH_3$$

At a constant reducing potential the silver iodide electrode would dissolve. Instead, by stepping from an oxidized to a reducing potential and back, much of the iodide is recaptured.

$$I^- + Ag \rightarrow AgI + e^-$$

The current is sampled near the end of the cathodic step. The applied potential and current is shown in Figure 1.

Complete selectivity for hypochlorous acid over monochloramine can be obtained by controlling the pH. The forward rate constant for the monochloramine-iodide reaction is smaller than for the hypochlorous acid-iodide reaction. At pH greater than 5 no measurable current is observed for monochloramine, while below pH 3, the current due to monochloramine is equivalent to the current for hypochlorous acid. The sensitivity for hypochlorous acid is invariant from pH 2.5 to 5.5. This facilitates the subtraction of the hypochlorous acid component from a mixture by measuring first at pH 5.5 and then at pH 3.

The apparatus consists of a flow injection system which allows selection of two separate background electrolytes. The timing, potentiostat, current following and data treatment are micro-processor controlled. The potential is continuously pulsed and the current sampled. When a chlorine containing sample is injected, the maximum in the cathodic current as it flows past the electrode is measured by the microprocessor and displayed in units of concentration.

## Measurement Conditions

| | |
|---|---|
| Pulse amplitude: | +0.05V to -0.20 V vs. SCE |
| Pulse width: | 100 milliseconds |
| Pulse frequency: | 1 pulse/150 milliseconds |
| Background electrolyte: | 0.5M $KNO_3$ + 0.05 M |
| | pH 3.0 or pH 5.5 buffer |

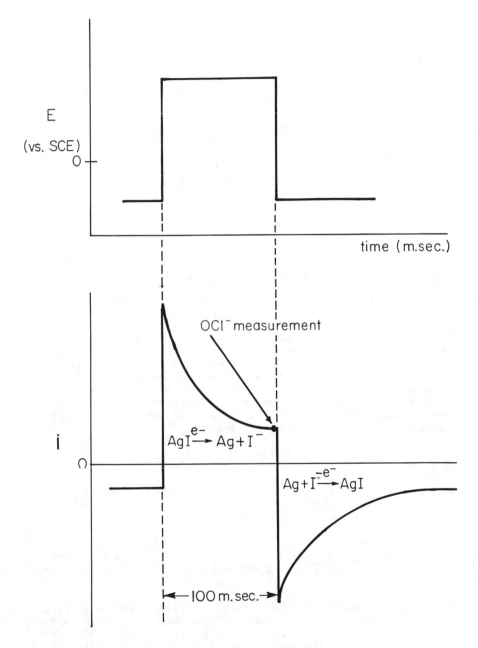

Figure 1.   Applied potential and current time profiles.

## Measurements Results

|  | HOCl | NH$_2$Cl |
|---|---|---|
| Linear range: | $\overline{<}10^{-3}$M | $\overline{<}10^{-4}$M |
| Lower detection limit | $2\times10^{-7}$M | $5\times10^{-7}$M |
| Sensitivity | 1.1nA/$\mu$M | 0.7mA/$\mu$M |

The speed and low detection limits (three times standard error of estimate of concentration) of the new technique have been applied to measuring the reaction rate between hypochlorous acid and ammonia at realistic concentrations and pH conditions. Weil and Morris (1949) have shown the reaction to follow the rate law in acidic or basic solution:

$$k_f = k_{obs} \left[ 1 + \frac{K_a}{[H^+]} \right] \left[ 1 + \frac{K_b}{[OH^-]} \right]$$

where $K_a = K_a$ for HOCl = 3.0 x $10^{-8}$
$K_b = K_b$ for NH$_3$ = 1.76 x $10^{-5}$

The rate was found to be maximum between pH 7 and 9. The published data to date show rate measurements only in acidic or basic solutions where the reaction is slower or at unrealistic ionic strengths. The silver iodide electrode reported here was used to measure the rate of reaction of HOCl upon addition of NH$_3$ at pH 8. Because of the very low concentrations which this electrode detects, the reaction was slow enough to allow complete mixing of reactants and yield a rate measurement at the ionic strengths, low concentrations, and moderate pH values of realistic natural water conditions. The rate constant measured was 3.3 ± 0.6 x $10^6$ L/mole/sec. In a review of the literature (Morris, 1983) an overall value of 4.0 x $10^6$ ± 25% ±/mole sec was reported. The two numbers agree within experimental errors.

## REFERENCES

Brooks, A. S., and G. L. Seegert. 1977. The effects of intermittent chlorination on rainbow trout and yellow perch. Trans. Am. Fish. Soc. 106:278286.

Weil, I., and J. C. Morris. 1949. Kinetic studies on the chloramines. 1. The rates of formation of monochloramine, N-chloroethylamine and N-chlorodimethylamine. J. Am. Chem. Soc. 71:1664.

Morris, J. C. and R. A. Isaac. 1983. A critical review of kinetic and thermodynamic constants for the aqueous chlorine-ammonia system, pp. 49-62. *In* R. L. Jolley, et al., ed., Water chlorination: Environmental impact and health effects, Vol. 4. Ann Arbor Science Publishers, Inc. Ann Arbor.

BIOCONCENTRATION OF CREOSOTE COMPOUNDS IN SNAILS
OBTAINED FROM PENSACOLA BAY, FLORIDA, NEAR
AN ONSHORE HAZARDOUS-WASTE SITE

Colleen E. Rostad and Wilfred E. Pereira,
U.S. Geological Survey, Denver, Colorado

Snails were collected from two areas offshore in Pensacola Bay
near an onshore hazardous-waste site. The hazardous-waste site
involved creosote contamination of ground water, resulting from
activities at a wood-preserving facility. Because concentrations of
creosote compounds in the water from Pensacola Bay may be too low
for detection, snails, which are detritus feeders, were chosen as a
preliminary indicator of bioconcentration. The snails, *Thais
haemostoma*, were deshelled and the tissue homogenized. The tissue
was then hydrolyzed for 2 hours with ethanolic potassium hydroxide.
After extraction with isooctane and further cleanup with dimethyl-
sulfoxide to eliminate cholesterol, the snail-tissue extracts were
analyzed by gas chromatography-mass spectrometry. Along with
naturally occurring compounds, the snail tissue contained large
concentrations of polycyclic aromatic compounds, such as naphthalene, phenanthrene, acridine, dibenzothiophene, dibenzofuran,
and benzo[a]pyrene. Many of these compounds are characteristic of
the creosote contamination associated with the onshore hazardous-
waste site. In the snail tissue, 23 compounds indicative of
creosote contamination were quantitated. The compounds selected
included four chemical classes: 13 polycyclic aromatic hydrocarbons (indane to benzo[a]pyrene); seven nitrogen heterocycles
(2,4-dimethylpyridine to acridinone); two sulfur heterocycles
(benzoand dibenzo-thiophene); and an oxygen heterocycle
(dibenzofuran). Concentrations of the selected compounds in the
snail tissue were higher at one site than the other site; however,
when the concentrations were normalized to phenanthrene, the
distributions of the compounds were similar, indicating a common
source. Based on the n-octanol-water partition coefficient, the
expected bioconcentration was compared to the actual concentration
of each compound. An absence of a correlation between the
bioconcentration and the n-octanol-water partition coefficient

*Chemical Quality of Water and the Hydrologic Cycle*, Robert C. Averett and Diane M. McKnight (Eds.) © 1987 Lewis Publishers, Inc.,
Chelsea, Michigan. Printed in the United States of America.

suggested an anthropogenic source. The low concentration and pattern of alkylated polycyclic aromatic hydrocarbons indicated that the source of these compounds was not petroleum. Many of the compounds studied (naphthalene versus 1- and 2-methylnaphthalene, acenaphthene versus biphenyl, phenanthrene versus anthracene), had similar n-octanol-water partition coefficients, yet the concentration of the compounds in the snail tissue varied considerably. Compounds with a n-octanol-water partition coefficient less than three were not bioconcentrated except for 2-quinolinone. The concentration profiles of these compounds in the snail tissue were more closely correlated with profiles from nearby onshore ground water near the source of the contamination and near the shore. The concentration profiles in the snails resemble the creosote-contaminated ground-water profile rather than that predicted by the n-octanol-water partition coefficient. The anomaly of the 2-quinolinone present in the snail tissue was the result of extremely high concentrations of 2-quinolinone present in the ground water. An evaluation of the snail-tissue extracts provided an excellent preliminary indication that the contaminated ground water was indeed impacting the food chain in Pensacola Bay.

SPATIAL AND TEMPORAL VARIATION IN CHEMICAL CHARACTERISTICS
OF GROUND WATER AND WATER IN WETLANDS OF THE COTTONWOOD
LAKE AREA, STUTSMAN COUNTY, NORTH DAKOTA, 1979-85

J. W. LaBaugh, U.S. Geological Survey, Denver, Colorado

Selected wetlands and the local contiguous ground-water system
in the Cottonwood Lake area, North Dakota have been the subject of
detailed hydrogeological and hydrochemical studies since 1979
(Winter and Carr, 1980; LaBaugh et al., 1987). The wetlands are in
an area in which average annual evaporation exceeds average annual
precipitation. Thus, water levels in wetlands and ground water
fluctuate considerably in response to year-to-year variation in
precipitation (Eisenlohr, 1972; Sloan, 1972) as well as seasonal
variation in precipitation (LaBaugh et al., 1987).

Two distinct types of wetlands were examined during the study;
seasonal and semipermanent. Seasonal wetlands contain water for
periods of weeks or months after snowmelt, eventually becoming
completely dry. Semipermanent wetlands usually contain water
throughout the year except in extremely dry periods, such as in the
early 1960's (Eisenlohr, 1972) and again in 1977 (Swanson, personal
communication, 1979), when the semipermanent wetlands were
completely dry in mid-summer.

LaBaugh et al. (1987) found that four wetlands under detailed
investigation in the study area had three different hydrologic
functions with respect to the local contiguous ground-water system.
One of the seasonal wetlands, wetland T8, recharged ground water,
while the other seasonal wetland, wetland T3, received input from
ground water on one side and lost water by output to ground water
on another side. Both semipermanent wetlands, wetlands P1 and P8
received ground-water discharge. An additional wetland, wetland
P11, was included in the investigations of the area in 1984. It
also is in an area of ground-water discharge but does not receive
discharge from the local flow system affecting wetlands P1 and P8.

Previous investigations in the region suggested there was a
link between ground-water flow systems and the relative salinity of
water in wetlands of the region (Sloan, 1972). However, details of
the relation between a wetland's position in the local ground-water

flow system and chemical characteristics of the wetland have only recently been determined (LaBaugh et al., 1987). Bicarbonate and sulfate water types occurred in ground water and the wetlands: bicarbonate waters and smaller values of specific conductance commonly were associated with areas of ground-water recharge, whereas sulfate waters and larger values of specific conductance commonly were associated with areas of ground-water discharge, or areas of reversals in ground-water flow.

There was no distinct relation between specific conductance and whether or not water was of a bicarbonate or sulfate type. Waters from wells immediately adjacent to wetland P1 were all a sulfate type even though wells located within 100 m of each other had values of specific conductance between 500 and 15,000 µS/cm. Water in adjacent wetlands had values between 100 and 3,500 µS/cm and wetland P11 had values as large as 13,500 µS/cm.

Between 1979-85 smaller values of specific conductance and concentrations of major ions in the wetlands were associated with spring snowmelt, loss of ice cover, and wet years. Larger concentrations were associated with ice cover formation, and dry years. Seasonal fluctuations in specific conductance were the smallest in wetland T8 and commonly in the range of less than 100 µS/cm. The largest seasonal fluctuations in specific conductance, in the range of thousands of µS/cm, occurred in wetlands P1 and P11. Differences between wet and dry years in the value of specific conductance was as much as 2,500 µS/cm in wetland P1, based on comparisons of values in early autumn prior to formation of ice cover. Temporal variation in chemical characteristics of ground water was not as readily associated with climatic variations as in the wetlands.

REFERENCES

Eisenlohr, W. S. 1972. Hydrologic investigations of prairie potholes in North Dakota, 1959-68. U.S. Geological Survey Professional Paper 585-A. U.S. Government Printing Office, Washington, D.C., 102 p.

LaBaugh, J. W., T. C. Winter, V. A. Adomaitis, and G. A. Swanson. 1987. Hydrology and chemistry of selected prairie wetlands in the Cottonwood Lake area, Stutsman County, North Dakota 1979-82. U.S. Geological Survey Professional Paper 1431 (in press).

Sloan, C. E. 1972. Ground-water hydrology of prairie potholes in North Dakota. U.S. Geological Survey Professional Paper 585-C. U.S. Government Printing Office, Washington, D.C., 28 p.

Swanson, G. A. U.S. Fish and Wildlife Service, Jamestown, North Dakota. 1979.

Winter, T. C. and M. R. Carr. 1980. Hydrologic setting of wetlands in the Cottonwood Lake area, Stutsman County, North Dakota. U.S. Geological Survey Water Resources Investigations Report 80-99, 42 p.

# MODIFICATIONS OF NUTRIENT TRANSPORT DURING FLOW THROUGH A NORTHERN FLORIDA RIVER WETLAND SYSTEM

John F. Elder, U.S. Geological Survey, Madison, Wisconsin

Research on wetland ecology in recent years has shown that wetland processes can significantly alter both the quantity and character of nutritive and contaminant substances as they flow through the wetland (Brinson, 1977; de la Cruz, 1979; Brinson et al., 1980). However, the nature of the alterations is highly variable in different systems and at different times. In some cases, the wetland acts as a net importer (Reichle, 1975; Mitsch et al., 1979); in some cases, it may act as a net exporter (Kowalczewski, 1978); and in still other cases, it may act as a combination of both (Peverly, 1982).

The Apalachicola River in northern Florida, a large subtropical river-wetland system, was the site of a study conducted by the U.S. Geological Survey to describe some of the effects of flood-plain processing on nutrient and detritus flow through the system. Flowing 170 km southward from Lake Seminole at the Georgia-Florida border (Fig. 1), the Apalachicola River is the downstream branch of a 50,000 km² watershed covering parts of three states. Its discharge varies from less than 300 m³/s to spring flood levels as high as 4,000 m³/s. The mean annual flow is 800 m³/s.

Approximately 15% (450 km²) of the Apalachicola basin consists of a relatively flat flood plain that supports dense stands of cypress, tupelo, and mixed bottomland hardwood trees. Nearly 400,000 tons of organic litterfall are generated annually by the flood-plain vegetation (Elder and Cairns, 1982).

Concentrations of various fractions of nitrogen, phosphorus, and organic carbon were monitored monthly for 19 months at seven sites in the Apalachicola river-wetland system (Fig. 1). Each of the sampling sites coincided with a continuous-recording streamflow-gaging site. Sites 1-4 were located on the main river channel and served to separate the subbasin into three drainage areas, each with measured inflow and outflow (Mattraw and Elder, 1984). Atmospheric precipitation collectors also were located at these sites. The other sites were located in the flood plain

*Chemical Quality of Water and the Hydrologic Cycle*, Robert C. Averett and Diane M. McKnight (Eds.) © 1987 Lewis Publishers, Inc., Chelsea, Michigan. Printed in the United States of America.

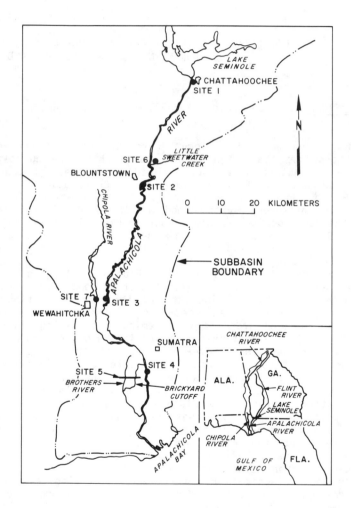

Figure 1.    Apalachicola-Chipola River subbasin, showing sampling sites; the entire Apalachicola-Chattahoochee-Flint River basin is shown in inset.

itself (site 5) and on tributaries of the river (sites 6 and 7). Ground-water flow was estimated from background data collected from wells throughout the subbasin.

The methods of analyses of the difference fractions were described in an earlier paper (Elder, 1985). These analyses yielded data for the following nitrogen and phosphorus fractions: dissolved nitrogen as nitrate and nitrite ($NO_3$-N), dissolved nitrogen as ammonia ($NH_4$-N), dissolved inorganic nitrogen (DIN=$NO_3$-N+$NH_4$-N), dissolved organic nitrogen (DON), particulate organic nitrogen (PON), total organic nitrogen (TON=DON+PON), total nitrogen (TN), soluble reactive phosphorus (SRP), dissolved phosphorus (DP), particulate phosphorus (PP), and total phosphorus (TP).

## Nitrogen: Phosphorus Ratios

The ratio of total nitrogen to total phosphorus (TN:TP) was nearly constant throughout the main channel, ranging from 12 to 15 at sites 1 through 5. The two tributary sites differed considerably from this value, both showing phosphorus-poor conditions with TN:TP ratios >20.

Concentrations of dissolved inorganic species are more significant biologically than are the total nutrient concentrations, because the readily-bioavailable pool is limited to nitrate and ammonia nitrogen and to SRP (Fitzgerald, 1972; Scavia, 1980). Autotrophic production is thus more likely to be dependent on the DIN:SRP ratio than on the TN:TP ratio. Unlike the latter, the DIN:SRP ratio showed a distinct tendency to increase in a downstream direction, reaching levels of about 40 near the mouth. An estuary that receives that type of nutrient load from the river is likely to be phosphorus limited. This is consistent with observations by Livingston and Loucks (1979) of phosphorus limitation in Apalachicola Bay.

## Flux of Different Species in the System

Mass-balance calculations were used to estimate nutrient input from various sources, including inflow at the river's headwaters, inflow from tributaries, ground-water discharge, precipitation, and surface-water runoff. In this system, where the hydrology is strongly streamflow dominated, the latter three sources contributed very little to the annual budget of most fractions. During the period of June 3, 1979, through June 2, 1980, there was negligible net change (difference between outflow at the mouth and the sum of inflows) in total nitrogen (Table 1). For total phosphorus, outflow exceeded inflow by 10%. Greater net changes were found for individual species. Ammonia, in particular, was depleted by more than 50% net uptake in the system. There was also a substantial net uptake of SRP. Error in measurement and analysis may account for some of the calculated residuals, but changes approaching 10% or more must be at least partially attributed to flood-plain processing.

Table 1.  Output Loads of Nutrient Fractions at Site 4 during the
Period June 3, 1979, through June 2, 1980, and Percent
Contributions from Various Sources to the Outputs.
Residuals are shown as either positive or negative,
indicating net export or net import, respectively,
during flow through the river-wetland system.

| | \multicolumn{6}{c}{Nutrient fraction} | | | | | |
| | TN | $NO_3$-N | $NH_4$-N | PON | TP | SRP |
|---|---|---|---|---|---|---|
| Total outflow (metric tons) | 21,400 | 8,200 | 830 | 4,500 | 1,650 | 230 |
| Contributions (percent of total inflow) | | | | | | |
| Headwater inflow | 81 | 87 | 119 | 80 | 81 | 92 |
| Major tributary inflow | 12 | 13 | 14 | 4 | 4 | 11 |
| Other inflow[1] | 5 | 3 | 20 | 5 | 5 | 11 |
| Residual | +2 | -3 | -53 | +11 | +10 | -14 |

[1]Sum of ground water, atmospheric precipitation, and surface-water
runoff.

The overall result of flood-plain processing in the
Apalachicola system is that there tends to be net uptake of
dissolved, inorganic nutrients and net release of particulate,
organic nutrients.  The nutrient load supplied to the estuary is
likely to be quantitatively similar to that entering the river at
its headwaters, but qualitatively it is quite different.  Thus, the
wetland functions more as a transformer than either a source or a
sink for nutrients.  The increase in particulate, organic material
during riverine transport creates a nutrient supply to the estuary
that favors growth of a community rich in detritivores.  Oyster and
shrimp populations, which are abundant and economically important
in Apalachicola Bay, depend on the detritus that is supplied by the
flood plain and transported by river flow.

REFERENCES

Brinson, M. M. 1977.  Decomposition and nutrient exchange of litter
in an alluvial swamp forest.  Ecology. 58:601-609.

Brinson, M. M., H. D. Bradshaw, R. N. Holmes, and J. B. Elkins, Jr.
1980.  Litterfall, stemflow, and throughfall nutrient fluxes in
an alluvial swamp forest.  Ecology. 61:827-835.

de la Cruz, A. A. 1979.  Production and transport of detritus in
wetlands, pp. 162-174.  *In* P. E. Greeson, J. R. Clark, Jr., and
J. E. Clark, eds., Wetland functions and values:  The state of
our understanding.  American Water Resources Assoc.,
Minneapolis.

Elder, J. F. 1985. Nitrogen and phosphorus speciation and flux in a large Florida river wetland system. Water Resources Research 21:724-732.

Elder, J. F., and D. J. Cairns. 1982. Production and decomposition of forest litterfall on the Apalachicola River flood plain, Florida. U.S. Geological Survey Water-Supply Paper 2196-B, p. 42.

Fitzgerald, G. P. 1972. Bioassay analysis of nutrient availability, pp. 147-169. *In* H. E. Allen and J. R. Kramer, eds., nutrients in natural waters. Wiley-Interscience, New York.

Livingston, R. J., and O. L. Loucks. 1979. Productivity, trophic interactions, and food-web relationships in wetlands and associated systems, pp. 101-119. *In* P. E. Greeson, J. R. Clark, Jr., and J. E. Clark, eds., Wetland functions and values: The state of our understanding. American Water Resources Assoc., Minneapolis.

Kowalczewski, A. 1978. Importance of a bordering wetland for chemical properties of lake water. Verhandlungen Internationale Vereinigung fur Theoretisch und Angewandte Limnologie. 20:2182-2185.

Mattraw, H. C., Jr., and J. F. Elder. 1984. Nutrient and detritus transport in the Apalachicola River, Florida. U.S. Geological Survey Water-Supply Paper 2196-C, p. 62.

Mitsch, W. J., C. L. Dorge, and J. R. Wiemhoff. 1979. Ecosystem dynamics and a phosphorus budget of an alluvial cypress swamp in southern Illinois. Ecology. 60:1116-1124.

Peverly, J. H. 1982. Stream transport of nutrients through a wetland. Journal of Environmental Quality. 11:38-43.

Reichle, D. E. 1975. Advances in ecosystem analysis. Bioscience. 25:257-264.

Scavia, D. 1980. Conceptual model of phosphorus cycling, nutrient cycling in the Great Lakes: A summarization of factors regulating the cycling of phosphorus. University of Michigan, Ann Arbor, Great Lakes Research Div., Special Report No. 83, pp. 119-140.

# INHIBITION OF AQUEOUS COPPER AND LEAD ADSORPTION ONTO GOETHITE BY DISSOLVED CARBONATE SPECIES

Kathleen S. Smith, U.S. Geological Survey, Denver, Colorado

Donald Langmuir, Colorado School of Mines, Golden, Colorado

Adsorption of aqueous copper and lead by goethite is inhibited by copper and lead carbonato complexing, which becomes important at pH's greater than the range 6.5 to 7 at a total dissolved carbonate ($C_T$ concentration of 0.01 $M$, and total metal concentrations of 0.1 to 0.01 $mM$. This reflects the relatively weak adsorption by goethite of neutral and anionic metal carbonato complexes in contrast to strong adsorption of free ions and/or cationic copper and lead hydroxo complexes. This study indicates that the mobilities of copper and lead in soil and ground-water systems are a function of the pH and $C_T$ content of those systems.

Trace-metal mobilities in natural water-rock systems are usually dominantly controlled by adsorption-desorption reactions (Jenne, 1968). This control is partially due to the often rapid equilibration times of sorption reactions as compared to the often longer equilibration times of water-mineral solubility reactions. Our laboratory study investigated copper and lead adsorption by goethite ($\alpha$-FeOOH) at elevated total carbonate concentrations. Goethite was chosen as the sorbent because it is often a major component of soils, sediments, and suspended material in streams and aquifers. Goethite has a high adsorptive capacity for heavy metals and is active in the attenuation and transport of these metals, especially at neutral to alkaline pH's. Adsorption inhibition onto goethite by dissolved carbonate species has previously been reported for the uranium system (Hsi, 1981; Tripathi, 1984; Hsi and Langmuir, 1985) and the plutonium system (Sanchez et al., 1985).

The synthetic goethite used in the adsorption experiments was prepared according to the method of Atkinson et al. (1967). Adsorption was studied in batch laboratory experiments using the goethite in suspension at a concentration of 1 g/L (31 m²/L). The pH was adjusted to between 5 and 9 using $HNO_3$ or KOH. Ionic

*Chemical Quality of Water and the Hydrologic Cycle*, Robert C. Averett and Diane M. McKnight (Eds.) © 1987 Lewis Publishers, Inc., Chelsea, Michigan. Printed in the United States of America.

strength (I) was maintained at 0.01 $M$ with $KNO_3$ for systems run under atmospheric conditions. For experiments performed under higher $C_T$ conditions, both $C_T$ and I were adjusted with $KHCO_3$. Experiments were run at initial total metal concentrations of $10^{-4}$, $10^{-5}$, and $10^{-6}$ $M$ for copper and $10^{-5}$, $10^{-6}$, and $10^{-7}$ $M$ for lead. These concentrations were chosen to simulate those found in natural systems. Rate studies indicated that adsorption attained equilibrium within 2 hours. In all adsorption experiments, solutions were prepared in 50-mL polycarbonate centrifuge tubes, tightly capped, and placed in a 25°C water bath for 4 hours. Occasional shaking of the tubes kept the goethite in suspension. The suspensions were centrifuged at 16,000 rpm for 20 min using a refrigerated centrifuge. The supernatant was removed in 15-mL aliquots, then transferred to acid-washed polyethylene bottles, and acidified with ULTREX nitric acid for total metal determination by atomic absorption spectrophotometry. Random samples were analyzed for iron to ensure that none of the goethite remained in suspension. The supernatant solutions were analyzed in random sequence to eliminate any systematic analytical errors. The pH was measured in supernatant solutions remaining in the centrifuge tubes. The pH measurements were found to be reproducible to within 0.05 pH units for a given adsorption run and 0.1 pH units between adsorption runs. The average standard deviation for copper by flame atomic absorption was ± 0.018 ppm, and by graphite furnace atomization was ± 0.66 ppb. The average standard deviation for lead by flame atomic absorption was ± 0.010 ppm, and by graphite furnace atomization was ± 2.6 ppb. The analytical limit of detection was 1 ppb for both copper and lead.

Data are plotted as fractional metal adsorption versus pH in Figures 1 and 2. The metal scales are logarithmic and are expressed as amount of metal remaining in solution after centrifugation. This presentation of adsorption isotherms emphasizes the lower metal concentrations where most natural interactions take place. Solid curves are free-drawn best fits of experimentally derived points. Data points of < 1 ppb are plotted on the 1 ppb line; therefore, there is no graphical distinction between sample values of 1 ppb and < 1 ppb. Under atmospheric conditions (a partial $CO_2$ pressure of $10^{-3.5}$ atm.), solutions at pH 7 and 9 correspond to $C_T = 10^{-4.2}$ and $10^{-2.3}$ $M$, respectively. Blank experiments performed without goethite indicated that dissolved metal concentrations in the adsorption experiments were controlled by adsorption reactions rather than solid precipitation reactions. Blank data are indicated by dashed lines (for atmospheric conditions) and separate points (for higher C conditions) in Figures 1 and 2.

Figures 1 and 2 show the experimental results for $10^{-4}$ $M$ (6,354 ppb) total-copper and $10^{-5}$ $M$ (2,072 ppb) total-lead systems, respectively, under atmospheric and above-atmospheric $C_T$ conditions. Both copper and lead are specifically adsorbed by goethite as the adsorption edge of these metals occurs at a pH below the point of zero charge (pH(PZC) = 7.6), the pH at which the net surface charge of the goethite substrate is zero. Results indicate that there is an inhibitory effect on copper and lead

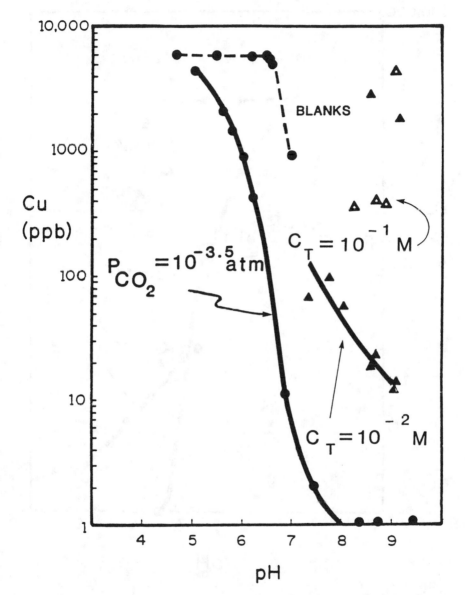

Figure 1. Adsorption of copper onto 1 g/L (31 m²/L) goethite in suspension for $10^{-4}$ M (6,354 ppb) total copper under atmospheric conditions ($P_{CO_2} = 10^{-3.5}$ atm) (circles), and for $C_T = 10^{-2}$ (closed triangles) and $C_T = 10^{-1}$ M (open triangles) with I = 0.01 and I = 0.1 M (as either $KNO_3$ or $KHCO_3$) at 25°C. Data from blank experiments performed without goethite are indicated by a dashed line for atmospheric conditions and by separate points for above-atmospheric C conditions.

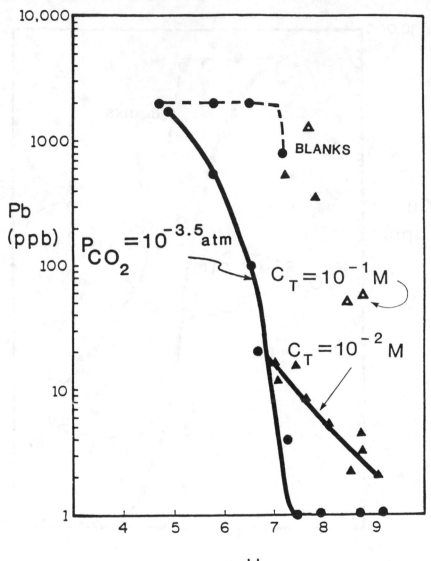

Figure 2.    Adsorption of lead onto 1 g/L (31 m²/L) goethite in suspension for $10^{-5}$ $M$ (2,072 ppb) total lead under atmospheric conditions ($P_{CO_2} = 10^{-3.5}$ atm) (circles), and for $C_T = 10^{-2}$ (closed triangles) and $C_T = 10^{-1}$ $M$ (open triangles) with I = 0.01 and I = 0.1 $M$ (as either $KNO_3$ or $KHCO_3$) at 25°C.    Data from blank experiments performed without goethite are indicated by a dashed line for atmospheric conditions and by separate points for above-atmospheric C conditions.

adsorption in the higher $C_T$ experiments. At $C_T = 10^{-2}$ $M$, about 80 ppb copper and 8 ppb lead remain in solution at the pH(PZC). This value represents only about 1.3% of the total copper present and less than 1% of the total lead present. However, these aqueous concentrations could be significant in natural settings. The degree of adsorption inhibition is proportional to $C_T$ and total-metal concentration. No noticeable adsorption inhibition was seen in the lower-metal-concentration experiments.

The adsorption inhibition of copper and lead is probably due to increased carbonato complexing of the metals. Figure 3 illustrates aqueous speciation for $10^{-4}$ $M$ total copper under atmospheric conditions (solid lines) and $C_T = 10^{-2}$ $M$ conditions (dashed lines). A comparison of the species distribution for these different $C_T$ conditions shows that there is both an increase in carbonato complexing at the expense of hydroxo complexing, and a downward pH shift of the carbonato complexes with increasing $C_T$. These combined effects place the dominant neutral carbonato complex in the pH range of greatest adsorption inhibition, and at a pH below completion of the adsorption edge seen in Figure 1. This neutral carbonato complex would be expected to be weakly adsorbed, if at all, compared to free metals and/or hydroxo complexes. Adsorption edges for the lower-metal-concentration experiments are shifted to lower pH regions; therefore, the adsorption edges are completed at a pH prior to the neutral-carbonato-complex region of dominance and no adsorption inhibition by dissolved carbonate species takes place. The arguments are similar for lead (Fig. 4).

Although these experiments were performed in the laboratory, the findings have implications to metal mobility in natural soil systems and aquifers. In neutral to alkaline (especially arid-climate) soils and in ground-water systems that contain calcite and dolomite, carbonato complexing of copper and lead can be expected to increase the mobilities of those metals. A paper is currently in preparation which presents a more quantitative discussion of copper- and lead-adsorption results modeled with the surface complexation site-binding treatment (Yates et al., 1974; Davis et al., 1978) as applied with the computer program MINEQL (Westall et al., 1976).

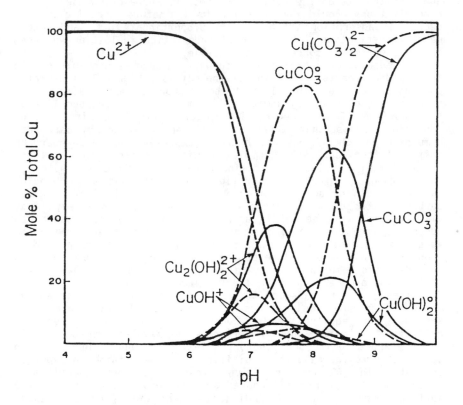

Figure 3.   Distribution diagram for copper species as a function of pH for the $Cu^{2+}$-$H_2O$-$CO_2$ system for $10^{-4}$ $M$ (6,354 ppb) total copper, $I = 0.01$ $M$ (as $KNO_3$ or $KHCO_3$), 25°C, and $P_{CO_2} = 10^{-3.5}$ atm. (solid lines) and $C_T = 10^{-2}$ $M$ (dashed lines).

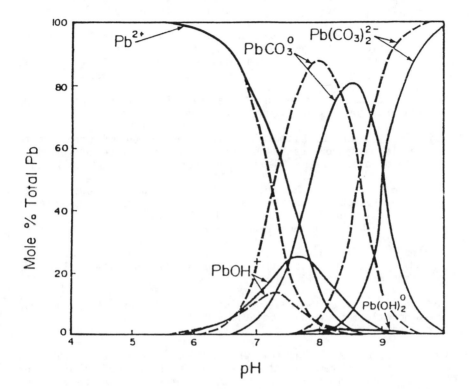

Figure 4.    Distribution diagram for lead species as a function of pH for the $Pb^{2+}$-$H_2O$-$CO_2$ system for $10^{-5}$ $M$ (2,072 ppb) total lead, I = 0.01 $M$ (as $KNO_3$ or $KHCO_3$), 25°C, and $P_{CO_2}$ = $10^{-3.5}$ atm. (solid lines) and $C_T$ = $10^{-2}$ $M$ (dashed lines).

REFERENCES

Atkinson, R. J., A. M. Posner, and J. P. Quirk. 1967. Adsorption of potential-determining ions at the ferric oxide-aqueous electrolyte interface. J. Phys. Chem. 71:550-558.

Davis, J. A., R. O. James, and J. O. Leckie. 1978. Surface ionization and complexation at the oxide/water interface. I. Computation of electrical double layer properties in simple electrolytes. J. Colloid Interface Sci. 63:480-499.

Hsi, C-K. D. 1981. Sorption of uranium (VI) by iron oxides. Ph.D. Dissertation. Colorado School of Mines. Golden.

Hsi, C-K. D. and D. Langmuir. 1985. Adsorption of uranyl onto ferric oxyhydroxides: Application of the surface complexation site-binding model. Geochim. Cosmochim. Acta. 49:1931-1941.

Jenne, E. A. 1968. Controls on Mn, Fe, Co, Ni, Cu, and Zn concentrations in soils and water. The significant role of hydrous Mn and Fe oxides, p. 337-387. *In* Trace inorganics in water, Adv. Chem. Ser. 73. Am. Chem. Soc., Washington, D.C.

Sanchez, A. L., J. W. Murray, and T. H. Sibley. 1985. The adsorption of plutonium IV and V on goethite. Geochim. Cosmochim. Acta. 49:2297-2307.

Tripathi, V. S. 1984. Uranium (VI) transport modeling: Geochemical data and submodels. Ph.D. Dissertation. Stanford University.

Westall, J. C., J. L. Zachary, and F. M. M. Morel. 1976. MINEQL, a computer program for the calculation of chemical equilibrium composition of aqueous systems. Water Quality Laboratory, Tech. Note 18. Department of Civil Engineering. Massachusetts Institute of Technology, Cambridge.

Yates, D. E., S. Levine, and T. W. Healy. 1974. Site-binding model of the electrical double layer at the oxide/water interface. Chem. Soc. Faraday Transactions I. 70:1807-1818.

ABSTRACT H

# AN EXAMINATION OF IRON OXYHYDROXIDE PHOTOCHEMISTRY AS A POSSIBLE SOURCE OF HYDROXYL RADICAL IN NATURAL WATERS

K. M. Cunningham and M. C. Goldberg, U.S. Geological Survey, Denver, Colorado

E. R. Weiner, University of Denver, Denver, Colorado

Recent laboratory and onsite investigations have established that photolysis of humic substances, $Fe^{+3}$ complexes with humic substances, and iron oxide minerals can yield reactive radical species like $ROO^{.}$ (organic peroxy radicals), $^{.}OH$ (hydroxyl radicals), and $^1O_2$ (singlet oxygen) (Roof, 1982; Cooper and Zika, 1983; Ross and Crosby, 1985). In the instance of iron oxide minerals and iron-humic complexes, the basic photoreaction is thought to be a charge transfer in the $Fe^{+3}$-ligand complex (either on the surface of the oxide or in solution) to yield a free radical residue of the ligand and $Fe^{+2}$. Such reactions have been postulated for a number of different ligands adsorbed on $Fe_2O_3$ and FeOOH (Sakata et al., 1984; Waite and Morel, 1984a; Finden et al., 1984; Cunningham et al., 1985; Bencala and McKnight, 1986). There is evidence that even $OH^-$ can be photo-oxidized by amorphous FeOOH, partially crystalline lepidocrocite (gamma-FeOOH), and crystalline hematite (alpha-$Fe_2O_3$) to yield $^{.}OH$ (Waite and Morel, 1984b; Haupt et al., 1984). However, two questions remain: (1) Is $^{.}OH$ generated primarily in a direct charge-transfer reaction between $Fe^{+3}$ and $OH^-$, or via the multistep Haber-Weiss mechanism for $Fe^{+2}$ oxidation (Sung and Morgan, 1980), an overall expression for which (in acidic solution) is,

$$3Fe^{+2} + O_2 + 8H_2O \longrightarrow 3Fe^{+3} + {}^{.}OH + 6H^+ \qquad (1)$$

and, (2) What effect does the degree of crystallinity of the iron oxide have on $^{.}OH$ generation? The answer to these questions has a bearing on the larger question of the significance of iron oxide photochemistry in natural systems.

We have photolyzed highly crystalline goethite, alpha-FeOOH, using near UV light (300-400 nm) in the presence of the organic

*Chemical Quality of Water and the Hydrologic Cycle*, Robert C. Averett and Diane M. McKnight (Eds.) © 1987 Lewis Publishers, Inc., Chelsea, Michigan. Printed in the United States of America.

anions succinate, oxalate, and benzoate. Benzoate was present as an ·OH scavenger. Our objective was to determine the relative importance of the direct photolysis and Haber-Weiss mechanisms for generating ·OH under conditions where the rate of $Fe^{+2}$ photogeneration can be varied by changing the organic composition of the suspension. Photolyzed suspensions contained 0.5 g/L goethite and one of the following three solutions: oxalate and benzoate (ob+g), succinate and benzoate (sb+g), and benzoate alone (b+g), each carboxylate having a concentration of 0.001 mol/L. Total ionic strength was about 0.01. The pH (5.5±0.1) and $[O_2]$ were maintained either chemically when purging with a $N_2/CO_2$ mixture, or manually when purging with air. $[Fe^{+2}]$ was determined colorimetrically (Tamura et al., 1974). In the presence of $O_2$, the product of the reaction of ·OH with benzoate is a mixture of ortho-, meta-, and para-hydroxy isomers (Matthews, 1984). The ortho-hydroxy isomer (salicylate) was selectively analyzed by a fluorometric method (Thommes and Leininger, 1958). Other experimental procedures have been previously described (Cunningham et al., 1985). The following types of experiments were performed:

Type I.     Photolyses of ob+g, sb+g, and b+g suspensions with $N_2/CO_2$ purging and analysis of $Fe^{+2}$.

Type II.    Photolyses of ob+g, sb+g, and b+g suspensions with air purging and analysis of salicylate.

Type III.   Dark reactions in which known aliquots of $Fe^{+2}$ were added to air-purged ob+g, sb+g, and b+g suspensions followed by analysis of salicylate, after allowing 50 minutes for the $Fe^{+2}$ aliquot to oxidize.

Type IV.    Addition of $Fe^{+2}$ to unphotolyzed, $N_2$-purged b+g suspensions and recovery of the $Fe^{+2}$ as a function of time to assess the importance of adsorption of $Fe^{+2}$ on the goethite surface.

Representative data for moles of $Fe^{+2}$ and moles of salicylate generated as a function of moles of absorbed photons (experiment types I and II) are shown in Figure 1 for the sb+g suspension. Ordinates for salicylate and $Fe^{+2}$ were chosen to give the data points as much overlap as possible for purposes of comparison. In all suspension types, the ratio of moles of salicylate produced to moles of $Fe^{+2}$ produced was constant throughout the photolysis. These ratios are listed in Table 1, column 1. Plots of moles of salicylate formed as a function of moles of $Fe^{+2}$ added (experiment type III) were close to linear for all suspensions. The slopes, representing the chemical yield of salicylate resulting from $Fe^{+2}$ oxidation, are summarized in Table 1, column 2. Initial quantum yields of salicylate and $Fe^{+2}$ production are listed in columns 3 and 4 of Table 1. These apparent yields increase by about 800 times in the series b+g $\longrightarrow$ sb+g $\longrightarrow$ ob+g. Experiments of type IV for the b+g suspension indicate that a rapid, but saturable, surface adsorption occurs that removes $Fe^{+2}$ from solution and competes strongly for $Fe^{+2}$ generated in the photolysis. Surface adsorption was also determined to be an important process in sb+g suspension, but not in ob+g suspension.

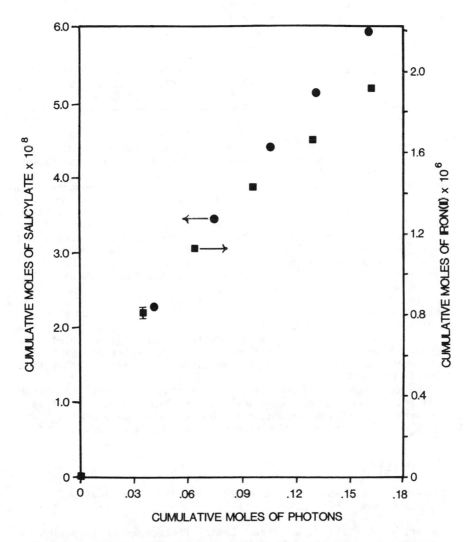

Figure 1.   Moles of salicylate and $Fe^{+2}$ produced as a function
of moles of photons absorbed for a sb+g suspension.

Table 1.    Product Yield Data from Photolysis and Dark Reactions in Carboxylate-Goethite Suspensions.

| Suspension type | Column 1 Salicylate/ $Fe^{+2}$ yield ratio in photolysis | Column 2 Salicylate chemical yield | Column 3 Initial quantum yield, salicylate | Column 4 Initial quantum yield, $Fe^{+2}$ |
|---|---|---|---|---|
| b+g  | 0.23(0.046)  | 0.031 | $3\times10^{-8}$ | $1\times10^{-7}(5\times10^{-7})$ |
| sb+g | 0.027(0.015) | 0.012 | $6\times10^{-7}$ | $2\times10^{-5}(3.6\times10^{-5})$ |
| ob+g | 0.21         | 0.55  | $2\times10^{-5}$ | $8\times10^{-5}$ |

Comparison of the data in Figure 1 and Table 1 enables an estimation of the relative contributions of the direct and Haber-Weiss routes to ·OH, once a correction has been made for the adsorption reaction in the b+g and sb+g suspensions. The corrected initial quantum yields and product ratios for the b+g and sb+g suspensions (values in parentheses in Table 1) are similar to their respective chemical yields in column 2. Chemical yields are due entirely to the Haber-Weiss mechanism for ligands like benzoate and succinate that bind relatively weakly to $Fe^{+3}$ (Sillen and Martell, 1971). Therefore, in sb+g and b+g suspensions, the similarity of the data in columns 1 and 2 indicates that the major source of ·OH is the Haber-Weiss mechanism.

In the instance of a strong-binding $Fe^{+3}$ ligand like oxalate, the large quantum and chemical yields in Table 1 indicate that a modification of reaction (1) is needed, perhaps including the perferryl ion, $FeO_2^{+2}$, as the hydroxylating agent (Dearden et al., 1968). Regardless of the nature of the hydroxylating agent, its source in the ob+g suspension is the oxidation of $Fe^{+2}$, rather than direct photolysis.

We conclude from these results that highly crystalline FeOOH will not generate significant amounts of ·OH by direct photolysis; however, significant amounts of ·OH will be generated by the Haber-Weiss mechanism in the presence of suitable ligands.

REFERENCES

Bencala, K. E., and D. M. McKnight. Written communication. U.S. Geological Survey (1986).

Cooper, W. J., and R. G. Zika. 1983. Photochemical formation of hydrogen peroxide in surface and ground waters exposed to sunlight. Science. 220:711-712.

Cunningham, K. M., M. C. Goldberg, and E. R. Weiner. 1985. The aqueous photolysis of ethylene glycol adsorbed on goethite. Photochem Photobiol. 41:409-416.

Dearden, M. B., C. R. E. Jefcoate, and J. R. Lindsay-Smith. 1968. Hydroxylation of aromatic compounds induced by the activation of oxygen, p. 260-278. *In* F. R. Mayo, ed., Ozone chemistry, photo-, singlet oxygen, and biochemical oxidations. Advances in Chemistry Series 77, Vol. III, American Chemical Society.

Finden, D. A. S., E. Tipping, G. H. M. Jaworski, and C. S. Reynolds. 1984. Light-induced reduction of natural iron(III) oxide and its relevance to phytoplankton. Nature. 309:783-784.

Haupt, J., J. Peretti, and R. Van Steenwinkel. 1984. Production of oxygen from water upon illumination by visible light of a solution of ferric oxide colloid. Nouv. J. Chim. 8:633-636.

Matthews, R. W. 1984. Hydroxylation reactions induced by near-ultraviolet photolysis of aqueous titanium dioxide suspensions. J. Chem. Soc., Far. Trans. I. 80:457-471.

Roof, A. A. M. 1982. Aquatic photochemistry, p. 43-72. *In* O. Hutzinger, ed., Handbook of environmental chemistry, Part 2B, Reactions and processes. Springer.

Ross, R. D., and D. G. Crosby. 1985. Photooxidant activity in natural waters. Environ. Toxicol. Chem. 4:773-778.

Sakata, T., T. Kawai, and K. Hashimoto. 1984. Heterogeneous photo-catalytic reactions of organic acids and water. New reactions paths besides the photo-kolbe reaction. J. Phys. Chem. 88:2344-2350.

Sillen, L. G., and A. E. Martell. 1971. Stability constants of metal-ion complexes. Special publication 25. The Chem. Soc.

Sung, W., and J. J. Morgan. 1980. Kinetics and product of ferrous iron oxygenation in aqueous systems. Environ. Sci. Technol. 14:561-568.

Tamura, H., K. Goto, T. Yotsuyanagi, and M. Nagayama. 1974. Spectrophotometric determination of iron(III) with 1,10-phenanthroline in the presence of large amounts of iron(III). Talanta. 21:314-318.

Thommes, G. A., and E. Leininger. 1958. Fluorometric determination of o- and m-hydroxybenzoic acids in mixtures. Anal. Chem. 30:1361-1363.

Waite, T. D., and F. M. M. Morel. 1984a. Photoreductive dissolution of colloidal iron oxide: Effect of citrate. J. Colloid Interface Sci. 102:121-137.

____. 1984b. Photoreductive dissolution of colloidal iron oxides in natural waters. Environ. Sci. Technol. 18:860-868.

ABSTRACT I

PROCESSES CONTROLLING THE VERTICAL DISTRIBUTION OF
SELECTED TRACE METALS IN FILSON CREEK BOG,
LAKE COUNTY, MINNESOTA

Katherine Walton Day and Clara S. E. Papp, U.S. Geological Survey,
Denver, Colorado

Lorraine H. Filipek, U.S. Geological Survey, Reston, Virginia

A small peat bog along Filson Creek in northeastern Minnesota
is separated from underlying Cu-Ni sulfide mineralized bedrock by
several meters of glacial debris. The processes controlling the
vertical distribution of trace metals in the bog were studied using
geochemical data from peat and pore water samples collected from
the bog, hydrologic measurements made in the bog, physical
characteristics of the peat and underlying sediments, and certain
(subjective) visual characteristics of these materials. An
important aim of the study was to determine whether metals in the
bog could be traced to mineralized bedrock in nearby outcrops
and/or to mineralized bedrock underlying the bog and glacial
debris. The results of the study showed that primary depositional
processes control the location of maximum concentrations of Fe, Mn,
and Co in the peat column, whereas diagenetic processes control the
location of the highest Cu concentrations. Possible sources of Cu
in the bog include: (1) Cu solubilized from surface rock outcrop
and brought into the bog in runoff, (2) Cu solubilized from bedrock
and sediments beneath the bog and transported into the bog in
upwelling groundwater, and (3) Cu solubilized and redistributed
from detrital mineral grains within the bog.
Chemical analyses of sectioned peat cores collected from the
bog in 1984 and 1985 revealed a stratification in the trace-element
content of the peat. The maximum Cu concentration occurs in a
shallow zone (above 40 cm depth) overlying a zone of maximum Fe,
Mn, and Co concentrations near the base of the peat (50-80 cm
depth) (Fig. 1). This pattern was observed in three peat cores
collected from one area in the bog.

*Chemical Quality of Water and the Hydrologic Cycle*, Robert C. Averett and Diane M. McKnight (Eds.) © 1987 Lewis Publishers, Inc.,
Chelsea, Michigan. Printed in the United States of America.

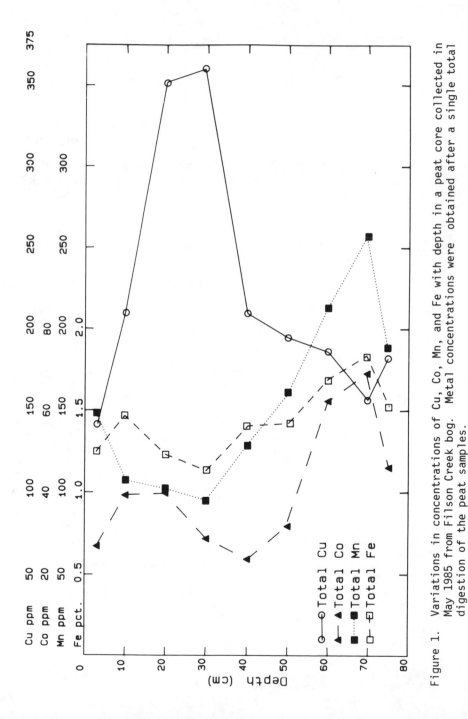

Figure 1.   Variations in concentrations of Cu, Co, Mn, and Fe with depth in a peat core collected in May 1985 from Filson Creek bog.  Metal concentrations were  obtained after a single total digestion of the peat samples.

A four-step sequential extraction procedure was applied to the peat samples to estimate the distribution of the metals among various geochemical phases. The four extractions were: (1) Na-pyrophosphate (pH = 9), which dissolves some organic material (primarily humic and fulvic acids); (2) hot hydroxylamine-hydrochloric acid (pH = 2), which dissolves Mn and amorphous Fe oxides, carbonates, acid-soluble sulfides, and metals sorbed to clay minerals; (3) hydrogen peroxide, which removes sulfides and some organic material; and finally (4) concentrated nitric, perchloric, and hydrofluoric acids, which dissolve the remaining residue. The residue is presumably composed of silicate and resistant oxide minerals, and highly resistant organic matter.

The sequential extraction results indicate that most of the Fe, Mn, and Co found in the deeper zone of the peat do not dissolve until the final "residual" extraction (Fig. 2), which is evidence that most of these metals are in silicate and resistant oxide minerals. The highest ash content (66 weight %) and highest visually estimated volume percent of mineral grains (15%) also occur in the deep zone of the peat. Both measurements correlate well with the total and residual contents of Fe, Mn, and Co, giving further evidence for a primary detrital source for these metals, especially near the base of the bog.

Variations of the Fe, Mn, and Co content in the pore water and in the non-residual dissolutions indicate that these elements were mobilized and redistributed from bedrock and sediment sources. However, the concentrations of these metals resulting from weathering and diagenetic redistribution are not as high as the concentrations of the metals in primary detrital mineral material in the deeper zone of the peat.

The sequential extraction results show that most of the Cu in the shallow high-Cu zone of the bog is soluble in the Na-pyrophosphate extraction (Fig. 3). These results suggest that the Cu in the shallow zone of the peat is associated with, and probably complexed by, organic matter.

The association of high Cu concentrations with organic matter in the shallow zone of the peat suggests that the Cu has been mobilized and removed from its source, redeposited, and concentrated with organic matter in the peat. The principal sources of soluble Cu for the bog are the bedrock and glacial sediments in the Filson Creek drainage basin and minerals contained in the peat. Minor amounts of metal enter the drainage basin in precipitation. Copper solubilized from sediment or bedrock sources outside the bog may enter the bog at the surface in runoff or from beneath by upward advection and/or diffusion. In addition, some Cu may be solubilized from detrital sediments within the peat.

The location of high total Cu concentrations in the upper portion of the peat deposit could be interpreted to mean that the Cu is introduced into the peat from the surface in runoff, and that the source of soluble Cu, therefore, is mineralized bedrock at or near the surface in the drainage basin. Bedrock outcrops do occur in the drainage basin. However, hydrologic measurements show evidence of seasonal reversals in the vertical hydrologic gradient which provides a mechanism for the advective transport to the bog of Cu solubilized from bedrock and sediments beneath the bog.

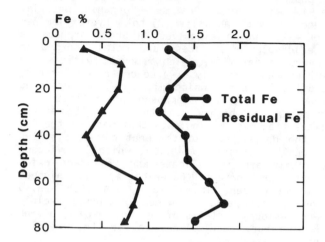

Figure 2.   Comparison of Fe in the total extraction with Fe in the
residual extraction.   Below 50 cm, most of the Fe occurs
in the residual extraction.   Manganese and Co data show
a similar pattern.

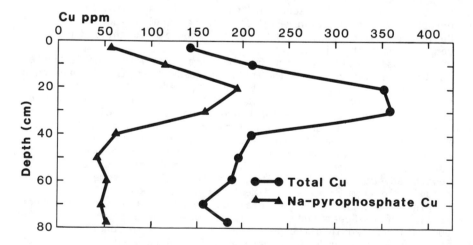

Figure 3.   Comparison of Cu in the total extraction with Cu in the
Na-pyrophosphate extraction.   Most Cu above 40-cm depth
occurs in the Na-pyrophosphate extraction.

The third possible source of Cu in the bog is detrital mineral material within the peat column. Indirect evidence, from the sequential extraction data, suggests that this source may have provided some soluble Cu to the bog sediment. Small amounts (less than 20%) of the total Cu concentration at each depth were removed by the residual extraction. Significantly larger proportions (ranging from 20-90%) of the total Fe, Mn, and Co concentrations in each sample were solubilized by the residual extraction. This observation implies either that insoluble silicate minerals are a minor source of Cu to the bog, or that a large proportion of silicate minerals containing Cu have been previously weathered out of the peat column.

A digenetic process has been proposed by Hallberg (1978) whereby Cu is concentrated in the shallow zone of an accumulating, subaqueous, sedimentary pile containing organic matter. In this model, Cu-organic complexes occur in the upper oxidized zone of the sediment above a deeper, reduced zone containing metal sulfide precipitates. Over time, as the oxidized sediments are buried and reduced, soluble Cu-organic complexes are transported upward, reprecipitate, and accumulate in the oxidized zone. In Filson Creek bog, some evidence suggests that increasingly reduced conditions occur with depth and that a sulfide zone may exist below 40 cm. These inferences are supported by: (1) decreasing concentrations of sulfate in bog pore water with depth; (2) a higher proportion of preserved wood and plant fragments with depth in the peat; and (3) an increase in the proportion of hydrogen-peroxide-extractable Fe and Cu below 40 cm. This evidence suggests that the conditions in Filson Creek bog are similar to those described in Hallberg's model. Therefore, we propose that a similar diagenetic process--involving the formation, upward transport by advection and/or diffusion, and precipitation of Cu-organic complexes in the oxidized zone--may be controlling the vertical distribution of Cu at the sites sampled in the bog.

In Hallberg's model, the implicit source of the soluble metals is the accumulating sediments. In Filson Creek bog, three possible sources of soluble Cu metal exist: surface runoff; upwelling groundwater; and detrital minerals within the peat. Diagenetic remobilization of Cu in the peat provides a mechanism whereby Cu from any of the three sources can be concentrated in the upper layers of the sediment with time rather than being buried.


REFERENCE

Hallberg, R. 1978. Metal-organic interaction at the redoxcline, p. 947-953. *In* W. E. Krumbein, ed., Environmental biogeochemistry and geomicrobiology. V. 3. Ann Arbor Science, Ann Arbor.

Zuhair Al-Shaieb, Department of Geology, Oklahoma State University, Stillwater, Oklahoma 74078

E. D. Andrews, U.S. Geological Survey, Water Resources Division, P.O. Box 25046, Mail Stop 413, Denver Federal Center, Denver, Colorado 80225

Robert C. Averett, U.S. Geological Survey, Water Resources Division, Mail Stop 432, 12201 Sunrise Valley Drive, Reston, Virginia 22092

Jill Baron, National Park Service, Water Resources Division, Natural Resources Ecology Laboratory, Colorado State University, Fort Collins, Colorado 80523

Marcia H. Bates, School of Civil Engineering, Oklahoma State University, Stillwater, Oklahoma 74078

Peter W. Bayley, Department of Geology, Oklahoma State University, Stillwater, Oklahoma 74078

Kenneth F. Bencala, U.S. Geological Survey, Water Resources Division, 345 Middlefield Road, Menlo Park, California 94025

Jon L. Bennet, Department of Chemistry, Colorado School of Mines, Golden Colorado 80401

Duane A. Benoit, U.S. Environmental Protection Agency, Environmental Research Laboratory-Duluth, 6201 Congdon Boulevard, Duluth, Minnesota 55804

Bert Bledsoe, R. S. Kerr Environmental Research Laboratory, P.O. Box 119, Ada, Oklahoma 74820

Owen Bricker, U.S. Geological Survey, Water Resources Division, Mail Stop 432, 12201 Sunrise Valley Drive, Reston, Virginia 22092

Myron H. Brooks, U.S. Geological Survey, P.O. Box 25046, Mail Stop 407, Denver Federal Center, Denver, Colorado 80225

Anthony R. Carlson, U.S. Environmental Protection Agency, Environmental Research Laboratory-Duluth, 6201 Congdon Boulevard, Duluth, Minnesota 55804

Kirkwood Cunningham, U.S. Geological Survey, P.O. Box 25046, Mail Stop 424, Denver Federal Center, Denver, Colorado 80225

Cliff N. Dahm, Department of Biology, University of New Mexico, Albuquerque, New Mexico 87131

John F. Elder, U.S. Geological Survey, 1815 University Avenue, Madison, Wisconsin 53705-4042

Lorraine H. Filipek, U.S. Geological Survey, Water Resources Division, Mail Stop 432, 12201 Sunrise Valley Drive, Reston, Virginia 22092

John R. Garbarino, U.S. Geological Survey, P.O. Box 25046, Mail Stop 407, Denver Federal Center, Denver, Colorado 80225

Marvin Goldberg, U.S. Geological Survey, P.O. Box 25046, Mail Stop 424, Denver Federal Center, Denver, Colorado 80225

Chad P. Gubala, School of Public and Environmental Affairs, Indiana University, Bloomington, Indiana 47405

John D. Hem, U.S. Geological Survey, 345 Middlefield Road, Mail Stop 427, Menlo Park, California 94025

Calvin O. Huber, Department of Chemistry, Center for Great Lakes Studies, University of Wisconsin-Milwaukee, Milwaukee, Wisconsin 53201

A. K. Huggins, School of Civil Engineering, Oklahoma State University, Stillwater, Oklahoma 74078

Jennifer L. Hughes, U.S. Geological Survey, 705 North Plaza, Room 224, Carson City, Nevada 89701

Don Kampbell, R. S. Kerr Environmental Research Laboratory, P.O. Box 119, Ada, Oklahoma 74820

J. Karleskint, School of Civil Engineering, Oklahoma State University, Stillwater, Oklahoma 74078

Douglas C. Kent, Department of Geology, Oklahoma State University, Stillwater, Oklahoma 74078

J. W. LaBaugh, U.S. Geological Survey, Mail Stop 413, Denver Federal Center, Denver, Colorado 80225

Donald Langmuir, Department of Chemistry, Colorado School of Mines, Golden, Colorado 80401

Roger W. Lee, U.S. Geological Survey, Water Resources Division, 75 Spring Street, Atlanta, Georgia 30303

Michael S. Lico, U.S. Geological Survey, 705 North Plaza, Room 224, Carson City, Nevada 89701

Carol J. Lind, U.S. Geological Survey, 345 Middlefield Road, Mail Stop 427, Menlo Park, California 94025

Donna C. Marron, U.S. Geological Survey, Water Resources Division, P.O. Box 25046, Mail Stop 413, Denver Federal Center, Denver, Colorado 80225

Vincent R. Mattson, U.S. Environmental Protection Agency, Environmental Research Laboratory-Duluth, 6201 Congdon Boulevard, Duluth, Minnesota 55804

Diane M. McKnight, U.S. Geological Survey, Water Resources Division, P.O. Box 25046, Mail Stop 407, Denver Federal Center, Denver, Colorado 80225

T. N. Morrison, Department of Chemistry, Center for Great Lakes Studies, University of Wisconsin-Milwaukee, Milwaukee, Wisconsin 53201

Clara S. E. Papp, U.S. Geological Survey, P.O. Box 25046, Mail Stop 973, Denver Federal Center, Denver, Colorado 80225

W. E. Periera, U.S. Geological Survey, Water Resources Division, P.O. Box 25046, Mail Stop 407, Denver Federal Center, Denver, Colorado 80225

John E. Poldoski, 517 Rose Street, Duluth, Minnesota 55803

Larry J. Puckett, U.S. Geological Survey, Water Resources Division, Mail Stop 432, 12201 Sunrise Valley Drive, Reston, Virginia 22092

Charles E. Robinson, U.S. Geological Survey, 345 Middlefield Road, Mail Stop 427, Menlo Park, California 94025

C. E. Rostad, U.S. Geological Survey, P.O. Box 25046, Mail Stop 407, Denver Federal Center, Denver, Colorado 80225

Leroy J. Schroder, U.S. Geological Survey, P.O. Box 25046, Mail Stop 407, Denver Federal Center, Denver, Colorado 80225

James R. Sedell, U.S. Department of Agriculture-Forest Service, Corvallis, Oregon 97331

Kathleen S. Smith, U.S. Geological Survey, P.O. Box 25046, Mail Stop 912, Denver Federal Center, Denver, Colorado 80225

J. R. Snider, Department of Atmospheric Science, University of Wyoming, Laramie, Wyoming 82071

M. Tate, School of Civil Engineering, Oklahoma State University, Stillwater, Oklahoma 74078

Howard E. Taylor, U.S. Geological Survey, P.O. Box 25046, Mail Stop 407, Denver Federal Center, Denver, Colorado 80225

Eleonora H. Trotter, Department of Biology, University of New Mexico, Albuquerque, New Mexico 87131

T. Ung, School of Civil Engineering, Oklahoma State University, Stillwater, Oklahoma 74078

David M. Updegraff, Department of Chemistry, Colorado School of Mines, Golden, Colorado 80401

David W. Vaden, Department of Geology, Oklahoma State University, Stillwater, Oklahoma 74078

G. Vali, Department of Atmospheric Science, University of Wyoming, Laramie, Wyoming 82071

Katherine Walton-Day, U.S. Geological Survey, P.O. Box 25046, Mail Stop 973, Denver Federal Center, Denver, Colorado 80225

E. R. Weiner, Department of Chemistry, University of Denver, 2199 S. University Blvd., Denver, Colorado 80208

Alan H. Welch, U.S. Geological Survey, 705 North Plaza, Room 224, Carson City, Nevada 89701

Jeffrey R. White, School of Public and Environmental Affairs, Indiana University, Bloomington, Indiana 47405

Timothy C. Willoughby, Goodson and Associates, Denver, Colorado, 80215

Barbara H. Wilson, Environmental and Ground Water Institute, University of Oklahoma, Norman, Oklahoma 73019

Y. Yang, School of Civil Engineering, Oklahoma State University, Stillwater, Oklahoma 74078

INDEX